Springer Theses

Recognizing Outstanding Ph.D. Research

Aims and Scope

The series "Springer Theses" brings together a selection of the very best Ph.D. theses from around the world and across the physical sciences. Nominated and endorsed by two recognized specialists, each published volume has been selected for its scientific excellence and the high impact of its contents for the pertinent field of research. For greater accessibility to non-specialists, the published versions include an extended introduction, as well as a foreword by the student's supervisor explaining the special relevance of the work for the field. As a whole, the series will provide a valuable resource both for newcomers to the research fields described, and for other scientists seeking detailed background information on special questions. Finally, it provides an accredited documentation of the valuable contributions made by today's younger generation of scientists.

Theses may be nominated for publication in this series by heads of department at internationally leading universities or institutes and should fulfill all of the following criteria

- They must be written in good English.
- The topic should fall within the confines of Chemistry, Physics, Earth Sciences, Engineering and related interdisciplinary fields such as Materials, Nanoscience, Chemical Engineering, Complex Systems and Biophysics.
- The work reported in the thesis must represent a significant scientific advance.
- If the thesis includes previously published material, permission to reproduce this must be gained from the respective copyright holder (a maximum 30% of the thesis should be a verbatim reproduction from the author's previous publications).
- They must have been examined and passed during the 12 months prior to nomination.
- Each thesis should include a foreword by the supervisor outlining the significance of its content.
- The theses should have a clearly defined structure including an introduction accessible to new PhD students and scientists not expert in the relevant field.

Indexed by zbMATH.

More information about this series at https://link.springer.com/bookseries/8790

Jennet Elizabeth Dickinson

ATLAS Measurements of the Higgs Boson Coupling to the Top Quark in the Higgs to Diphoton Decay Channel

Doctoral Thesis accepted by
University of California, Berkeley, USA

Springer

Author
Dr. Jennet Elizabeth Dickinson
Particle Physics Division
Fermi National Accelerator
Laboratory
Batavia, IL, USA

Supervisor
Prof. Marjorie D. Shapiro
Physics Department
University of California
Berkeley, CA, USA

ISSN 2190-5053 ISSN 2190-5061 (electronic)
Springer Theses
ISBN 978-3-030-86367-8 ISBN 978-3-030-86368-5 (eBook)
https://doi.org/10.1007/978-3-030-86368-5

This Springer imprint is published by the registered company Springer Nature Switzerland AG
The registered company address is: Gewerbestrasse 11, 6330 Cham, Switzerland

Supervisor's Foreword

The realization that the masses of the fundamental particles can be generated dynamically is a triumph of modern physics. This mechanism is known as electroweak symmetry breaking (EWSB). In the standard model (SM), the quarks, charged leptons, and W and Z bosons obtain their masses through their interactions with the Higgs field, whose fluctuations give rise to a single scalar particle: the Higgs boson. Beyond the SM theories incorporate electroweak symmetry breaking but often predict a richer phenomenology where the single fundamental Higgs is replaced by multiple scalar or pseudoscalar bosons or where the Higgs boson(s) can be a composite particle or resonance.

The existence of a spin-0 particle with mass approximately 125 GeV is now well established. But it is not yet established that this particle is the SM Higgs. Furthermore, the existence of a Higgs candidate leads to a new, more detailed set of questions: Does the Higgs couple to dark matter? Does the Higgs boson play a role generating neutrino masses? Are there additional Higgs particles to be found?

This thesis represents an important, early step in the campaign to use the Higgs boson as a tool for discovery. It presents the observation of production of the Higgs in association with a top quark pair ($t\bar{t}$). Because the top quark is the heaviest particle in the SM, this measurement is particularly sensitive to the Higgs mechanism. Furthermore, by separating the observed signal into a number of categories, these data also provide constraints on a potential "CP-violating" term that might result in slightly different couplings of the Higgs to matter and antimatter.

This work focuses on the Higgs decay $H \to \gamma\gamma$. The branching ratio for this mode is small ($\approx 0.2\%$), but because the photon energy and direction can be well measured in the ATLAS detector, the Higgs mass is observed as a narrow peak in the diphoton mass spectrum. In addition, the "sideband" regions above and below the mass peak provide excellent control regions for characterizing the relevant backgrounds.

Unlike the "clean" peak in diphoton mass, the top quark pair is a messy system. The top quark (antiquark) decays to a W boson and a b quark (antiquark), and the W subsequently decays either to a charged lepton and a neutrino or to a quark–antiquark pair. Identifying this variety of decay modes, as well as separating the $t\bar{t}$ signal from background relies on detailed machine learning algorithms implemented using boosted decision trees (BDTs). Inputs to these BDTs are the momenta of the jets of particles that serve as proxies for the quarks, the leptons, and the photons observed in each event. Jets themselves are characterized by the probability that they contained a *b* quark. This *b*-jet identification is also performed using machine learning techniques.

To increase the statistical power of the measurement, events are separated into multiple categories with different signal-to-background ratios, and the number of Higgs signal events in each category is determined from the diphoton invariant mass distributions. The fitted yields in each category are then combined, taking into account correlated experimental and theoretical systematic uncertainties, and the resulting yield is used to measure the $t\bar{t}H$ production cross section. An alternative categorization focused on observables sensitive to the CP structure of the Higgs coupling is also presented. Using this categorization, a fully CP odd Higgs-top coupling is excluded at the 3.9σ level, and the CP mixing angle is constrained. A by-product of the analysis is an upper bound on the cross section for single-top-plus-Higgs production of 11.6 times the SM prediction.

Jennet Dickinson's thesis is an impressive piece of work. As a graduate student, she played a central role in the first 5σ observation of $t\bar{t}H$ production and spearheaded the analysis that constrained the CP structure of the Higgs. It has been a pleasure to work with her and to observe her growth into a fully developed scientist.

Berkeley, CA, USA
May 2021

Prof. Marjorie D. Shapiro

Abstract

During Run 2 of the Large Hadron Collider, the ATLAS experiment recorded proton–proton collision events with a center of mass energy of 13 TeV, the highest energy ever achieved in a collider. Analysis of this dataset, which corresponds to an integrated luminosity of 139 fb^{-1}, has provided new opportunities for precision measurements of the Higgs boson. The interaction between the Higgs boson and the top quark, the heaviest known particle, is a fundamental probe of the Higgs mechanism, which describes how Standard Model particles obtain mass.

The coupling of the Higgs boson to the top quark can be directly probed through the production of a Higgs boson in association with a top–antitop quark pair ($t\bar{t}H$) at the LHC. A statistical combination of ATLAS searches in multiple Higgs boson decay channels using 79.8 fb^{-1} of the Run 2 dataset yielded the first observation of $t\bar{t}H$ production at the level of 6.3σ.

The Higgs to diphoton ($\gamma\gamma$) decay channel is among the most sensitive for $t\bar{t}H$ measurements due to the excellent diphoton mass resolution of the ATLAS detector and the clean signature of this decay. A powerful multivariate analysis was developed to select $t\bar{t}H(\gamma\gamma)$ events based on the momenta of jets, photons, and leptons in the final state. The $t\bar{t}H$ cross section times $H \rightarrow \gamma\gamma$ branching fraction was measured to be $\sigma_{t\bar{t}H} \times B_{\gamma\gamma} = 1.59^{+0.43}_{-0.39}$ fb using the full Run 2 ATLAS dataset. Top quarks were reconstructed with high probability in selected $t\bar{t}H(\gamma\gamma)$ events.

Following the observation of the $t\bar{t}H$ process, a direct measurement of the CP properties of the Higgs-top interaction was carried out in the $H \rightarrow \gamma\gamma$ decay channel. A CP-sensitive multivariate categorization was developed using reconstructed top quark variables, yielding an observed $t\bar{t}H$ significance of 5.2σ and an upper limit on the cross section of single-top-plus-Higgs (tH) production of 11.6 times the SM expectation. The observed data excludes a fully CP odd Higgs-top coupling at the level of 3.9σ, and the CP mixing angle is constrained to $|\alpha| > 43°$ at 95% confidence level.

Acknowledgments

The work included in this thesis could not have been accomplished without substantial support from colleagues, friends, and family, including many who have assumed more than one of these roles.

My educational journey has been a long one, and I am fortunate to have found tremendous mentors at every step of the way. Rest assured that I will do my best to pay it forward. I am grateful to you all: to the educators at St. Catherine's for guiding me toward a scientific career; to those at Columbia, especially Georgia, for giving me my first taste of research work; and to my ATLAS collaborators around the globe, especially the group at Lawrence Berkeley National Laboratory. Thank you especially to Haichen, not only for the scientific guidance, but also for pushing me to do more than I knew I could. And to Marjorie, for your constant counsel and support, and for setting an example of the type of scientist I aspire to become.

I would like to thank the friends who helped me to stay (mostly) sane throughout my doctoral work. Sam, Jeffrey, Kimber, Megan, and countless more: I have learned that there is no one place, be it Richmond, New York, Berkeley, or Geneva, where you can be close to everyone. I will miss you when I am far away and celebrate when we are together again.

And finally, and most importantly, my family has been a constant source of support and encouragement throughout my life. Thank you for instilling in me the deep love of learning without which I would not be here today. I love you, Mom and Dad.

Contents

1 Introduction ... 1
References ... 4

2 Phenomenology ... 5
2.1 Theory of Strong Interactions ... 5
2.1.1 Factorization of Proton–Proton Collisions ... 7
2.2 Theory of Electroweak Interactions ... 9
2.2.1 The Higgs Mechanism ... 10
2.2.2 Fermion Masses and Interactions ... 11
2.3 Higgs Boson Physics ... 13
2.3.1 Production Mechanisms ... 13
2.3.2 Decay Channels ... 16
2.3.3 Coupling Measurement Framework ... 19
2.4 Beyond the Standard Model Higgs ... 22
2.4.1 CP Violation in the Higgs Sector ... 22
References ... 25

3 The ATLAS Experiment ... 29
3.1 The Large Hadron Collider ... 29
3.1.1 Luminosity ... 31
3.2 The ATLAS Detector ... 31
3.2.1 Inner Detector ... 33
3.2.2 Calorimeters ... 35
3.2.3 Muon Spectrometer ... 39
3.2.4 Trigger ... 41
References ... 43

4 Object Reconstruction ... 45
4.1 Tracks and Vertices ... 45
4.2 Photons ... 46
4.2.1 Photon Identification ... 47
4.2.2 Photon Isolation ... 48
4.2.3 Primary Vertex Selection in Diphoton Events ... 48

4.3 Electrons ... 49
4.3.1 Electron Identification ... 49
4.3.2 Electron Isolation ... 50
4.4 Muons ... 50
4.4.1 Muon Identification ... 50
4.4.2 Muon Isolation ... 51
4.5 Hadronic Jets ... 51
4.5.1 Flavor Tagging ... 52
4.6 Missing Transverse Energy ... 53
References ... 53

5 Data and Monte Carlo Samples ... 55
5.1 Data: LHC Run 2 ... 55
5.1.1 Data Control Samples ... 57
5.2 Simulated Samples ... 59
5.2.1 SM Higgs Boson Samples ... 60
5.2.2 BSM Higgs Samples ... 61
5.2.3 Background Samples ... 63
References ... 63

6 Statistical Model ... 65
6.1 Motivation for Categorization ... 65
6.1.1 A Short Proof ... 66
6.2 Signal and Background Shapes ... 67
6.2.1 Signal Model ... 67
6.2.2 Background Model ... 68
6.3 Likelihood Model ... 69
6.3.1 Treatment of Systematics ... 71
6.3.2 Generation of Asimov Datasets ... 72
6.4 Sources of Systematic Uncertainty ... 73
6.4.1 Theoretical Uncertainties ... 74
6.4.2 Experimental Uncertainties ... 76
6.5 Weighting Data for Presentation ... 77
References ... 78

7 Machine Learning Techniques ... 81
7.1 Multivariate Analysis ... 81
7.1.1 Classification ... 82
7.2 Boosted Decision Trees ... 84
7.2.1 Boosting ... 86
7.2.2 Multiclass BDT ... 87
7.3 Hyper-Parameter Optimization ... 88
7.3.1 Grid Scan ... 88
7.3.2 Gaussian Processes Minimization ... 89
References ... 89

8 Reconstruction of Top Decays 91
8.1 Reconstruction Strategy 93
8.1.1 Hadronic Channel 93
8.1.2 Leptonic Channel 94
8.2 Training of the Top Reco BDT 95
8.2.1 Constructing Signal and Background Samples 95
8.2.2 Training Variables 97
8.3 Performance 101
8.3.1 Hadronic Channel 101
8.3.2 Leptonic Channel 103
References 106

9 Selection of $t\bar{t}H(\gamma\gamma)$ Events 107
9.1 Poisson Number-Counting Significance 107
9.2 Categorization of Events 109
9.2.1 Hadronic Channel 111
9.2.2 Leptonic Channel 118
9.3 Decomposition of Continuum Background 121
9.3.1 Hadronic Channel 122
9.3.2 Leptonic Channel 124
References 125

10 First Observation of $t\bar{t}H$ Production 127
10.1 Combination of Higgs Decay Channels 127
10.1.1 Input Analyses 127
10.1.2 Systematic Uncertainties 129
10.1.3 Results of Combined Analysis 130
10.2 $t\bar{t}H$ in the Diphoton Decay Channel 133
10.2.1 Signal and Background Shapes 133
10.2.2 Systematic Uncertainties 134
10.2.3 Results 135
References 139

11 CP-Sensitive Categorization of $t\bar{t}H(\gamma\gamma)$ Events 141
11.1 Number-Counting Limit on CP Mixing 141
11.2 Sensitive Observables 143
11.3 Multivariate Categorization Strategy 145
11.3.1 Hadronic Channel 147
11.3.2 Leptonic Channel 149
11.4 Determination of Category Boundaries 153
References 162

12 Measurement of CP Properties in $t\bar{t}H$ and tH 163
12.1 Treatment of Loop Processes 163
12.1.1 Kinematics of Gluon Fusion 164
12.1.2 Effective Loop Couplings 165
12.2 Signal and Background Model 166

12.2.1 Signal and Background Shapes 166
12.2.2 Signal Yield Dependence on Mixing Angle 167
12.3 Systematic Uncertainties 171
12.4 Results 173
12.4.1 Sensitivity to SM $t\bar{t}H$ and tH 175
12.4.2 Constraints on CP Mixing 177
References 182

13 Higgs Couplings in $H \to \gamma\gamma$ 183
13.1 Event Categorization 183
13.1.1 Leptonic VH 185
13.1.2 Vector Boson Fusion and Hadronic VH 186
13.1.3 Gluon Fusion 188
13.2 Signal and Background 188
13.3 Systematic Uncertainties 189
13.4 Results (79.8 fb^{-1}) 191
13.4.1 Higgs Cross Section by Production Mode 192
13.4.2 Measurement of STXS 196
References 198

14 Conclusions 199
References 200

Appendix A: Search for TeV-Scale Gravity 201

Appendix B: Multijet Modeling with MG5_aMC@NLO 209

Appendix C: Radiation Damage in the IBL 215

Chapter 1
Introduction

The Standard Model of Particle Physics (SM) has enjoyed widespread experimental success over the last fifty years: its predictions have been confirmed by many experiments studying a large range of energy scales, with few exceptions. The list of fundamental particles included in the Standard Model is complete to the best of our knowledge. These particles are shown in Fig. 1.1, and an overview of their basic properties is presented in Table 1.1.

In addition to a list of fundamental particles, the Standard Model also provides a set of rules governing particle interactions. The Standard Model includes Strong, Electromagnetic, and Weak interactions, which are breifly described in Chap. 2. The well-known Gravitational interaction is not part the Standard Model, and its contribution is typically neglected in high energy physics experiments: the effects of gravity are more than 10^{30} times weaker than SM interactions. Many Beyond the Standard Model (BSM) theories seek to include gravity, and some theories predict signatures that are accessible to current collider experiments [2, 3]. Appendix A includes a search for such signatures.

The final Standard Model particle to be experimentally observed was the Higgs boson, which was discovered in 2012 by the ATLAS and CMS experiments [4, 5] during the first run of the Large Hadron Collider (LHC), which delivered a large dataset of proton–proton collisions with a center of mass energy $\sqrt{s}$ of 7–8 TeV. In the second run of the LHC (2015–2018), an even larger dataset and higher center of mass energy ($\sqrt{s} = 13$ TeV) have facilitated the observation of new Higgs boson production and decay modes, as well as improved measurements of Higgs couplings to fermions and bosons.

The coupling of the Higgs boson to the top quark, the heaviest particle in the Standard Model, is a fundamental probe of the nature of the Higgs boson. The measurement of this coupling is a sensitive test of the Higgs Mechanism, which describes how Standard Model particles obtain mass through their interaction with the Higgs field. Prior to LHC Run 2, the ATLAS and CMS collaborations performed indirect

J. E. Dickinson, *ATLAS Measurements of the Higgs Boson Coupling to the Top Quark in the Higgs to Diphoton Decay Channel*, Springer Theses,
https://doi.org/10.1007/978-3-030-86368-5_1

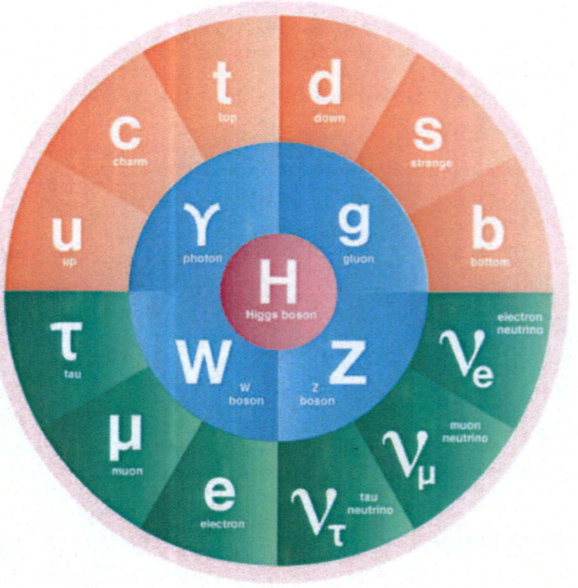

Fig. 1.1 Constituents of the standard model of particle physics [1]

Table 1.1 Summary of properties of the fundamental particles in the standard model, from the particle data group [8]

Name	Symbol	Mass	Electric charge	Spin	Antiparticle
Bosons					
Higgs Boson	H	125.09 GeV	0	0	H
W Boson	W^+	80.379 GeV	+1	1	W^-
Z Boson	Z	91.188 GeV	0	1	Z
Photon	γ	0	0	1	γ
Gluon	g	0	0	1	g
Fermions					
Quarks					
Up	u	2.16 MeV	+2/3	1/2	$\bar{u}$
Down	d	4.67 MeV	−1/3	1/2	$\bar{d}$
Strange	s	93 MeV	−1/3	1/2	$\bar{s}$
Charm	c	1.27 GeV	+2/3	1/2	$\bar{c}$
Bottom	b	4.18 GeV	−1/3	1/2	$\bar{b}$
Top	t	172.9 GeV	+2/3	1/2	$\bar{t}$
Leptons					
Electron	e or e^-	511 keV	−1	1/2	$\bar{e}$ or e^+
Muon	μ or μ^-	105 MeV	−1	1/2	$\bar{\mu}$ or μ^+
Tau	τ or τ^-	1.78 GeV	−1	1/2	$\bar{\tau}$ or τ^+
Electron Neutrino	ν_e	< 2 eV	0	1/2	$\bar{\nu}_e$ $(? = \nu_e)$
Muon Neutrino	ν_μ	< 2 eV	0	1/2	$\bar{\nu}_\mu$ $(? = \nu_\mu)$
Tau Neutrino	ν_τ	< 2 eV	0	1/2	$\bar{\nu}_\tau$ $(? = \nu_\tau)$

measurements of this coupling [6]. However, these measurements were predicated on assumptions that no BSM physics processes contribute to Higgs boson production. LHC Run 2 presents an opportunity to probe the Higgs-top coupling directly through Higgs boson production in association with a top-antitop quark pair ($t\bar{t}H$). This process is expected to be rare: it makes up only 1% of the Standard Model Higgs boson cross section at $\sqrt{s} = 13$ TeV.

This Thesis presents a series of measurements of $t\bar{t}H$ production using the ATLAS detector at the LHC. The first observation of $t\bar{t}H$ production is reported, which establishes a direct measurement of the interaction between the Higgs boson and quarks. Measurements of combined and single channel $t\bar{t}H$ cross sections are described, and a study of the charge-parity (CP) symmetry of the Higgs-top interaction is presented.

The measurements presented in this work focus on $t\bar{t}H$ production in conjunction with the Higgs boson decay to a pair of photons ($H \rightarrow \gamma\gamma$), which has a low branching fraction but clean signature in the ATLAS detector. $H \rightarrow \gamma\gamma$ measurements are performed by fitting to the diphoton invariant mass spectrum in data

$$m_{\gamma\gamma} = \sqrt{2E_{\gamma 1}E_{\gamma 2}(1 - \cos\theta_{\gamma\gamma})} \tag{1.1}$$

where $E_{\gamma 1}$ and $E_{\gamma 2}$ represent the energies of two photons separated by an angle $\theta_{\gamma\gamma}$. Due to the high diphoton mass resolution in ATLAS, the Higgs boson signal appears as a narrow peak in this distribution. By using a fit function composed of a smoothly falling background distribution plus a signal peak, the background for $H \rightarrow \gamma\gamma$ processes is estimated in a fully data-driven way.

This Thesis is organized as follows: Chap. 2 reviews the phenomenology of collider physics, including a detailed look at Higgs production in proton–proton collisions. Chapter 3 describes the experimental setup, including the Large Hadron Collider and the ATLAS detector. Chapter 4 describes the reconstruction of particles from their signatures in the ATLAS detector. The data and simulated samples used to study $t\bar{t}H$ production in conjunction with the diphoton decay ($t\bar{t}H(\gamma\gamma)$) are detailed in Chap. 5, and the statistical tools required to perform the fit to $m_{\gamma\gamma}$ are presented in Chap. 6. Chapter 7 gives a brief introduction to Machine Learning and the multivariate techniques used in this work. In Chap. 8, a novel method for reconstruction of top quark decays using Machine Learning is presented. This technique is applied extensively in later Chapters to calculate variables dependent on top quark kinematics. The development of a multivariate analysis selection sensitive to $t\bar{t}H$ production in the $H \rightarrow \gamma\gamma$ decay channel is outlined in Chap. 9. Chapter 10 presents the first observation of $t\bar{t}H$ production in ATLAS, including measurements of combined and single channel cross sections. Chapters 11 and 12 present a search for violation of charge-parity (CP) symmetry in the interaction between the Higgs boson and the top quark, which could contribute to the observed asymmetry between matter and anti-matter in the universe [7]. Chapter 11 describes the development of multivariate analysis categories sensitive to the CP properties of $t\bar{t}H$ and tH production. Chapter 12 presents the results of this study, including stringent limits on CP mixing. Finally, Chap. 13 describes the optimization of analysis categories targeting $H \rightarrow \gamma\gamma$ events in the three Higgs boson production modes with highest cross section: gluon fusion, vector boson fusion, and vector boson associated production. Cross sections in each production mode and in different kinematic regions are obtained by a simultaneous fit to all $H \rightarrow \gamma\gamma$ categories, including $t\bar{t}H$. Conclusions and final thoughts are included in Chap. 14. Appendixes A–C contain short summaries of additional ATLAS work by the author of this Thesis.

References

1. Riesselmannl K (2015) The standard model of particle physics. In: Symmetry
2. Arkani-Hamed N, Dimopoulos S, Dvali GR (1998) The Hierarchy problem and new dimensions at a millimeter. Phys Lett B 429:263–272. https://doi.org/10.1016/S0370-2693(98)00466-3. arXiv: hep-ph/9803315
3. Randall L, Sundrum R (1999) A Large mass hierarchy from a small extra dimension. Phys Rev Lett 83:3370–3373. https://doi.org/10.1103/PhysRevLett.83.3370 arXiv: hep-ph/9905221
4. Aad G et al (2012) Observation of a new particle in the search for the standard model Higgs boson with the ATLAS detector at the LHC. Phys Lett B 716:1–29. https://doi.org/10.1016/j.physletb.2012.08.020. arXiv: 1207.7214 [hep-ex]
5. Chatrchyan S et al (2012) Observation of a New Boson at a Mass of 125 GeV with the CMS experiment at the LHC. Phys Lett B 716:30–61. https://doi.org/10.1016/j.physletb.2012.08.021. arXiv: 1207.7235 [hep-ex]
6. Aad G et al (2016) Measurements of the Higgs boson production and decay rates and constraints on its couplings from a combined ATLAS and CMS analysis of the LHC pp collision data at $\sqrt{s} = 7$ and 8 TeV. JHEP 08:045. https://doi.org/10.1007/JHEP08(2016)045. arXiv: 1606.02266 [hep-ex]
7. Sakharov AD (1991) Violation of CP in variance, C asymmetry, and baryon asymmetry of the universe. Phys USP 34(5):392–393. https://doi.org/10.1070/PU1991v034n05ABEH002497
8. Tanabashi M et al (2018) Review of particle physics. Phys Rev D 98:030001. https://doi.org/10.1103/PhysRevD.98.030001. https://link.aps.org/doi/10.1103/PhysRevD.98.030001

Chapter 2
Phenomenology

The primary goal of high energy collider experiments is to test the predictions of the Standard Model. Direct searches for Beyond the Standard Model (BSM) physics exploit the high center of mass energy reached by colliders to produce new particles and processes. Indirect measurements rely on the high statistics of collider data to make precision measurements of Standard Model interactions in order to identify deviations that could arise from new physics via loop processes.

This Chapter presents the relevant theory for calculating Standard Model and Beyond the Standard Model predictions. Section 2.1 focuses on the strong interaction and the framework used to perform difficult calculations. The theoretical necessity of the Higgs boson is its role in electroweak symmetry breaking is discussed in Sect. 2.2. Production and decay modes of the Higgs boson are described in detail in Sect. 2.3. Finally, Beyond the Standard Model Higgs scenarios, including the introduction of CP violation in the top Yukawa coupling, are addressed in Sect. 2.4.

2.1 Theory of Strong Interactions

In order to study physics at the highest energies and smallest distance scales, collider experiments focus on **hard scatter** interactions, in which there is a large momentum transfer (Q^2) between two incident particles. During the course of a single proton–proton event, many interactions occur with widely varying Q^2. This presents a major challenge for calculating properties of proton–proton events due to the changing behavior of the strong interaction at different energy scales.

Strong interactions between quarks and gluons are described by the non-Abelian $SU(3)_{\text{color}}$ gauge theory known as **quantum chromodynamics (QCD)**. The rate, or **cross section** σ, of any strong interaction can be expanded as a power series in the strong force coupling α_S:

J. E. Dickinson, *ATLAS Measurements of the Higgs Boson Coupling to the Top Quark in the Higgs to Diphoton Decay Channel*, Springer Theses,
https://doi.org/10.1007/978-3-030-86368-5_2

$$\sigma = \sum_{i}^{\infty} c_i \alpha_S^i \tag{2.1}$$

For small enough α_S, this expansion is **perturbative**: the contribution of each subsequent term is small compared to the previous terms. Summing over a finite number of terms therefore gives a good approximation of σ, and high order terms can be safely neglected. If the highest order term included in a calculation of σ is linear in α_S, the calculation is called "leading order" (LO). If the sum is completed up to the quadratic, cubic, or quartic term in α_S, the calculation is called "next-to-leading order" (NLO), "next-to-next-to-leading order" (NNLO), or "next-to-next-to-next-to-leading order" (N^3LO).

The primary computational challenges in collider physics arise from a unique property of QCD: interactions at low Q^2 (long distance) are stronger than interactions at high Q^2 (short distance). The running of the strong coupling (that is, its dependence on a **renormalization scale** μ_R) is shown in Fig. 2.1 for $\mu_R^2 = Q^2$. As the renormalization scale is varied, each term in the perturbative expansion in Eq. 2.1 changes, though the infinite sum remains unchanged. Typical values of μ_R^2 used in calculations are $\mathcal{O}(Q^2)$, which ensure that higher order terms remain small.

Due to the self-interaction of gluons, the strong coupling α_S grows large at small Q^2: this results in the **confinement** of quarks and gluons into $SU(3)_{\text{color}}$-singlet states (hadrons). However, at high Q^2 the strong coupling is very small and quarks are not confined to bound states: this is known as **asymptotic freedom**.

This behavior of the strong interaction contrasts strikingly with quantum electrodynamics, for example, where interaction rates fall nearly to zero for particles that are separated by very long distances. This difference in the behavior of these forces is a result of different running of the couplings: the electroweak coupling decreases with Q^2, and its magnitude varies by less than 10% over the range 1 GeV $< Q^2 <$ 10^4 GeV [1].

In order to perform perturbative calculations, the smallness of the coupling is imperative. For this reason, calculations of collider phenomena are factorized into two components [3–5]:

1. Physics at short distance (high Q^2, perturbative in QCD)
2. Physics at long distance (low Q^2, non-perturbative in QCD)

The **factorization scale** μ_F separates these two regimes. Above the factorization scale, α_S is sufficiently small that cross sections can be calculated at fixed order in perturbation theory. Below the factorization scale, predictions rely heavily on data and phenomenological models of low-energy processes. Like the renormalization scale, μ_F^2 is typically taken to be $\mathcal{O}(Q^2)$.

Cross sections and kinematic distributions calculated at high enough order in perturbation theory are independent of the renormalization and factorization scales. However, small residual dependence on μ_R and μ_F are expected from finite perturbative expansions: these effects are considered as systematic uncertainties.

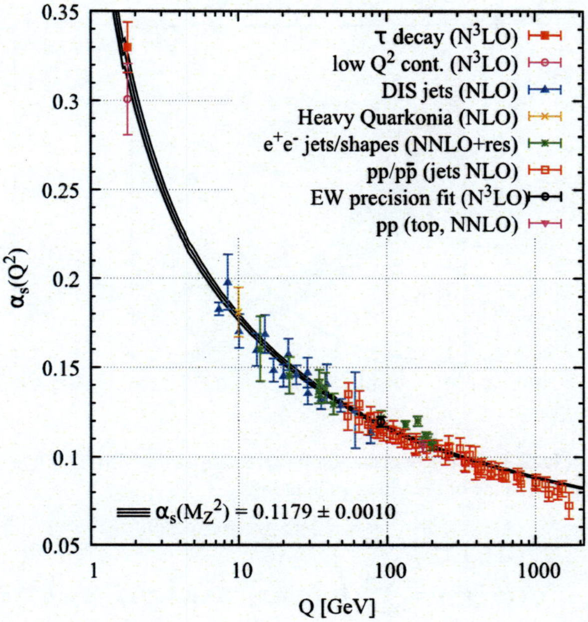

Fig. 2.1 Measurements of the running of the strong coupling collected by the particle data group [2]

2.1.1 Factorization of Proton–Proton Collisions

In proton–proton collisions, one constituent parton (quark, antiquark, gluon) of each incoming proton participates in the hard scatter interaction. The remaining partons contribute low-energy hadronic activity to the final state of the event: this is referred to as the **underlying event (UE)**. Figure 2.2 shows a diagram of a proton–proton event containing a hard scatter interaction of cross section $\hat{\sigma}$.

The incoming proton constituents and their momenta are described by **parton distribution functions (PDFs)**, which are convolved with parton–parton hard scatter cross sections $\hat{\sigma}$ to obtain the full proton–proton cross section σ:

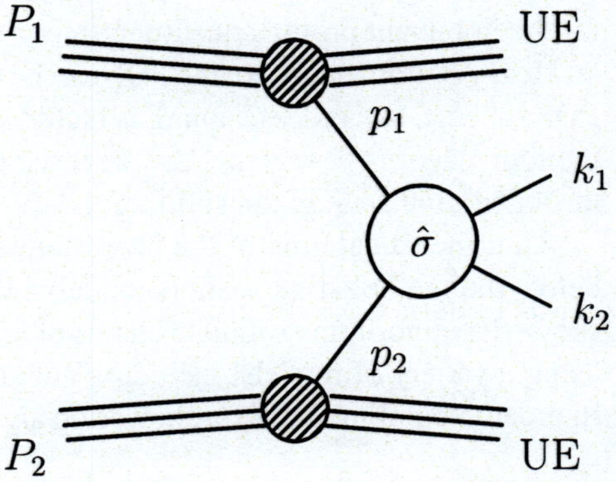

Fig. 2.2 Feynman diagram of an example proton–proton event containing a hard scatter interaction of cross section $\hat{\sigma}$

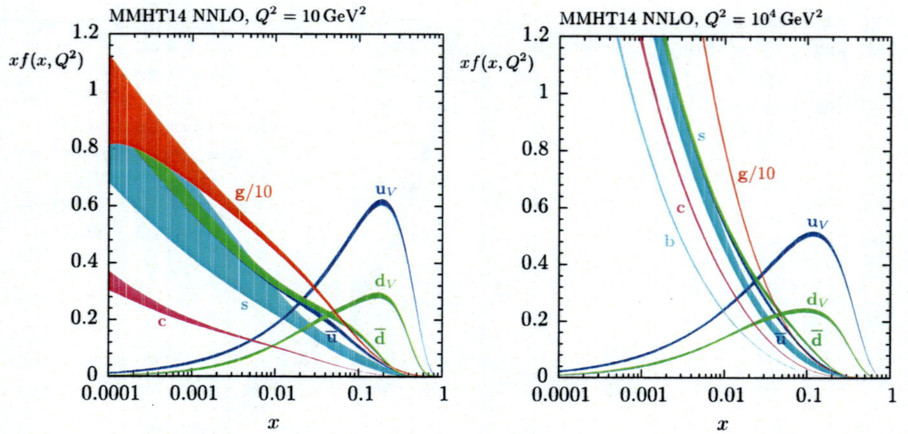

Fig. 2.3 Parton distribution functions from the MMH14 PDF set [7], at two values of Q^2. Note the dominance of the gluon PDF at low x

$$\sigma(P_1, P_2) = \sum_{i,j} \int dx_1 dx_2 f_i(x_1, \mu_F^2) f_j(x_2, \mu_F^2) \hat{\sigma}_{ij}(p_1, p_2, \alpha_S(\mu_F^2), Q^2/\mu_F^2) \tag{2.2}$$

where i and j represent the different types of partons. The cross section of the hard scatter interaction $\hat{\sigma}$ is calculated to fixed order in perturbation theory for all combinations of incoming partons i, j.

The PDFs f_i depend on:

- the **Björken** x such that $\overrightarrow{p} = x\overrightarrow{P}$ where p is the parton momentum and P the proton momentum
- the factorization scale μ_F

Measurements of parton distribution functions are primarily based on data from Deep Inelastic Scattering, with additional input data from neutrino scattering and hadron collider measurements [5, 6]. Examples of parton distribution functions at two values of $\mu_F^2 = Q^2$ are shown in Fig. 2.3. At high values of Q^2, the gluon is by far the dominant proton constituent.

Heavy bottom quarks in the PDF can be treated in two ways: in the five-flavor (5F) PDF scheme, a massless b quark is included among the constituents of the proton; in the four-flavor (4F) scheme, the presence of b quarks in the proton is possible only through initial state gluon splitting.

After the calculation of the hard scatter cross section, additional QCD radiation below the factorization scale is modeled by **parton showering** algorithms. Parton showering algorithms simulate chains of low-energy and small-angle QCD radiation using a variety of models, including angular ordered [8] and dipole [9] showering. In order to combine a parton showering algorithm with a hard scatter matrix element calculation, matching and merging algorithms are used to ensure the calculations are consistent and no diagrams are over-counted [10].

During parton showering, quarks and gluons lose energy and pass from the regime of high Q^2 to that of confinement. During this transition, known as **hadronization** [11], partons form $SU(3)_{\text{color}}$-singlet states, resulting in a columnated **jet** of hadrons. Multiple models are available for simulating hadronization and color flow in hadronic final states: the most widely used are the string [12] and cluster [13] models.

Because the underlying event, parton shower, and hadronization (UEPS) are low Q^2 processes that are non-perturbative in QCD, parameters of these models are typically tuned to data rather than predicted *a priori*. These **tunes** are applied extensively in predictions and simulations [14]. Differences in the predictions from multiple UEPS models are taken as a source of systematic uncertainty.

For a discussion and comparison of different models of strong interactions in Monte Carlo, see Appendix B.

2.2 Theory of Electroweak Interactions

The Glashow-Weinberg-Salam theory unifies the weak and electromagnetic forces under the gauge group $SU(2)_L \times U(1)_Y$ [15–17]. $SU(2)_L$ is the weak isospin group, and its generators in the spinor representation are the Pauli matrices:

$$\sigma_1 = \begin{pmatrix} 0 & 1 \\ 1 & 0 \end{pmatrix};\ \sigma_2 = \begin{pmatrix} 0 & -i \\ i & 0 \end{pmatrix};\ \sigma_3 = \begin{pmatrix} 1 & 0 \\ 0 & -1 \end{pmatrix} \tag{2.3}$$

Weak isospin I_3 is the eigenvalue of the operator σ_3. The generator of $U(1)_Y$, the Abelian hypercharge group, is simply denoted 1, and its eigenvalue hypercharge by Y.

A scalar field ϕ is introduced that transforms as a doublet under $SU(2)_L$ and is symmetric under gauge transformations of the form

$$\phi \to e^{i\alpha^a \sigma^a/2} e^{i\beta/2} \phi \tag{2.4}$$

for all values of $\alpha^1, \alpha^2, \alpha^3$ and β. The electroweak Lagrangian is constructed from a kinetic term and a potential term for ϕ:

$$\mathcal{L} = |\mathcal{D}_\mu \phi|^2 + V(\phi) \tag{2.5}$$

where $\mathcal{D}_\mu$ is the electroweak covariant derivative. To ensure invariance of the kinetic term under the gauge transformation in Eq. 2.4, the covariant derivative must take the form

$$\mathcal{D}_\mu \phi = (\partial_\mu - i\frac{g'}{2} B_\mu - i\frac{g}{2} W_\mu^j \sigma_j)\phi \tag{2.6}$$

where $W^{1,2,3}$ and B are the field strengths of the gauge groups $SU(2)_L$ and $U(1)_Y$, respectively. Expanding Eq. 2.5 in terms of field tensors and making use of Pauli matrix identities, the Lagrangian becomes

$$\mathcal{L} = |\partial_\mu \phi|^2 + \frac{g'^2}{2} B_\mu B^\mu |\phi|^2 + \frac{g^2}{2} W^j_\mu W^\mu_j |\phi|^2 + \frac{g'g}{4} B_\mu W^{k\mu} (\phi^\dagger \sigma_k \phi + (\sigma_k \phi)^\dagger \phi) + V(\phi) \tag{2.7}$$

The following substitutions simplify the interpretation:

$$\begin{aligned} W^\pm_\mu &= \frac{W^1_\mu \pm i W^2_\mu}{\sqrt{2}} \\ Z_\mu &= \frac{g' B_\mu - g W^3_\mu}{\sqrt{g'^2 + g^2}} \\ A_\mu &= \frac{g' B_\mu + g W^3_\mu}{\sqrt{g'^2 + g^2}} \end{aligned} \tag{2.8}$$

The transformed field strengths W^+_μ, W^-_μ and Z_μ correspond to the W and Z bosons, and A_μ corresponds to the electromagnetic field and has eigenvalue $Q = I_3 + Y$.

The simplified Lagrangian becomes

$$\mathcal{L} = |\partial_\mu \phi|^2 + \frac{g'^2 + g^2}{2} Z_\mu Z^\mu |\phi|^2 + \frac{g^2}{2} W^+_\mu W^{-\mu} |\phi|^2 + V(\phi) \tag{2.9}$$

2.2.1 *The Higgs Mechanism*

The theory of weak interactions thus far contains only massless gauge bosons. However, the masses of the W and Z bosons have been measured to be substantial (80.379 GeV and 91.188 GeV, respectively [2]). The introduction of mass for the W and Z bosons while maintaining the gauge invariance of $\mathcal{L}$ is accomplished through **electroweak symmetry breaking**.

The above gauge symmetry is spontaneously broken by means of the Higgs Mechanism [18, 19]. The potential for the doublet ϕ is taken to be

$$V(\phi) = -\mu^2 \phi^\dagger \phi + \frac{\lambda}{2} (\phi^\dagger \phi)^2 \tag{2.10}$$

For $\mu^2 > 0$, this potential has a minimum, or **vacuum expectation value**, at $\phi = (0, v)$ where $v = \sqrt{\mu^2/\lambda}$. The shape of this potential is shown in Fig. 2.4.

The field ϕ can be expressed as oscillations about this constant minimum:

$$\phi = (0, v + H) \tag{2.11}$$

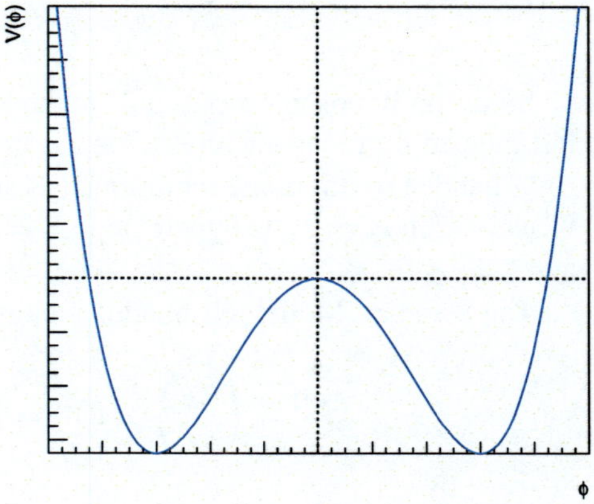

Fig. 2.4 Functional form of the potential of the scalar field ϕ

where H represents the Higgs boson. Performing this substitution, the Lagrangian becomes

$$\mathcal{L} = (\partial_\mu H)^2 + \frac{g'^2 + g^2}{2} Z_\mu Z^\mu (v + H)^2 + g^2 W^+_\mu W^{-\mu} (v + H)^2 + V(v + H) \tag{2.12}$$

Thus mass terms for the W and Z bosons have been obtained ($m_W = \frac{1}{2} v g$ and $m_Z = \frac{1}{2} v \sqrt{g'^2 + g^2}$) in addition to interactions between these particles and the Higgs field H. One gauge boson, the photon, remains massless and has no tree-level interactions with the Higgs boson.

The final term in Eq. 2.12 gives a mass term for the Higgs boson ($m_H = \sqrt{2\lambda} v$), as well as interactions between the Higgs boson and itself. This Higgs boson self-coupling is a parameter of great interest, and experimental searches for the Higgs self-interaction are underway [20, 21]. However, unless this coupling is enhanced by Beyond the Standard Model physics, there is no sensitivity to the Standard Model Higgs boson self-interaction at the LHC during Run 2 or even Run 3 [22, 23].

2.2.2 Fermion Masses and Interactions

The Lagrangian for a fermion field ψ can be written

$$\mathcal{L}_\psi = \bar{\psi} i \not{\mathcal{D}} \psi = \bar{\psi} i \gamma^\mu \mathcal{D}_\mu \psi \tag{2.13}$$

where the co-variant derivative $\mathcal{D}_\mu$ is that defined in Eq. 2.6. This can be broken into separate terms for the left- and right-handed components of the field:

$$\bar{\psi} i \not{D} \psi = \bar{\psi}_L i \not{D} \psi_L + \bar{\psi}_R i \not{D} \psi_R \tag{2.14}$$

Since the W boson couples only to the left-handed components of a fermion field, left-handed fermions are represented as doublets in the $SU(2)_L$ gauge group, while right-handed fermions are represented as singlets. Thus, right-handed fermions have weak isospin $I_3 = 0$ and hypercharge $Y = Q$, and left-handed fermions have $I_3 = \frac{1}{2}$ and $Y = Q \pm \frac{1}{2}$.

The Standard Model left-handed lepton doublets are

$$E_L^{(1)} = \begin{pmatrix} \nu_{eL} \\ e_L \end{pmatrix} \; ; \; E_L^{(2)} = \begin{pmatrix} \nu_{\mu L} \\ \mu_L \end{pmatrix} \; ; \; E_L^{(3)} = \begin{pmatrix} \nu_{\tau L} \\ \tau_L \end{pmatrix} \tag{2.15}$$

and the left-handed quark doublets are

$$Q_L^{(1)} = \begin{pmatrix} u_L \\ d_L \end{pmatrix} \; ; \; Q_L^{(2)} = \begin{pmatrix} c_L \\ s_L \end{pmatrix} \; ; \; Q_L^{(3)} = \begin{pmatrix} t_L \\ b_L \end{pmatrix} \tag{2.16}$$

The Lagrangian for a single generation of fermions (for simplicity) can be written

$$\mathcal{L} = \bar{E}_L i \not{D} E_L + \bar{e}_R i \not{D} e_R + \bar{Q}_L i \not{D} Q_L + \bar{u}_R i \not{D} u_R + \bar{d}_R i \not{D} d_R \tag{2.17}$$

Expansion of the covariant derivative gives kinetic terms for each fermion, as well as interactions between the fermions and gauge bosons.

While the mass terms for the W and Z bosons result directly from the Higgs mechanism, no fermion mass terms are explicitly required by electroweak symmetry breaking. However, mass terms can be naturally added with the form

$$\mathcal{L}_e = \lambda_e \bar{E}_{Li} \phi e_{Rj} \tag{2.18}$$

where ϕ is the Higgs doublet and λ_e is the (real-valued) electron Yukawa coupling. Continuing with the first generation leptons as an example, this becomes:

$$\mathcal{L}_e = \frac{-\lambda_e}{\sqrt{2}} \left[(\bar{\nu}_{eL}, \bar{e}_L) \begin{pmatrix} 0 \\ v + H \end{pmatrix} e_R + \bar{e}_R \, (0, v + H) \begin{pmatrix} \nu_{eL} \\ e_L \end{pmatrix} \right] \tag{2.19}$$

This yields both a mass term for the electron and a Higgs boson coupling to the electron:

$$\mathcal{L}_e = \frac{-v\lambda_e}{\sqrt{2}} \bar{e}_L e_R + \frac{-\lambda_e}{\sqrt{2}} H \bar{e}_L e_R \tag{2.20}$$

The electron mass obtained is $m_e = \lambda_e v/\sqrt{2}$, and the strength of the Higgs boson coupling is proportional to m_e. The same procedure can be used to generate mass terms and Higgs boson coupling terms for other charged leptons and down-type quarks. The gauge-invariant transformation $\phi \to -i\sigma_2\phi^*$ gives Yukawa terms of a slightly different form, which generate mass terms for up-type quarks (and Dirac neutrinos):

$$\mathcal{L}_u = -i\lambda_u \bar{Q}_L \sigma_2 \phi^* u_R \tag{2.21}$$

In the case of the top quark, which is the primary interest of this work, the mass and coupling terms are completely analogous to the case of the electron:

$$\mathcal{L}_t = \frac{-v\lambda_t}{\sqrt{2}} \bar{t}_L t_R + \frac{-\lambda_t}{\sqrt{2}} H \bar{t}_L t_R \tag{2.22}$$

with the top mass proportional to the Yukawa coupling via $m_t = \lambda_t v/\sqrt{2}$. Thus for the top quark (and for all fermions), the mass is not predicted directly by the Standard Model. Instead, the Standard Model provides an explicit prediction of the relationship between the mass of fermion f and its Yukawa coupling λ_f.

2.3 Higgs Boson Physics

It was demonstrated in the previous Section that particles with higher mass have stronger couplings to the Higgs boson. This has been confirmed by all measurements of Higgs boson couplings to date [24, 25]. Higgs boson production is therefore dominated by interactions between the Higgs boson and the heaviest Standard Model particles: the top quark, W boson, and Z boson.

2.3.1 Production Mechanisms

A Higgs boson can be produced in many ways in proton–proton collisions. Cross sections of the dominant Higgs boson production processes are shown in Fig. 2.5 as a function of proton–proton center of mass energy. The best available calculation of the cross sections at $\sqrt{s} = 13$ TeV is reported in Table 2.1. See [26] for Higgs boson production cross sections at lower center of mass energies.

In proton–proton collisions, the Higgs boson production cross section is dominated by gluon fusion through a quark loop (ggF, or ggH). In the Standard Model, this loop is dominated by the contribution from the top quark due to its high mass. In BSM models, massive particles that are charged under $SU(3)_{\text{color}}$ can also contribute to the rate of ggF production.

Sizable contributions to the total Higgs boson cross section also arise from vector boson fusion (VBF) and vector boson associated production, where a Higgs is radiated off of a W or Z boson ($qq/qg \rightarrow VH$ or simply VH, also known as Higgs-strahlung). Feynman diagrams of these three processes are shown in Fig. 2.6.

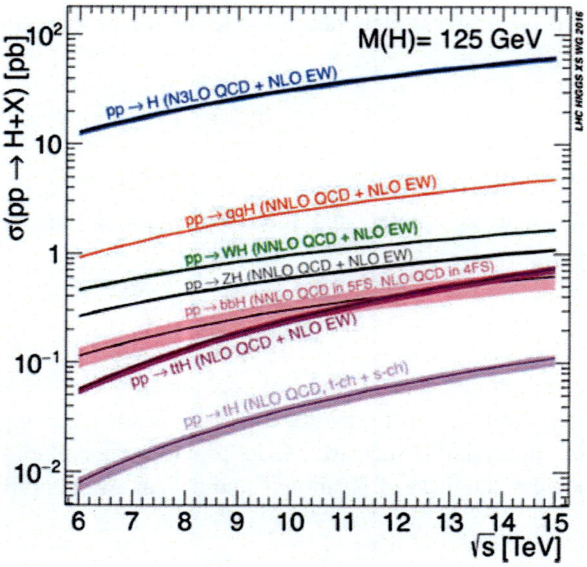

Fig. 2.5 Standard Model cross section as a function of proton–proton center of mass energy for different Higgs boson production modes [27]

Table 2.1 Most accurate available cross sections for Standard Model Higgs boson production, calculated at $m_H = 125.09$ GeV [27]

Process	$\sigma(\sqrt{s} = 13$ TeV) [pb]	Order of calculation
ggF	48.52	$\mathrm{N^3LO(QCD)}$ + NLO(EW)
VBF	3.779	NNLO(QCD) + NLO(EW)
WH	1.369	NNLO(QCD) + NLO(EW)
$qq/qg \rightarrow ZH$	0.7612	NNLO(QCD) + NLO(EW)
$gg \rightarrow ZH$	0.1227	NLO + NLL(QCD)
$t\bar{t}H$	0.5065	NLO(QCD) + NLO(EW)
$b\bar{b}H$	0.4863	5FS NNLO or 4FS NLO(QCD)
$tHjb$	0.07426	4FS NLO(QCD)
tWH	0.01517	5FS NLO(QCD)

An additional small contribution to the ZH final state comes from gluon-initiated production ($gg \rightarrow ZH$). Feynman diagrams for this process are shown in Fig. 2.7.

The primary focus of this work is a rare process in which a Higgs boson is produced in association with a top-antitop quark pair ($t\bar{t}H$). The lowest order Feynman diagrams for $t\bar{t}H$ production are shown in Fig. 2.8. The $t\bar{t}H$ cross section comprises only 1% of the Standard Model Higgs boson cross section at $\sqrt{s} = 13$ TeV.

A Higgs boson is produced in association with a bottom-antibottom pair through diagrams analogous to those shown in Fig. 2.8 for $t\bar{t}H$. At $\sqrt{s} = 13$ TeV, the $b\bar{b}H$ cross section is similar to that of $t\bar{t}H$.

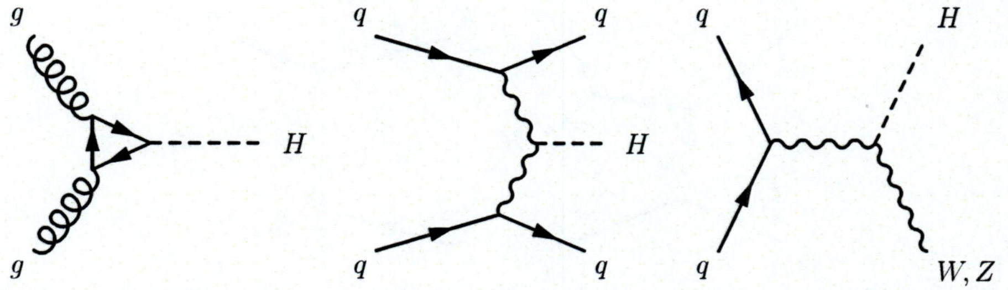

Fig. 2.6 Lowest order Feynman diagrams of the three Higgs boson production modes with highest cross section in 13 TeV proton–proton collisions. From left to right: gluon–gluon fusion, vector boson fusion, vector boson associated production

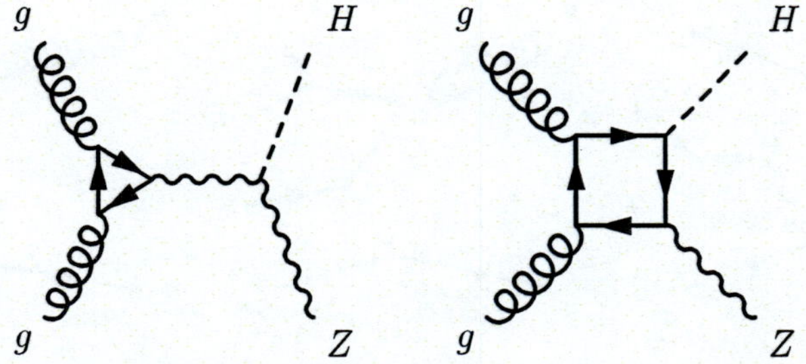

Fig. 2.7 Lowest order Feynman diagrams of the $gg \to ZH$ process

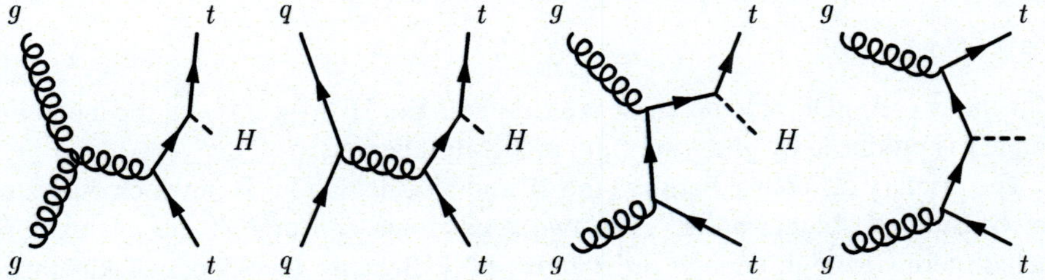

Fig. 2.8 Lowest order Feynman diagrams of the $t\bar{t}H$ process. Additional diagrams can be obtained by allowing a Higgs boson to radiate from the antitop, or by crossing the lines of initial gluons in t-channel diagrams

Higgs boson production in association with a single top quark (tH) is also predicted by the Standard Model, but with much lower cross section than $t\bar{t}H$ due to destructive interference between Higgs-top and Higgs-W diagrams. The sum of all tH processes is displayed in light purple in Fig. 2.5. Many models of BSM physics, including those discussed in Sect. 2.4.1, predict a large enhancement of tH cross sections: any observation of this process during LHC Run 2 would be a clear sign of new physics.

Figure 2.9 shows the lowest order diagrams for the tHj process as calculated five-flavor (5F) PDF scheme, which includes massless b quarks among the constituents of the proton. Figure 2.10 shows the leading diagrams for the tHj process in the

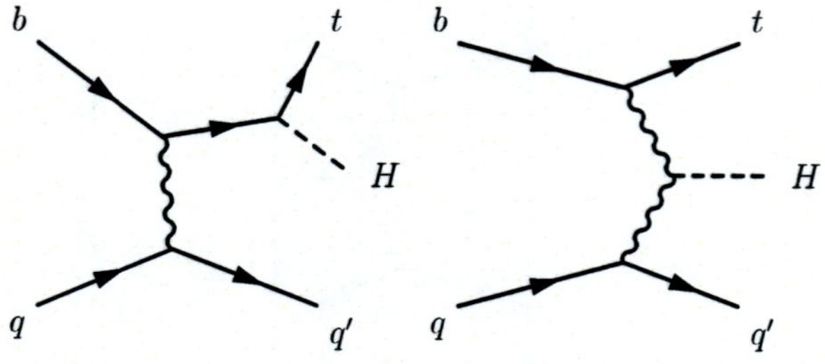

Fig. 2.9 Lowest order Feynman diagrams of the tHj process in the five-flavor PDF scheme. Additional diagrams can be obtained by reversing the fermion lines, or by crossing the lines of initial particles

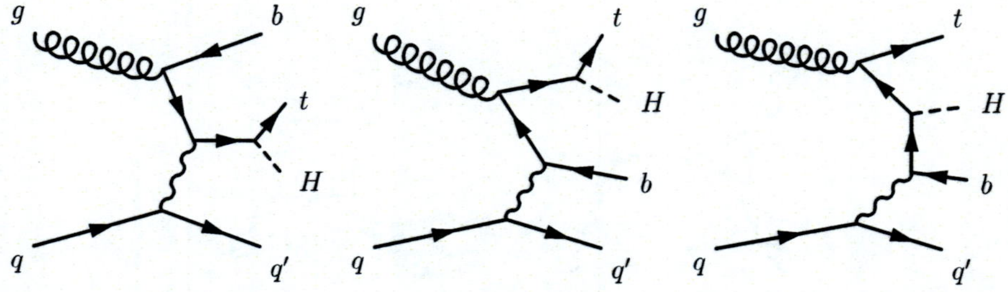

Fig. 2.10 Lowest order Feynman diagrams of the $tHjb$ process in a four-flavor PDF scheme. Additional diagrams can be obtained by allowing a Higgs boson to radiate from the W boson, by reversing the fermion lines, or by crossing the lines of initial particles

four-flavor (4F) PDF scheme: these are similar to the 5F diagrams, but the incoming b quark is produced by gluon splitting in the 4F scheme.

The primary difference between the 4F and 5F calculations is in which diagrams are considered "higher-order." The two predictions converge if the calculations include terms to high enough order in α_S [28]. Because the shapes of kinematic variables such as Higgs boson p_T and bottom quark p_T are described with smaller uncertainties in the 4F scheme, the tHj process is calculated in the 4F scheme and referred to as $tHjb$ for the remainder of this work. The small contribution from quark-initiated s-channel tHj production is neglected.

The leading diagrams for tWH production are shown in Fig. 2.11. At leading order, this process only occurs in the 5F PDF scheme. At next-to-leading order (NLO), there is interference between the tWH and $t\bar{t}H$ processes [29], as demonstrated in Fig. 2.12. The treatment of this interference in simulation is discussed in Chap. 5.

2.3.2 Decay Channels

Higgs decays to massive gauge bosons and fermions occur at tree level, as shown in Fig. 2.13. Higgs decays to massless particles, including a pair of gluons or photons, occur through loop diagrams and are suppressed accordingly.

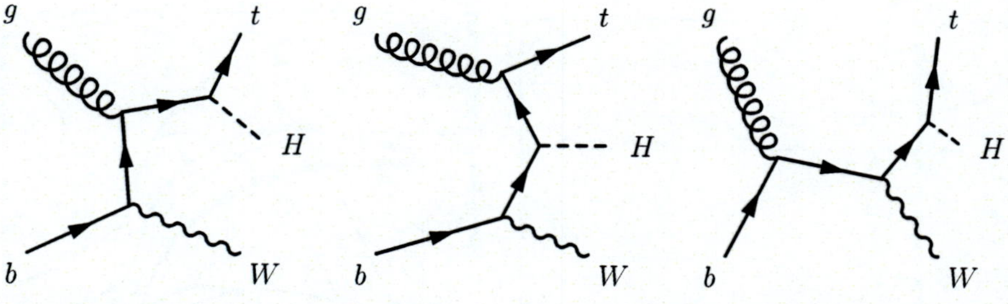

Fig. 2.11 Lowest order Feynman diagrams of the tWH process. These diagrams can only occur in a five-flavor PDF scheme. Additional diagrams can be obtained by allowing a Higgs boson to radiate from the W boson, by reversing the fermion lines, or by crossing the lines of initial particles in t-channel diagrams

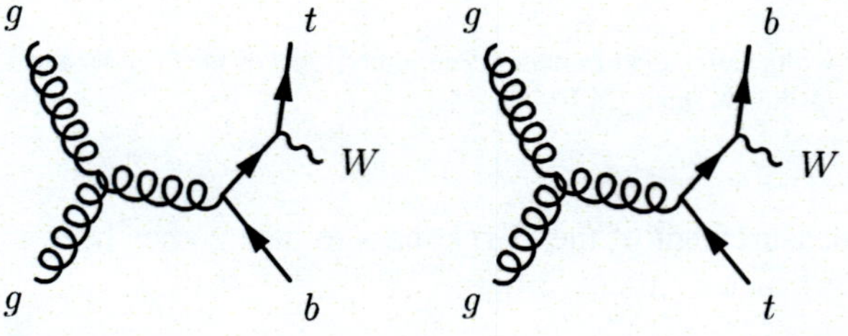

Fig. 2.12 NLO Feynman diagrams of tW production (left) and $t\bar{t}$ production (right) that produce interference. These are extended to the tWH and $t\bar{t}H$ case by adding Higgs-strahlung from the t or W lines

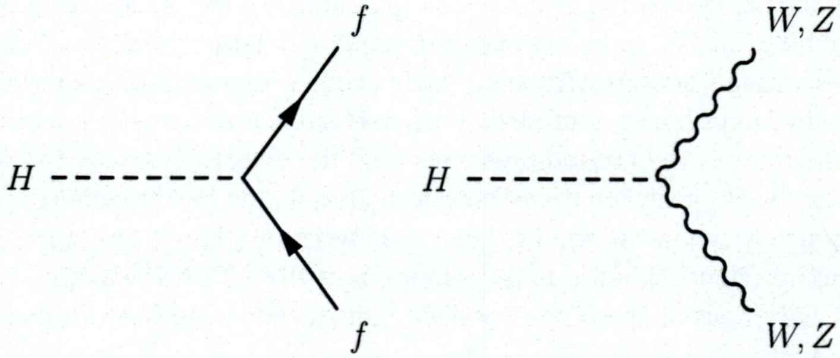

Fig. 2.13 Tree-level diagrams of Higgs boson decays to fermions (left) and vector bosons (right)

The total width of the SM Higgs boson Γ_H is shown in Fig. 2.14 as a function of the Higgs boson mass m_H. Figure 2.14 also shows the branching fractions B of the SM Higgs boson as a function of the Higgs boson mass, where

$$B(H \to X) = \frac{\Gamma(H \to X)}{\Gamma_H} \tag{2.23}$$

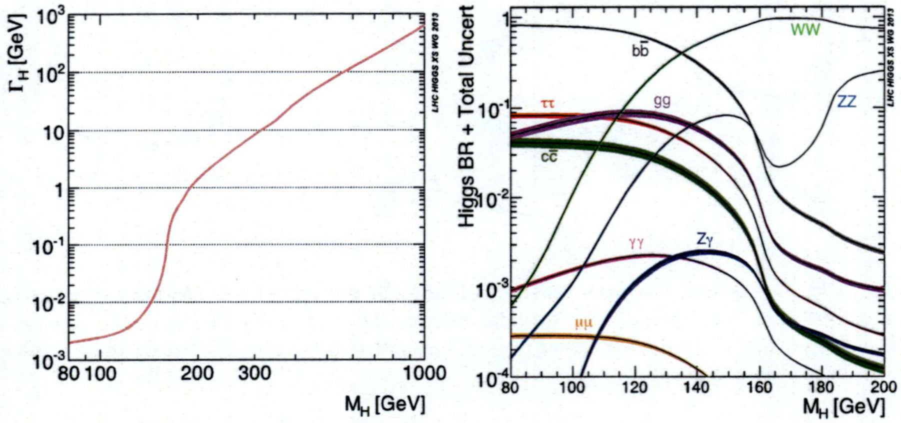

Fig. 2.14 Total width (left) and branching fractions (right) of the Standard Model Higgs boson as a function of Higgs boson mass [27]

The best measurement of the Higgs mass to date comes from the ATLAS/CMS combined Run 1 measurement [30]:

$$m_H = 125.09 \pm 0.21\,(\text{stat}) \pm 0.11\,(\text{syst})\,\text{GeV} \tag{2.24}$$

Table 2.2 lists the branching fractions of the SM Higgs boson at this value of m_H. Note that the decay of a Higgs boson to top quarks is energetically forbidden.

The number of Higgs boson events produced is larger in decay channels with higher branching fractions. However, when seeking to maximize experimental sensitivity to the Higgs boson, there are additional factors that must be considered. These include the rate of background processes, and the detector's ability to identify and reconstruct the Higgs boson decay products. Despite the low branching fraction, the $H \rightarrow \gamma\gamma$ decay channel is among the most powerful for Higgs boson measurements due to the excellent diphoton mass resolution of the ATLAS detector. The bulk of this work will focus on the $H \rightarrow \gamma\gamma$ decay, and its advantages are further discussed in later Chapters.

The lowest order diagrams contributing to the $H \rightarrow \gamma\gamma$ decay are shown in Fig. 2.15. In the Standard Model, this loop is dominated by the Higgs coupling to the W boson, and the decay through a quark loop provides some small sensitivity to the top Yukawa coupling. Destructive interference between the Higgs-top and Higgs-W diagrams suppresses the rate. In BSM models, massive particles that are electrically charged can also contribute to the rate of the $H \rightarrow \gamma\gamma$ decay.

Table 2.2 Branching fractions of the Standard Model Higgs boson [27]

Decay mode	Branching fraction (%)
$H \to bb$	57.5 ± 1.9
$H \to WW$	21.6 ± 0.9
$H \to gg$	8.56 ± 0.86
$H \to \tau\tau$	6.30 ± 0.36
$H \to cc$	2.90 ± 0.35
$H \to ZZ$	2.67 ± 0.11
$H \to \gamma\gamma$	0.228 ± 0.011
$H \to Z\gamma$	0.155 ± 0.014
$H \to \mu\mu$	0.022 ± 0.001

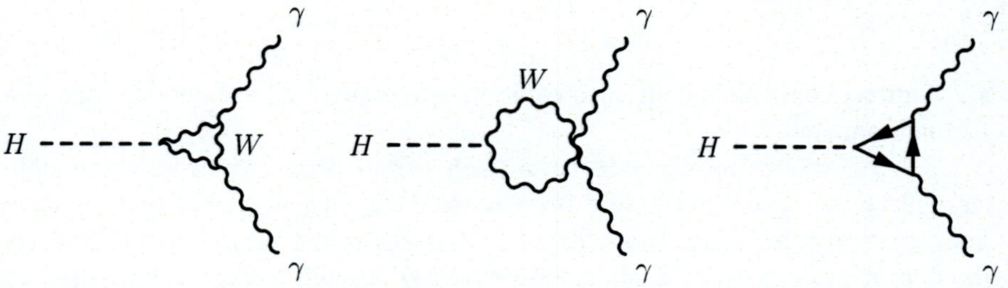

Fig. 2.15 Lowest order Feynman diagrams contributing to the $H \to \gamma\gamma$ decay. The fermion loop in the right-most diagram is dominated by the contribution from the top quark

2.3.3 *Coupling Measurement Framework*

When measuring Higgs boson couplings, it is convenient to refer to the ratio to the Standard Model prediction rather than the absolute coupling:

$$\kappa_i = \frac{\lambda_i^{\text{meas}}}{\lambda_i^{SM}} \tag{2.25}$$

for particle type i. This κ parameter is called the **coupling strength**, and measurements of κ that differ from one indicate deviations from the Standard Model prediction. This convention is known as the κ framework [31].

Table 2.3 shows the SM Higgs boson production cross sections calculated in terms of the coupling strengths κ at $\sqrt{s} = 13$ TeV (see [32] for comparable parameterizations at lower center of mass energies). The cross section of $t\bar{t}H$ production is dependent only on the top coupling strength κ_t, and therefore provides a direct probe of the top Yukawa coupling. The cross sections of $tHjb$ and tWH production depend on both the Higgs-top (κ_t) and Higgs-W (κ_W) couplings. The cross term ($\sim \kappa_t \kappa_W$) represents the interference, which provides sensitivity to the relative sign of κ_t and

Table 2.3 Production modes of the Standard Model Higgs boson in 13 TeV proton–proton collisions and their parameterization in the κ framework [27]

Production mode	Effective modifier	Resolved modifier
$\sigma(ggF)$	κ_g^2	$1.04\kappa_t^2 + 0.002\kappa_b^2 - 0.04\kappa_t\kappa_b$
$\sigma(VBF)$	–	$0.73\kappa_W^2 + 0.27\kappa_Z^2$
$\sigma(qq/qg \to WH)$	–	κ_W^2
$\sigma(qq/qg \to ZH)$	–	κ_Z^2
$\sigma(gg \to ZH)$	–	$2.46\kappa_Z^2 + 0.46\kappa_t^2 - 1.90\kappa_Z\kappa_t$
$\sigma(t\bar{t}H)$	–	κ_t^2
$\sigma(b\bar{b}H)$	–	κ_b^2
$\sigma(tHjb)$	–	$2.63\kappa_t^2 + 3.58\kappa_W^2 - 5.21\kappa_t\kappa_W$
$\sigma(tWH)$	–	$2.91\kappa_t^2 + 2.31\kappa_W^2 - 4.22\kappa_t\kappa_W$

κ_W. Indirect constraints on κ_t also come from top quark contributions to the gluon fusion loop.

Table 2.2 shows the dependence of each partial decay width on the coupling strengths κ at $m_H = 125.09$ GeV. Because the decay of a Higgs boson to top quarks is energetically forbidden, none of the SM Higgs boson decay modes provides a direct probe of κ_t. However, Higgs boson decays through loop processes, including $H \to \gamma\gamma$, have some indirect sensitivity to κ_t.

The resolved coupling modifiers in Tables 2.3 and 2.4 are calculated assuming the Standard Model. However, the presence of new particles in loop processes would introduce additional terms to these modifiers. In order to account for potential new physics in loop processes, **effective coupling strengths** are often preferred.

For loop-induced production processes such as gluon fusion, the effective coupling strength is defined by taking the ratio of the measured to Standard Model cross section:

$$\kappa_g^2 = \frac{\sigma_{ggF}^{\text{meas}}}{\sigma_{ggF}^{SM}} \tag{2.26}$$

The effective coupling strengths for loop decays are defined in an analogous way from the decay width:

$$\kappa_\gamma^2 = \frac{\Gamma_{\gamma\gamma}^{\text{meas}}}{\Gamma_{\gamma\gamma}^{SM}} \tag{2.27}$$

Because these effective couplings are not explicitly dependent on each particle contributing to the loop, the effective parameterization is more model independent than the resolved parameterization. Depending on the assumptions made in a given statistical analysis, resolved and effective loop couplings can both be useful. See Chap. 12 for an example of an analysis that uses both.

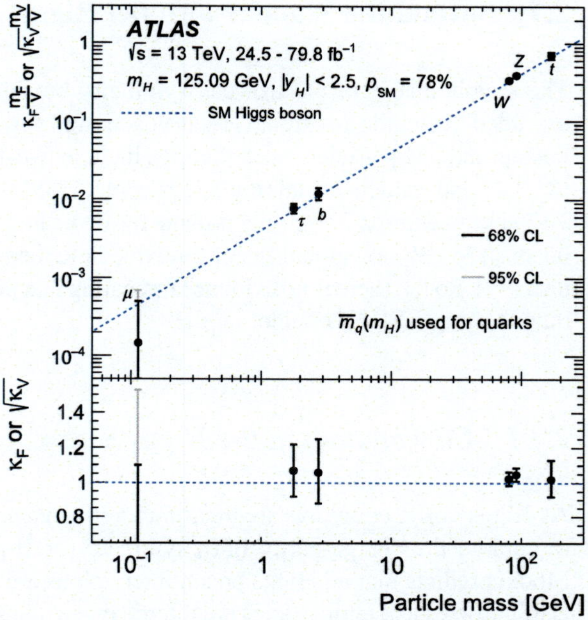

Fig. 2.16 Predicted and measured values of Higgs boson couplings to Standard Model particles and their relationship to particle mass [24]. Note that the vertical axis differs for bosons and fermions

Table 2.4 Decay modes of the Standard Model Higgs boson and their parameterization in the κ framework [27]

Decay mode	Effective modifier	Resolved modifier
$H \to bb$	–	κ_b^2
$H \to WW$	–	κ_W^2
$H \to gg$	κ_g^2	$1.11\kappa_t^2 + 0.01\kappa_b^2 - 0.12\,\kappa_b\kappa_t$
$H \to \tau\tau$	–	κ_τ^2
$H \to cc$	–	κ_c^2
$H \to ZZ$	–	κ_Z^2
$H \to \gamma\gamma$	κ_γ^2	$1.59\kappa_W^2 + 0.07\,\kappa_t^2 - 0.67\,\kappa_W\kappa_t$
$H \to Z\gamma$	$\kappa_{Z\gamma}^2$	$1.12\kappa_W^2 - 0.12\,\kappa_W\kappa_t$
$H \to \mu\mu$	–	κ_μ^2

The κ framework is widely used for measurements of Higgs boson couplings. Figure 2.16 shows the measured value of each κ parameter as a function of particle mass. Note that the vertical axis of this Figure differs for bosons and fermions. No significant deviation from the Standard Model expectation (represented as a dashed blue line) has yet been observed.

2.4 Beyond the Standard Model Higgs

The simple theory of electroweak symmetry breaking outlined in Sect. 2.2 can be extended to include additional Higgs bosons and new interactions between the Higgs bosons and SM particles. Such theories include two-Higgs-doublet models (2HDM) [33, 34] and models containing Higgs triplets [35].

These scenarios are being probed by dedicated searches for additional Higgs bosons [36–38] and global fits of effective field theory (EFT) parameters to collider data [39]. So far, there is no evidence indicating that the Higgs sector is more complex than the single SU(2) doublet model.

2.4.1 CP Violation in the Higgs Sector

As Higgs boson couplings are measured with greater precision, it becomes possible to explore the charge-parity (CP) symmetry of Higgs interactions. The Standard Model predicts that all Higgs boson interactions are CP-conserving (CP even): that is, if particles and antiparticles are interchanged and spatial coordinates are inverted, all Higgs boson interactions are unchanged. The measurement of even a small CP-violating component in any Higgs boson coupling would constitute a discovery of physics beyond the Standard Model and would potentially address one of the key open questions in particle physics today: the understanding of the matter-antimatter imbalance in the Universe.

Measurements of Higgs couplings to vector bosons have constrained any CP violating component to be small [40–42]. However, CP-violating contributions to these bosonic couplings can only enter via higher order operators that are suppressed by powers of $1/\Lambda^2$, where Λ is the energy scale of new physics in the EFT.

The CP properties of Higgs-fermion interactions remain largely unexplored, and CP violating terms can enter at the same order as the CP even terms of the SM. These measurements are therefore expected to give higher sensitivity than measurements of Higgs couplings to vector bosons.

As the largest Yukawa coupling, the Higgs-top coupling is a natural choice for testing CP properties. Measurements of electric dipole moments have provided indirect constraints on the CP properties of the Higgs-top interaction [43], though these rely on assumptions of SM Yukawa couplings for light fermions such as the electron. In addition, hadron collider measurements of ggF production and $H \to \gamma\gamma$ decay at $\sqrt{s} = 8$ TeV have provided some indirect constraints [44], but again the results rely on an assumption of SM contributions to these loops. With LHC data at $\sqrt{s} =$ 13 TeV, the CP properties of the Higgs-top interaction can be probed directly using the $t\bar{t}H$ and tH processes.

The Higgs Characterization (HC) model [45] provides a framework for describing different couplings, including scalar, pseudo-scalar, or some combination thereof. Within this model, the effective Lagrangian that describes the top Yukawa coupling can be expressed as:

$$\mathcal{L} = -\frac{m_t}{v}(\kappa_t \cos(\alpha)\bar{t}t + i\kappa_t \sin(\alpha)\bar{t}\gamma_5 t)H \tag{2.28}$$

where α is the CP mixing angle. The Standard Model coupling corresponds to $\alpha = 0°$ and $\kappa_t = 1$, while a fully CP odd coupling is realized when $\alpha = 90°$. In this parameterization, κ_t is restricted to be ≥ 0 and α is in the range $(-180°, 180°)$.

The dependence of the $t\bar{t}H$ and tH cross sections on the model parameters κ_t and α can be calculated from the Lagrangian in Eq. 2.28. The $t\bar{t}H$ cross section takes the form

$$\sigma_{t\bar{t}H} = A\kappa_t^2 c_\alpha^2 + B\kappa_t^2 s_\alpha^2 + E\kappa_t^2 s_\alpha c_\alpha \tag{2.29}$$

where s_α and c_α represent the sine and cosine of α, respectively. The coefficient A corresponds to the CP even contribution to the cross section, and B corresponds to the CP odd contribution. The interference between CP even and CP odd is represented by coefficient E, which is expected to be small.

For $tHjb$ and tWH, the interference between the top quark and W boson couplings must also be taken into account. The cross section of both $tHjb$ and tWH take the same form (though with different values of the coefficients):

$$\sigma_{tH} = A\kappa_t^2 c_\alpha^2 + B\kappa_t^2 s_\alpha^2 + C\kappa_t c_\alpha + D\kappa_t s_\alpha + E\kappa_t^2 s_\alpha c_\alpha + F \tag{2.30}$$

The coefficients A, B, and E have the same meanings as in the $t\bar{t}H$ case. The Higgs coupling to the W boson is assumed to obey the Standard Model ($\kappa_W = 1$). The coefficient C corresponds to the interference between the CP even top-Higgs and W-Higgs couplings, D to the interference between the CP odd top-Higgs and W-Higgs couplings, and F to the contribution of the pure W-Higgs coupling.

Using the HC model and its implementation in the MG5_aMC@NLO Monte Carlo generator [46, 47], the cross sections for $t\bar{t}H$, $tHjb$, and tWH are calculated at multiple values of κ_t and α. The NLO inclusive cross sections calculated by MG5_aMC@NLO are shown in Table 2.5. These cross sections (normalized to the Standard Model prediction) are shown in Fig. 2.17 as a function of α with κ_t fixed to unity. The parameterizations of the loop processes ggH (κ_g) and $H \to \gamma\gamma$ (κ_γ) in terms of α are taken from [44].

The cross section of $t\bar{t}H$ decreases as the CP odd component becomes larger and is minimized at $\alpha = 90°$. However, the Standard Model point represents a global minimum of the cross section for both tH processes: the rate can only increase with changes in α. Note that at $\alpha = 180°$, the sign of the interference term between the top-Higgs and W-Higgs couplings is flipped with respect to the SM. This switch from destructive to constructive interference accounts for the large change in tH rate at $\alpha = 180°$.

Table 2.5 NLO inclusive cross sections from MG5_aMC@NLO for the $t\bar{t}H$, $tHjb$, and tWH processes in different CP scenarios

Parameters		$\sigma(\sqrt{s} = 13$ TeV) [pb]		
κ_t	α	$t\bar{t}H$	$tHjb$	tWH
1	0° (SM)	0.458	0.0606	0.0167
1	15°	0.443	0.0636	0.0184
1	30°	0.396	0.0743	0.0231
1	45°	0.329	0.0958	0.0308
1	60°	0.265	0.1374	0.0422
1	75°	0.217	0.1970	0.0563
1	90°	0.199	0.2707	0.0726
−1	0°	0.458	0.6971	0.1490
0.5	0°	0.1145	0.0939	0.0148
2	0°	1.832	0.2406	0.0924
2	45°	–	0.2055	0.1095

For $\kappa_t = 2$ and $\alpha = 45°$, no $t\bar{t}H$ cross section is calculated

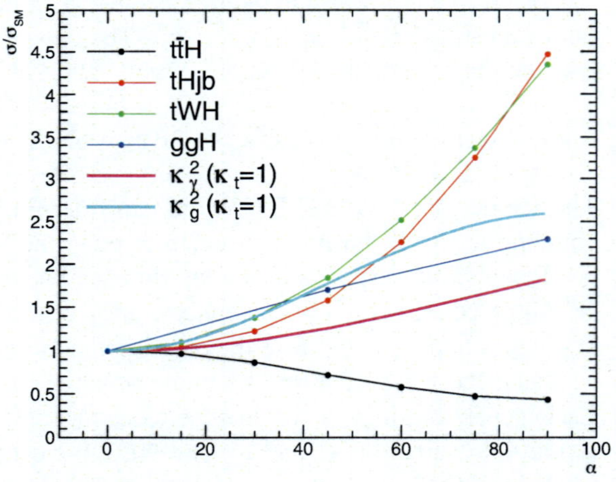

Fig. 2.17 NLO $t\bar{t}H$ and tH cross sections as a function of the CP mixing angle α (for $\kappa_t = 1$), normalized to the Standard Model prediction. The dependence of κ_g and κ_γ on α is taken from [44]

In addition to its impact on overall cross sections, the introduction of a CP odd component to the Higgs-top coupling also alters the shapes of sensitive kinematic observables in $t\bar{t}H$ and tH. Such observables are discussed at length in Chapter 11, which presents a multivariate analysis strategy that exploits the shape dependence of kinematic distributions on the CP mixing angle α.

References

1. Mele S (2006) Measurements of the running of the electromagnetic coupling at LEP'. In: Proceedings, 26th International symposium on physics in collision (PIC 2006): Buzios, Brazil, July 6-9, 2006 arXiv: hep-ex/0610037 [hep-ex]. http://www.slac.stanford.edu/econf/C060706/pdf/0610037.pdf
2. Tanabashi M et al (2018) Review of particle physics'. Phys. Rev. D 98(3):030001. https://doi.org/10.1103/PhysRevD.98.030001. https://link.aps.org/doi/10.1103/PhysRevD.98.03000
3. Collins JC, Soper DE, Sterman GF (1989) Factorization of hard processes in QCD. 5:1–91. https://doi.org/10.1142/9789814503266_0001. arXiv: hep-ph/0409313
4. Buckley A et al (2011) General-purpose event generators for LHC physics. Phys Rept 504:145–233. https://doi.org/10.1016/j.physrep.2011.03.005. arXiv: 1101.2599 [hep-ph]
5. Keith Ellis R, James Stirling W, Webber BR (1996) QCD and collider physics. Camb Monogr Part Phys Nucl Phys Cosmol 8:1–435
6. Buckley A et al (2015) LHAPDF6: parton density access in the LHC precision era. Eur Phys J C75:132. https://doi.org/10.1140/epjc/s10052-015-3318-8. arXiv: 1412.7420 [hep-ph]
7. Harland-Lang LA et al (2015) Parton distributions in the LHC era: MMHT 2014 PDFs. Eur Phys J C75(5):204. https://doi.org/10.1140/epjc/s10052-015-3397-6. arXiv: 1412.3989 [hep-ph]
8. Corcella G et al (2000) HERWIG 6: an event generator for Hadron emission reactions with interfering Gluons (including supersymmetric processes). In: JHEP 01.hep- ph/0011363. CAVENDISH-HEP-99-03. CERN-TH-2000-284. RAL-TR-2000-048, 010. 91 p. https://doi.org/10.1088/1126-6708/2001/01/010. https://cds.cern.ch/record/478828
9. Catani Stefano et al (2002) The Dipole formalism for next-to-leading order QCD calculations with massive partons. Nucl Phys B 627:189–265. https://doi.org/10.1016/S0550-3213(02)00098-6 arXiv: hep-ph/0201036
10. Alwall J et al (2008) Comparative study of various algorithms for the merging of parton showers and matrix elements in hadronic collisions. Eur Phys J C 53:473–500. https://doi.org/10.1140/epjc/s10052-007-0490-5. arXiv: 0706.2569 [hep-ph]
11. Webber BR (1994) Hadronization. In: hep-ph/9411384. CAVENDISH-HEP-94-17. https://cds.cern.ch/record/272660
12. Andersson Bo et al (1983) Parton fragmentation and string dynamics. Phys Rept 97:31–145. https://doi.org/10.1016/0370-1573(83)90080-7
13. Field Richard D, Wolfram Stephen (1983) A QCD model for e+ e- annihilation. Nucl Phys B 213:65–84. https://doi.org/10.1016/0550-3213(83)90175-X
14. ATLAS Pythia 8 tunes to 7 TeV datas. Tech. rep. ATL-PHYS-PUB-2014-021. Geneva: CERN, 2014. https://cds.cern.ch/record/1966419
15. Glashow SL (1980) Towards a unified theory: threads in a tapestry. Rev Mod Phys 52:539–543. https://doi.org/10.1103/RevModPhys.52.539
16. Weinberg Steven (1980) Conceptual foundations of the unified theory of weak and electromagnetic interactions. Rev Mod Phys 52:515–523. https://doi.org/10.1103/RevModPhys.52.515
17. Salam Abdus (1980) Gauge unification of fundamental forces. Rev Mod Phys 52:525–538. https://doi.org/10.1103/RevModPhys.52.525
18. Higgs PW (1964) Broken symmetries and the masses of gauge Bosons. Phys Rev Lett 13(16):508–509. https://doi.org/10.1103/PhysRevLett.13.508. https://link.aps.org/doi/10.1103/PhysRevLett.13.508
19. Englert F, Brout R (1964) Broken symmetry and the mass of gauge vector mesons. Phys Rev Lett 13(9): 321–323. https://doi.org/10.1103/PhysRevLett.13.321. https://link.aps.org/doi/10.1103/PhysRevLett.13.321
20. Aad G et al (2020) Combination of searches for Higgs boson pairs in *pp* collisions at $\sqrt{s} = 13$ TeV with the ATLAS detector. Phys Lett B 800:135103. https://doi.org/10.1016/j.physletb.2019.135103. arXiv: 1906.02025 [hep-ex]

21. Sirunyan AM et al (2019) Combination of searches for Higgs boson pair production in proton-proton collisions at $\sqrt{s} = 13$ TeV". Phys Rev Lett 122(12):121803. https://doi.org/10.1103/PhysRevLett.122.121803. arXiv: 1811.09689 [hep-ex]
22. Measurement prospects of the pair production and self-coupling of the Higgs boson with the ATLAS experiment at the HL-LHC. Tech. rep. ATL-PHYS-PUB-2018-053. Geneva: CERN, 2018. https://cds.cern.ch/record/2652727
23. Sensitivity projections for Higgs boson properties measurements at the HL-LHC. Tech. rep. CMS-PAS-FTR-18-011. Geneva: CERN, 2018. https://cds.cern.ch/record/2647699
24. Aad G et al (2020) Combined measurements of Higgs boson production and decay using up to 80 fb^{-1} of proton-proton collision data at $\sqrt{s} = 13$ TeV collected with the ATLAS experiment". In: Phys. Rev. D 101(1):012002. https://doi.org/10.1103/PhysRevD.101.012002. arXiv: 1909.02845 [hep-ex]
25. Combined Higgs boson production and decay measurements with up to 137 fb^{-1} of proton-proton collision data at $\sqrt{s} = 13$ TeV. Tech. rep. CMS-PAS-HIG-19-005. Geneva: CERN, 2020. https://cds.cern.ch/record/2706103
26. Andersen JR et al (2013) Handbook of LHC Higgs cross sections: 3. Higgs properties". In: Heinemeyer S et al (ed) https://doi.org/10.5170/CERN-2013-004. arXiv: 1307.1347 [hep-ph]
27. LHC Higgs Cross Section Working Group. Handbook of LHC Higgs Cross Sections: 4. Deciphering the Nature of the Higgs Sector. Tech. rep. FERMILAB/CERN, 2017. https://cds.cern.ch/record/2227475
28. Demartin F et al (2015) Higgs production in association with a single top quark at the LHC. Eur Phys J C75(6):267. https://doi.org/10.1140/epjc/s10052-015-3475-9. arXiv: 1504.00611 [hep-ph]
29. Demartin F et al (2017) tWH associated production at the LHC. Eur Phys J C77(1):34. https://doi.org/10.1140/epjc/s10052-017-4601-7. arXiv: 1607.05862 [hep-ph]
30. Aad G et al (2015) Combined measurement of the Higgs Boson mass in pp Collisions at vs = 7 and 8 TeV with the ATLAS and CMS experiments. Phys Rev Lett 114:191803. https://doi.org/10.1103/PhysRevLett.114.191803. arXiv: 1503.07589 [hep-ex]
31. David A et al (2012) LHC HXSWG interim recommendations to explore the coupling structure of a Higgs-like particle. arXiv: 1209.0040 [hep-ph]
32. Aad G et al (2016) Measurements of the Higgs boson production and decay rates and constraints on its couplings from a combined ATLAS and CMS analysis of the LHC pp collision data at $\sqrt{s}$ =7 and 8 TeV. JHEP 08:045. https://doi.org/10.1007/JHEP08(2016)045. arXiv: 1606.02266 [hep-ex]
33. Branco GC et al (2012) Theory and phenomenology of two-Higgs-doublet models. Phys Rept 516:1–102. https://doi.org/10.1016/j.physrep.2012.02.002. arXiv: 1106.0034 [hep-ph]
34. Fayet P (1977) Spontaneously broken supersymmetric theories of weak, electromagnetic and strong interactions. Phys Lett B 69: 489. https://doi.org/10.1016/0370-2693(77)90852-8
35. Arhrib A et al (2011) The Higgs potential in the Type II seesaw model. Phys Rev D84:095005. https://doi.org/10.1103/PhysRevD.84.095005. arXiv: 1105.1925 [hep-ph]
36. Aad G et al (2020) Search for heavy Higgs bosons decaying into two tau leptons with the ATLAS detector using pp collisions at $\sqrt{s}$ =13 TeV. arXiv: 2002.12223 [hep-ex]
37. Aaboud M et al (2019) Search for doubly charged scalar bosons decaying into same-sign W boson pairs with the ATLAS detector. Eur Phys J C79(1):58. https://doi.org/10.1140/epjc/s10052-018-6500-y. arXiv: 1808.01899 [hep-ex]
38. Aaboud M et al (2018) Search for charged Higgs bosons decaying into top and bottom quarks at $\sqrt{s} = 13$ TeV with the ATLAS detector. JHEP 11:085. https://doi.org/10.1007/JHEP11(2018)085. arXiv: 1808.03599 [hep-ex]
39. Constraints on an effective Lagrangian from the combined $H \to ZZ* \to 4\ell and H \to \gamma\gamma$ channels using 36.1 fb^{-1} of $\sqrt{s} = 13$ TeV pp collision data collected with the ATLAS detector. Tech. rep. ATL-PHYS-PUB-2017-018. Geneva: CERN, 2017. https://cds.cern.ch/record/2293084
40. Aad G et al (2016) Test of CP invariance in vector-boson fusion production of the Higgs boson using the optimal observable method in the ditau decay channel with the ATLAS detector.

Eur Phys J C76(12):658. https://doi.org/10.1140/epjc/s10052-016-4499-5. arXiv: 1602.04516 [hep-ex]
41. Khachatryan V et al (2016) Combined search for anomalous pseudoscalar HVV couplings in VH(H $\rightarrow b\bar{b}$) production and H $\rightarrow$ VV decay. Phys Lett B759:672–696. https://doi.org/10.1016/j.physletb.2016.06.004. arXiv: 1602.04305 [hep-ex]
42. CMS Collaboration (2019) Measurements of the Higgs boson width and anomalous HVV couplings from on-shell and off-shell production in the four-lepton final state. arXiv: 1901.00174 [hep-ex]
43. Brod J, Haisch U, Zupan J (2013) Constraints on CP-violating Higgs couplings to the third generation. JHEP 11:180. https://doi.org/10.1007/JHEP11(2013)180. arXiv: 1310.1385 [hep-ph]
44. Ellis J et al (2014) Disentangling Higgs-top couplings in associated production. JHEP 04: 004. https://doi.org/10.1007/JHEP04(2014)004. arXiv: 1312.5736 [hep-ph]
45. Demartin F et al (2014) Higgs characterisation at NLO in QCD: CP properties of the top-quark Yukawa interaction. Eur Phys J C74(9):3065. https://doi.org/10.1140/epjc/s10052-014-3065-2. arXiv: 1407.5089 [hep-ph]
46. Alwall J et al (2014) The automated computation of tree-level and next-to-leading order differential cross sections, and their matching to parton shower simulations. JHEP 07:079. https://doi.org/10.1007/JHEP07(2014)079. arXiv: 1405.0301 [hep-ph]
47. Frixione S, Webber BR (2002) Matching NLO QCD computations and parton shower simulations. JHEP 06:029. arXiv: hep-ph/0204244

Chapter 3
The ATLAS Experiment

Experimental high energy physics in the twenty-first century requires collaboration on an unprecedented scale. The design and construction of the world's highest-energy collider, a complex detector the size of the White House, and novel technologies that withstand immense radiation are ambitious projects that necessitate the formation of large international collaborations.

The ATLAS collaboration is made up of over 3000 scientists from 38 nations, all of whom have contributed to the experimental setup described in this Chapter. The first Section describes the accelerator itself, the Large Hadron Collider (LHC). The second describes the layout and function of the general-purpose ATLAS detector and its subsystems.

3.1 The Large Hadron Collider

The **Large Hadron Collider (LHC)** [1, 2] is a high energy particle accelerator located at the *Organisation Européenne pour la Recherche Nucléaire* (CERN) outside of Geneva, Switzerland. The LHC occupies a circular tunnel with a circumference of 27 km, spanning the border between Switzerland and neighboring France. The LHC began collecting proton–proton collision data in the fall of 2009. During the first run of the LHC (2009–2012), protons were collided with center of mass energy $\sqrt{s} = 7$ and 8 TeV. During the second run of the LHC (2015–2018), the center of mass energy was increased to $\sqrt{s} = 13$ TeV, the highest energy ever achieved by a collider.

J. E. Dickinson, *ATLAS Measurements of the Higgs Boson Coupling to the Top Quark in the Higgs to Diphoton Decay Channel*, Springer Theses,
https://doi.org/10.1007/978-3-030-86368-5_3

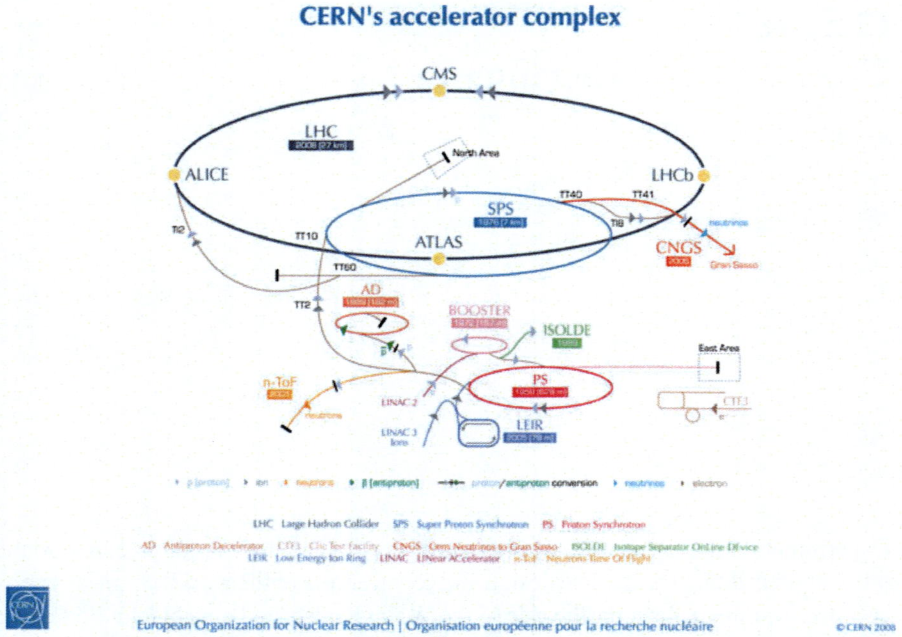

Fig. 3.1 Diagram of the CERN collider complex [8]

LHC protons originate from a bottle of hydrogen gas. Once the electrons have been stripped away using an electric field, Linear Accelerator 2 (LINAC2) [3] accelerates the protons using radiofrequency (RF) cavities, which use oscillating electromagnetic fields to accelerate protons to an energy of 50 MeV. The timing of a proton's arrival at a given point in the RF cavity determines how much it will be accelerated (or decelerated): this results in a regular pattern of distinct **proton bunches** [4].

From LINAC2, the protons are injected into three successive synchrotron systems: the Synchrotron Booster (BOOSTER) [5], which accelerates the protons to 1.4 GeV; the Proton Synchrotron (PS) [6], which accelerates the protons to 25 GeV; and the Super Proton Synchrotron (SPS) [7], which accelerates the protons to 450 GeV. Finally, the protons are injected into the LHC. The layout of the CERN collider complex, including each of these injectors, is shown in Fig. 3.1.

The LHC itself is composed of eight straight segments, each about half a kilometer long, where RF cavities accelerate protons to 6.5 TeV in bunches separated by 25 ns (50 ns in Run 1). The straight segments are connected by eight rounded arcs, which each contain three superconducting dipole magnets (for bending the particle trajectories) and one superconducting quadrupole magnet (for focusing) [1]. Two beams of protons circulate through the LHC in opposite directions, passing through twin bores in the dipole magnets. The beams cross at four points: the locations of the ATLAS [9], CMS [10], LHCb [11], and ALICE [12] experiments.

3.1.1 Luminosity

For a given process P, the expected rate of interactions at the LHC is given by

$$\frac{dN_P}{dt} = \sigma_P \times \mathcal{L} \tag{3.1}$$

where σ_P is the cross section for process P, and $\mathcal{L}$ is the **instantaneous luminosity** delivered by the machine. For two beams of identical Gaussian profile colliding head-on, the instantaneous luminosity is

$$\mathcal{L} = \frac{n_p^2 n_b f}{4\pi \sigma_x \sigma_y} \tag{3.2}$$

where n_p is the number of protons per bunch, n_b is the number of bunches, f is the revolution frequency, and σ_x and σ_y are the beam resolutions in the plane transverse to the beam. Additional corrections are required to account for beams crossing at slight angles, correlations between σ_x and σ_y, and other effects [13]. The ATLAS detector was designed for a maximum instantaneous luminosity of 10^{34} cm^{-2}s^{-1}.

The total number of interactions of process P produced at the LHC during time T is

$$N_P = \sigma_P \times \int_0^T \mathcal{L} dt' = \sigma_P \times \mathcal{L}_{\text{int}} \tag{3.3}$$

The **integrated luminosity** $\mathcal{L}_{\text{int}}$ is typically measured in **inverse femtobarns (fb^{-1})**. The integrated luminosity delivered to ATLAS during LHC Run 2 was measured by the LUCID-2 detector, which is located around the LHC beam pipe on either side of the ATLAS detector [14].

3.2 The ATLAS Detector

The ATLAS detector is a general-purpose particle detector located 100 m underground on the LHC ring. The cylindrical detector geometry uses the LHC beam pipe as its axis, and the forward-backward symmetry gives nearly 4π coverage in solid angle. Figure 3.2 shows a diagram of the ATLAS detector, which extends 44 m long and 25 m high. It is estimated that the detector (including both sensitive area and structural components) weighs 7000 tons.

The detector is described using a right-handed coordinate system with origin corresponding to the proton–proton collision point at the center of the detector. The direction of the incoming proton beams defines the z-axis, the x-axis points towards the center of the LHC ring, and the y-axis points upwards.

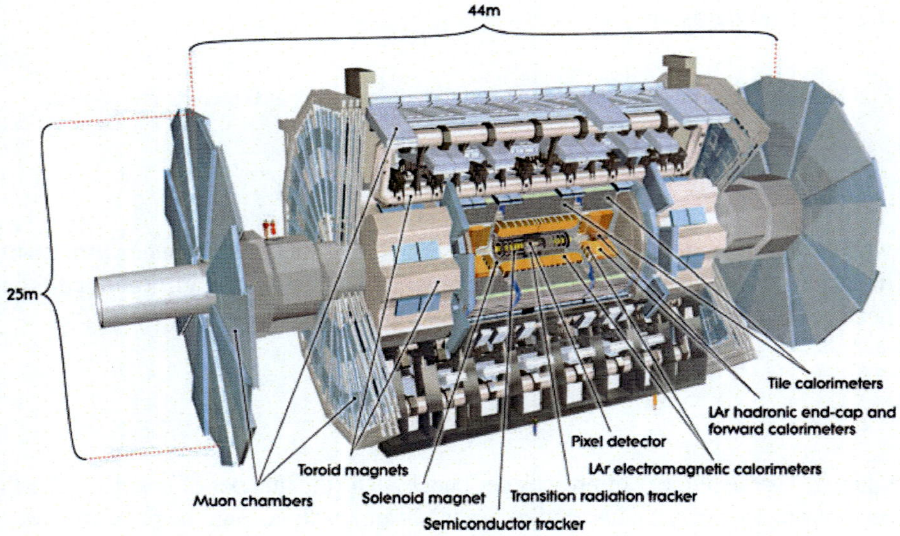

Fig. 3.2 Diagram of the ATLAS detector [15]

Due to the azimuthal symmetry of the detector, cylindrical coordinates r and ϕ (azimuthal angle around the z-axis) provide a convenient description. Each particle's momentum is projected onto the transverse plane to obtain the **transverse momentum**:

$$p_T \equiv \sqrt{p_x^2 + p_y^2} \tag{3.4}$$

The **pseudorapidity** is defined in terms of the polar angle θ:

$$\eta \equiv -\ln\left[\tan\left(\frac{\theta}{2}\right)\right] \tag{3.5}$$

For a particle with mass much smaller than its momentum ($m \ll |\mathbf{p}|$), the pseudorapidity approaches the rapidity y. In this limit, pseudorapidity differences are invariant under Lorentz boosts in the $\hat{z}$ direction. Angular distance is measured in terms of $\Delta R = \sqrt{(\Delta\eta)^2 + (\Delta\phi)^2}$.

The following sections detail the design and purpose of each detector system, from inner- to outermost: the Inner Detector, the Calorimeters, and the Muon Spectrometer. Each system is composed of a **barrel** region, with layers arranged on concentric cylinders around the z-axis, and a disk-shaped **endcap**, with layers oriented perpendicular to the beam axis at high $|z|$.

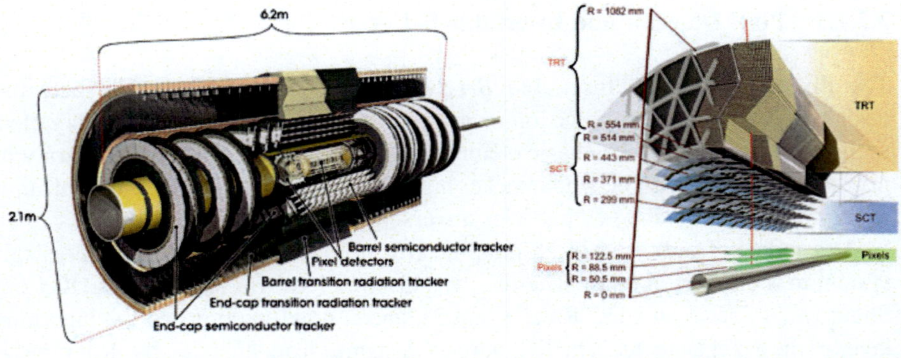

(a) Inner Detector (Full) (b) Inner Detector (Barrel)

Fig. 3.3 Diagram of the ATLAS Inner Detector layout [16]. (Left) the full Inner Detector, including end-caps. (Right) cross section of the barrel region of the ID. The Pixel Detector is shown in green, the SCT in blue, and the TRT in orange. The IBL is missing from this figure, but is positioned at $r = 33.25$ mm

Table 3.1 Extent of the sensitive region of each ATLAS inner detector subsystem [9]

System	Barrel			End-cap		
	Layers	r (mm)	$\|z\|$ (mm)	Disk layers	r (mm)	$\|z\|$ (mm)
IBL	1	33.25	0–330.15	–	–	–
Pixel	3	50.5–122.5	0–400.5	2×5	88.8–149.6	495–650
SCT	4	299–514	0–749	2×9	275–560	810–2797
TRT	73	563–1066	0–712	160	644–1004	848–2710

3.2.1 Inner Detector

The ATLAS Inner Detector (ID) [16] is composed of the subsystems closest to the collision point, which provide precise measurements of particle position and momentum needed for tracking. The subsystems include (from inner- to outermost) the Insertable B-Layer (IBL), the Pixel Detector, the Semiconductor Tracker (SCT), and the Transition Radiation Tracker (TRT).

A superconducting solenoid magnet encloses the full ID and immerses all subsystems in a 2 T magnetic field oriented in the $\hat{z}$ direction. The resulting curvature of charged particle trajectories in the transverse plane provides measurements of particle momenta.

Figure 3.3a shows the full layout of the Inner Detector (including end-caps), and Fig. 3.3b shows the radial position of each barrel layer. The IBL is missing from this figure, but is positioned at $r = 33.25$ mm. The extent of each ID subsystem is summarized in Table 3.1: all subsystems have coverage in the range $|\eta| < 2.5$.

3.2.1.1 Pixel Detector and Insertable B-Layer

The Pixel detector and Insertable B-Layer (IBL) are highly granular detectors designed to provide high-precision measurements of particle trajectories very close to the interaction point. The basic element of these detectors is a silicon sensor with pixelated readout, or **pixel**, which makes high-resolution measurements of ionization energy deposited by charged particles traversing the sensor.

At a radius of only $r = 33.25$ mm, the IBL [17, 18] is the closest ATLAS subsystem to the collision point. The IBL was inserted into ATLAS during LHC Long Shutdown 1 (between LHC Runs 1 and 2), and it now functions as the innermost layer of the Pixel Detector. The IBL derives its names from the critical role it plays in b-tagging, which relies heavily on identification secondary vertices from the decay of B-hadrons (see Sect. 4.5.1). Due to its proximity to the interaction point, the IBL is among the ATLAS subsystems most susceptible to radiation damage. See Appendix C for studies of radiation effects in the IBL.

The Pixel Detector [19] is composed of three cylindrical barrel layers (not including the IBL) and ten disk layers (five on each side) of silicon pixels. The barrel layers are named the B-layer, Layer 1, and Layer 2, with the B-layer being the innermost (like the IBL, named for its important role in b-tagging). Each barrel layer is tiled with pixels for two-dimensional segmentation in the $\hat{\phi} \times \hat{z}$ plane. Each pixel module is composed of 61,440 individual pixels and 16 readout chips. The resolution of a pixel is 10 μm in $r\phi$ and 115 μm in z in the barrel layers, and 10 μm in $r\phi$ and 115 μm in r in the endcap disks.

3.2.1.2 Semiconductor Tracker

Like the IBL and Pixel detector, the Semiconductor Tracker (SCT) [20] relies on silicon semiconductor technology. Each silicon microstrip detector is composed of 768 readout strips, each with a width of 80 μm. In the barrel region, SCT modules are formed from four single-sided silicon microstrip detectors: two are wire-bonded end to end, forming strips that are 12 cm long. Two such pairs are then positioned back-to-back, with one side oriented parallel to the beam direction and the other offset from the z axis by an angle of 40 mrad. This small angle offset allows for measurement of the z coordinate. The SCT barrel is composed of four layers of these modules.

In the endcap, SCT modules are similarly constructed from strip detectors positioned back-to-back with an angular offset of 40 mrad. However, the strips in the endcap are oriented radially and taper to smaller pitch at the origin. Each endcap consists of nine disk layers.

The SCT has $r\phi$ resolution of 17 μm in the barrel and endcap regions. The z coordinate is calculated from the $r\phi$ and stereo measurements (details are available in [21]). The resolution in z is 580 μm in the barrel, and the resolution on r is 580 μm in the endcap.

3.2.1.3 Transition Radiation Tracker

The Transition Radiation Tracker (TRT) [22] is composed of straw drift tubes filled with xenon-based gas and a central cathode wire. In the barrel region, the TRT straws are positioned parallel to the beam axis and extend to $|\eta| < 0.7$. In the end-cap wheels, TRT straws are positioned radially from the z axis. Each straw is 4 mm in diameter and provides 130 μm resolution in $r\phi$ (but none in z).

The TRT straws allow for continuous tracking, typically providing about 36 space points per charged particle track. The large number of space points compensates for the low resolution compared to the silicon detectors, and TRT inputs make a significant contribution to measurements of charged particle curvature (and hence momentum). Precise measurements in this outer region of the ID are also useful for identifying photon conversions.

The TRT derives its name from its ability to measure transition-radiation photons. The amount of transition radiation measured in the TRT provides discrimination between electrons and charged hadrons (pions).

3.2.2 Calorimeters

The ATLAS calorimeters [23–26] serve to measure both electromagnetic showers and hadronic jets. The electromagnetic showers spawned by photons and electrons are typically contained in the innermost calorimeter, the liquid argon electromagnetic calorimeter (ECAL). Because hadrons travel through more material than electrons or photons, the hadronic calorimeters make up a larger portion of ATLAS' volume than the ECAL: these are the Tile Calorimeter, Hadronic Endcap Calorimeters (HECs), and Forward Calorimeters (FCAL).

Different materials are used in each calorimeter in order to remain robust against radiation damage in all η regions. The layout of the ATLAS calorimeter systems is shown in Fig. 3.4. The number of layers and η coverage of each subsystem are presented Table 3.2.

3.2.2.1 Electromagnetic Calorimeter

The liquid argon Electromagnetic Calorimeter (ECAL) [23, 24] is a sampling calorimeter composed of alternating layers of lead (absorber) and liquid argon (active material). The lead and liquid argon are layered in an accordion structure in order to provide full coverage in ϕ. A pre-sampler detector is located just upstream of the ECAL in order to provide corrections for losses due to material in the ID.

In the barrel region, the ECAL extends from 1.5 m $< r <$ 1.97 m, which corresponds to 24 radiation lengths (X_0) at $\eta = 0$. The ECAL barrel is divided into three layers at different radii. In the central region, the layers are segmented into projective $\eta \times \phi$ regions of sizes 0.003×0.1, 0.025×0.025, and 0.5×0.025 in

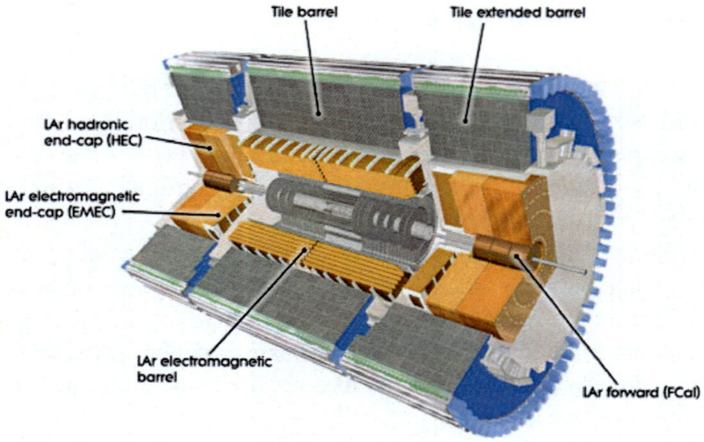

Fig. 3.4 Diagram of the layout of the ATLAS calorimeter systems [16]

Table 3.2 Extent of each ATLAS calorimeter system [9]

System	Barrel		End-cap	
	Layers	$\|\eta\|$ coverage	Layers	$\|\eta\|$ coverage
Presampler	1	< 1.52	1	1.5–1.8
ECAL	3 (2)	< 1.35 (1.35–1.475)	2 × 2 (3)	1.375–3.2 (1.5–2.5)
Tile	3	< 1.0	–	–
Tile Extended	2 × 3	0.8–1.7	–	–
HEC	–	–	2 × 4	1.5–3.2
FCAL	–	–	2 × 3	3.1–4.9

the first, second, and third ECAL layer respectively. The segmentation of the ECAL barrel becomes coarser for $|\eta| > 1.4$. Due to its fine granularity in η, the first ECAL layer is also referred to as the ECAL strips layer. It provides critical input for the determination of photon vertices and the identification of π^0 hadrons (see Sect. 4.2). Figure 3.5 shows a diagram of the ECAL layout in the barrel and the segmentation of each layer.

Each ECAL endcap is composed of two concentric wheels centered at and perpendicular to the z-axis. The total endcap extends radially from 302 mm $< r <$ 2077 mm and longitudinally from 3.7 m $< |z| <$ 4.2 m. The height and angle of the accordion folds in the endcap vary with radius in order to ensure uniform coverage in ϕ.

Though the ECAL provides full coverage in the transition between barrel and endcap ($1.37 < |\eta| < 1.52$, or the "crack"), the large density of material upstream from the ECAL in this region leads to degraded performance. Photon and electron candidates in this region are vetoed.

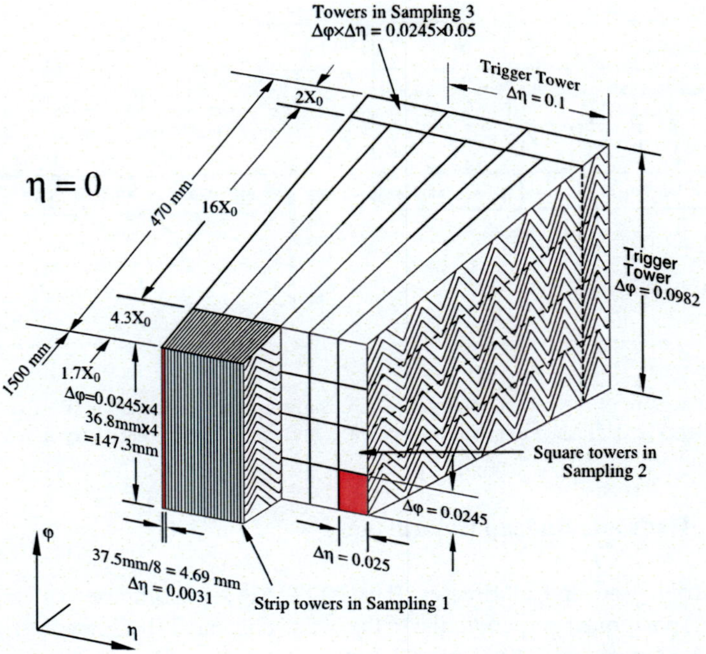

Fig. 3.5 Diagram of a barrel segment of the ATLAS electromagnetic calorimeter [16]

3.2.2.2 Tile Calorimeter

The hadronic Tile Calorimeter [25, 26] is a collection of sampling calorimeters using alternating layers of 3 mm-thick plastic scintillator tiles (active material) and 14 mm-thick iron plates (absorber). The calorimeters are the Tile Barrel ($|z| < 2.82$ m), which is divided into two halves, and the Tile Extended Barrels ($3.2\,\text{m} < |z| < 6.11\,\text{m}$). Each half of the Tile Barrel and each Tile Extended Barrel is composed of 64 modules positioned symmetrically in ϕ. Each module is composed of readout cells with the divisions in η giving a projective geometry centered at the detector origin (see Fig. 3.6). The readout cells are connected by wavelength-shifting fibers to photomultiplier tubes (PMTs) located at outer radial edge of the module, where the scintillation light is collected.

For good resolution of high energy jets, the Tile Calorimeter must capture the full radial extent of each hadronic shower to avoid overflow, or punch through, of jet activity into the muon system. The Tile Calorimeter extends from $r = 2.28$ m to $r = 4.25$ m, which corresponds to 9.7 nuclear interaction lengths (λ) at $\eta = 0$.

Calibration of the Tile Calorimeter is performed using laser-based methods, using ^{137}Cs source currents, or by injecting a pulse of known magnitude into the readout electronics. Details on these calibration techniques are available in [26].

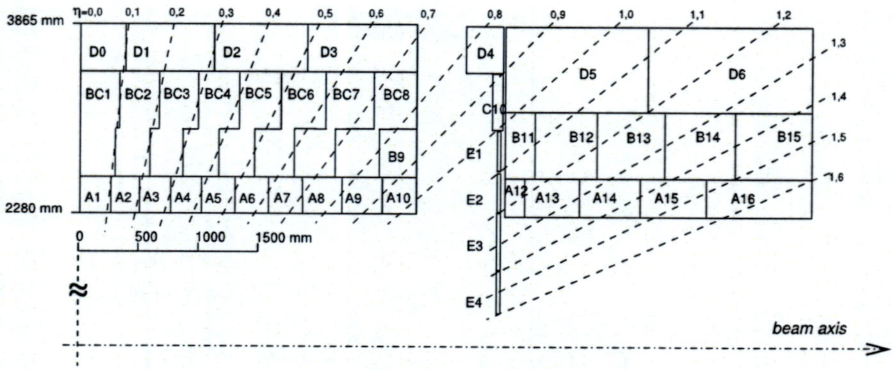

Fig. 3.6 Diagram of the cell geometry of a Tile Barrel module (left) and a Tile Extended Barrel module (right) [26]. Dashed lines show constant η, and the projective geometry in η is evident

3.2.2.3 Hadronic Endcap Calorimeters

The Hadronic Endcap Calorimeters (HECs) [23, 24] are composed of copper plates (absorber) positioned perpendicular to the beam axis. Each HEC extends from 4.26 m $< |z| <$ 6.12 m and radially from 475 mm $< r <$ 2.03 m. The gaps between copper plates are filled with liquid argon (active material), and three electrodes divide each endcap into four sampling layers at different $|z|$. In the two layers closest to the detector origin, the copper plates are 25 mm thick, compared to 50 mm thick in the outer two layers.

3.2.2.4 Forward Calorimeters

The Forward Calorimeters (FCAL) [23, 24] extend from 4.7 m $< |z| <$ 5.15 m and radially from 72 mm $< r <$ 455 mm, covering the range $3.1 < |\eta| < 4.9$. In this region, particle fluxes are high, and the FCAL is designed to be robust against radiation damage.

The FCAL consists of three layers on each side of the detector origin. The innermost layer on each side uses copper as the metal absorber, and the outer two use tungsten. The metal is used to create a matrix of grounded tubes with axes oriented parallel to the beam direction. Each tube contains a concentric cathode rod, and the gaps between each rod and the inner tube edge are filled with liquid argon (active material).

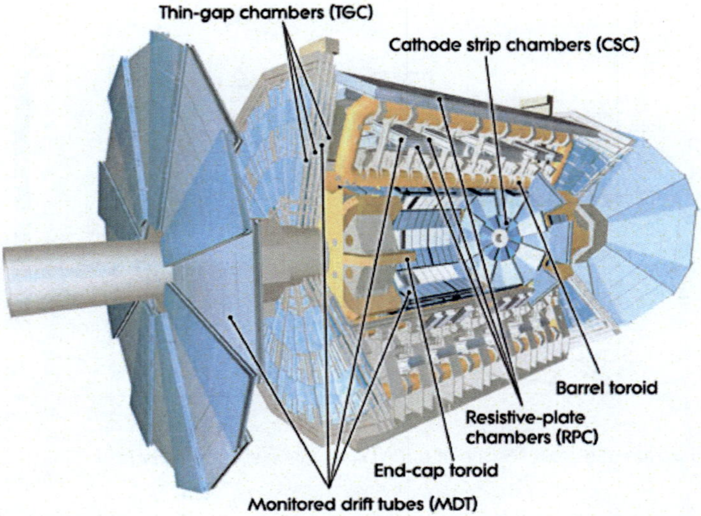

Fig. 3.7 Diagram of the ATLAS muon spectrometer [16]. The toroid magnets are shown in orange

3.2.3 *Muon Spectrometer*

The largest and outermost subsystem of ATLAS is the Muon Spectrometer (MS) [27, 28], shown in Fig. 3.7.

Three superconducting air-core toroid magnets, one barrel toroid and two endcap toroids, deflect charged particles towards the $\pm\hat{z}$ direction. Each toroid consists of 8 coils positioned at equal distance in ϕ. The magnetic field generated by the barrel toroid is dominant in the region $|\eta| < 1.4$, and the magnetic field generated by the endcap toroid is dominant in the region $1.6 < |\eta| < 2.7$. The region $1.4 < |\eta| < 1.6$, where both barrel and endcap toroids contribute to the magnetic field, is known as the transition region.

3.2.3.1 Monitored Drift Tubes

In most η regions, Monitored Drift Tubes (MDTs) dominate the precision measurement of muon tracks and momenta. A single MDT consists of a 15 mm-radius aluminum tube and a central cathode wire, with the region in between filled with argon gas.

In the barrel region, MDTs are positioned in three layers ("stations") and divided into sixteen azimuthal regions. The cylindrical barrel stations are centered around the z-axis at $r = 5$, 7.5, and 10 m. The innermost barrel station is composed of two stacks of four MDT monolayers each, separated by a support structure. The outer two barrel stations are structured similarly, but with only three MDT monolayers on either side of the support. The axis of each drift tube is oriented in the $\hat{\phi}$ direction,

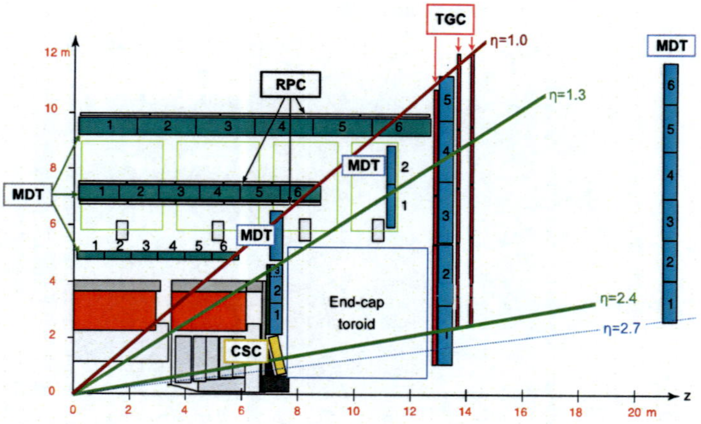

Fig. 3.8 Diagram of a cross section of the ATLAS muon spectrometer [29]

providing a projection of particle tracks in the bending plane of the magnetic field ($\hat{r} \times \hat{z}$). The resolution of a single MDT is 80 μm, but higher resolution is achieved by stacking monolayers of MDTs.

In each endcap, the MDTs are arranged in four disk-shaped stations centered at and perpendicular to the beam axis (see Fig. 3.8). The innermost station is composed of two stacks of four MDT monolayers, separated by a support structure. The other stations use only three MDT monolayers on either side of the support. Like the barrel, the endcap is divided into sixteen azimuthal regions and the axis of each drift tube is oriented in the $\hat{\phi}$ direction.

3.2.3.2 Cathode Strip Chambers

While the MDTs perform well in more central regions, their comparatively large radius and the high pressure requirements of argon gas make them ill-suited for regions of very high particle flux. Instead, Cathode Strip Chambers (CSCs) make precision measurements of muon momenta in the very forward region ($2 < |\eta| < 2.7$).

The CSCs are multi-wire proportional chambers, which measure particle trajectories by detecting charge induced on a segmented cathode readout strip by the avalanche formed on the anode wire. The wires are oriented in the $\hat{\phi}$ direction, providing a projection of particle tracks in the bending plane of the magnetic field ($\hat{r} \times \hat{z}$). The resolution of these measurements is about 50 μm.

Each CSC chamber is composed of two layers of four wire planes, separated by a support structure. The chambers are arranged in two pairs of rings at $z = \pm 7$ m, each divided into eight azimuthal segments. Each chamber is angled so that it is perpendicular to a straight particle trajectory starting at the detector origin. The CSCs are shown in yellow in Fig. 3.8.

3.2.3.3 Trigger Chambers

Additional subsystems of the muon spectrometer with lower resolution and faster readout are used for fast muon identification in the trigger system (Sect. 3.2.4): these include Resistive Plate Chambers (RPCs) in the barrel and Thin Gap Chambers (TGCs) in the endcaps. The location of these chambers is indicated in Fig. 3.8. As the studies in this Thesis do not rely on muon-based triggers, the description of these subsystems is left to [27].

3.2.4 Trigger

Due to the 25 ns spacing of LHC proton bunches, collisions in ATLAS occur at a rate of 40 MHz. An average event has a size of $\mathcal{O}(1\text{ MB})$ [30], and the resulting data rate is a whopping 1 TB/s. This is far too much data to store and process economically, and the signals from most ATLAS systems require longer than 25 ns to be fully read out. The ATLAS trigger system is responsible for making quick decisions about which collisions to read out and save for analysis, reducing the rate to < 1 kHz.

The trigger system is provided with a list of conditions, known as the **trigger menu**, that specify what combinations of physics objects are interesting enough to be saved. The trigger system first constructs a quick summary of the physics content of each event, then selects those events satisfying one or more of the criteria on the trigger menu. The first step of trigger selection, Level 1, is implemented directly in the detector hardware. The following steps, known as the High Level Trigger, are implemented in software.

The trigger used for analyses in this work requires two photons, with the leading (subleading) photon $p_T > 35$ (25) GeV. The performance of this trigger is reviewed in Chapter 5.

3.2.4.1 Level 1

The first step in the trigger, Level 1 (L1), is implemented at the hardware level and considers inputs from the muon and calorimeter systems. At L1, only regions of the detector where there is activity, called **Regions of Interest (ROI)**, are considered. This allows for processing of only a fraction of ATLAS' readout channels at L1.

To speed up calculations, L1 triggers do not use the full granularity of the calorimeters. Instead, energy deposits are summed over cells in projective regions called **calorimeter towers**, which have an angular size of $\Delta\eta \times \Delta\phi = 0.1 \times 0.1$ in the barrel region. Muon candidates are identified by coincidences in the dedicated Trigger Chambers, which have low resolution but fast readout compared to the MDTs and CSCs. A topological trigger, which calculates geometric or kinematic relationships between L1 trigger objects, was also added for LHC Run 2 [31].

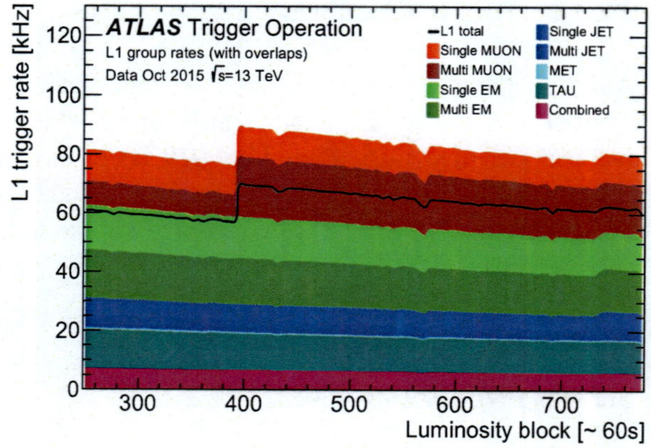

Fig. 3.9 Rate of collisions passing the Level 1 Trigger during a 13 TeV run in 2015 [31]. The total (black) is less than the stacked total due to overlap between different triggers

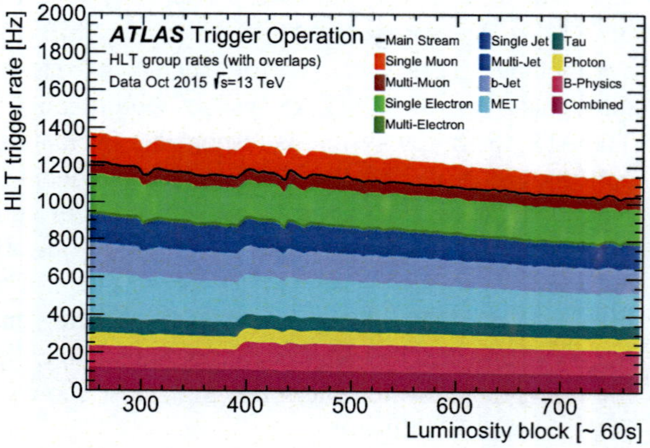

Fig. 3.10 Rate of collisions passing the High Level Trigger during a 13 TeV run in 2015 [31]. The total (black) is less than the stacked total due to overlap between different triggers

The rate of collisions passing L1 triggers is shown in Fig. 3.9 during a 13 TeV run in 2015. The rate is reduced to < 100 kHz after L1 selection. The diphoton trigger used in later Chapters falls under the "Multi EM" group (dark green).

3.2.4.2 High Level Trigger

The High Level Trigger (HLT) processes only those events passing L1 triggers. This system is software-based and uses information from all sub-detectors. Tracking information from the ID is added to each event, and information in the MDTs near any L1 muon candidates is also included. Energy deposits in the calorimeters are read out with the full detector granularity at HLT, and these deposits are clustered together using more complex algorithms [31].

The rate of collisions passing HLT triggers is shown in Fig. 3.10 during a 13 TeV run in 2015. The rate is reduced to < 1 kHz after HLT selection. The diphoton trigger used in later Chapters falls under the "Photon" group (yellow).

References

1. Pettersson TS, Lef'evre P (1995) The Large Hadron Collider: conceptual design. Tech. rep. CERN-AC-95-05-LHC. https://cds.cern.ch/record/291782
2. Sim Brüning O et al (2004) LHC design report. CERN yellow reports: monographs. Geneva: CERN. https://doi.org/10.5170/CERN-2004-003-V-1. https://cds.cern.ch/record/782076
3. Linear accelerator 2. In: (2012). https://cds.cern.ch/record/1997427
4. Radiofrequency cavities. In: (2012). https://cds.cern.ch/record/1997424
5. The Proton Synchrotron Booster. In: (2012). https://cds.cern.ch/record/1997372
6. The Proton Synchrotron. In: (2012). https://cds.cern.ch/record/1997189
7. The Super Proton Synchrotron. In: (2012). https://cds.cern.ch/record/1997188
8. Lefèvre C (2008) The CERN accelerator complex. Complexe des accélérateurs du CERN. https://cds.cern.ch/record/1260465
9. Aad G et al (2008) The ATLAS experiment at the CERN large hadron collider. JINST 3:S08003. https://doi.org/10.1088/1748-0221/3/08/S08003
10. Chatrchyan S et al (2008) The CMS experiment at the CERN LHC. JINST 3:S08004. https://doi.org/10.1088/1748-0221/3/08/S08004
11. Augusto AA Jr et al (2008) The LHCb detector at the LHC. JINST 3:S08005. https://doi.org/10.1088/1748-0221/3/08/S08005
12. Aamodt K et al (2008) The ALICE experiment at the CERN LHC. JINST 3:S08002. https://doi.org/10.1088/1748-0221/3/08/S08002
13. Herr W, Muratori B (2006) Concept of luminosity. https://doi.org/10.5170/CERN-2006-002.361. https://cds.cern.ch/record/941318
14. Avoni G et al (2018) The new LUCID-2 detector for luminosity measurement and monitoring in ATLAS. JINST 13(07):P07017. https://doi.org/10.1088/1748-0221/13/07/P07017
15. Pequenao J (2008) Computer generated image of the whole ATLAS detector. https://cds.cern.ch/record/1095924
16. Aad G et al (2010) The ATLAS inner detector commissioning and calibration. Eur Phys J C 70:787–821. https://doi.org/10.1140/epjc/s10052-010-1366-7. arXiv: 1004.5293 [physics.ins-det]
17. ATLAS Insertable B-Layer Technical Design Report. Tech. rep. 2010. https://cds.cern.ch/record/1291633
18. Abbott B et al (2018) Production and Integration of the ATLAS insertable B-layer. JINST 13(05):T05008. https://doi.org/10.1088/1748-0221/13/05/T05008. arXiv: 1803.00844 [physics.ins-det]
19. Aad G et al (2008) ATLAS pixel detector electronics and sensors. JINST 3:P07007. https://doi.org/10.1088/1748-0221/3/07/P07007
20. Ahmad A et al (2007) The Silicon microstrip sensors of the ATLAS semiconductor tracker. Nucl Instrum Meth A 578:98–118. https://doi.org/10.1016/j.nima.2007.04.157
21. ATLAS inner detector: Technical Design Report, 1. Technical Design Report ATLAS. Geneva: CERN, 1997. https://cds.cern.ch/record/331063
22. Akesson T et al (2004) Status of design and construction of the transition radiation tracker (TRT) for the ATLAS experiment at the LHC. Nucl Instrum Meth A 522:131–145. https://doi.org/10.1016/j.nima.2004.01.033
23. ATLAS liquid-argon calorimeter: Technical Design Report. Technical Design Report ATLAS. Geneva: CERN, 1996. https://cds.cern.ch/record/331061
24. Aad G et al (2010) Readiness of the ATLAS Liquid Argon Calorimeter for LHC Collisions. Eur Phys J C 70:723–753. https://doi.org/10.1140/epjc/s10052-010-1354-y. arXiv: 0912.2642 [physics.ins-det]
25. ATLAS tile calorimeter: Technical Design Report. Technical Design Report ATLAS. Geneva: CERN, 1996. https://cds.cern.ch/record/331062
26. Aad G et al (2010) Readiness of the ATLAS Tile Calorimeter for LHC collisions. Eur Phys J C 70:1193–1236. https://doi.org/10.1140/epjc/s10052-010-1508-y. arXiv: 1007.5423 [physics.ins-det]

27. ATLAS muon spectrometer: Technical Design Report. Technical Design Report AT- LAS. Geneva: CERN, 1997. https://cds.cern.ch/record/331068
28. Aad G et al (2010) Commissioning of the ATLAS Muon Spectrometer with Cosmic Rays. Eur Phys J C 70:875–916. https://doi.org/10.1140/epjc/s10052-010-1415-2. arXiv: 1006.4384 [physics.ins-det]
29. Aad G et al (2015) Performance of the ATLAS muon trigger in pp collisions at $\sqrt{s} = 8$ TeV. Eur Phys J C 75:120. https://doi.org/10.1140/epjc/s10052-015-3325-9. arXiv: 1408.3179 [hep-ex]
30. Aad G et al (1849) Performance of the ATLAS Trigger System in 2010. Eur Phys J C 72:1849. https://doi.org/10.1140/epjc/s10052-011-1849-1. arXiv: 1110.1530 [hep-ex]
31. Aaboud M et al (2017) Performance of the ATLAS Trigger System in 2015. Eur Phys J C 77(5):317. https://doi.org/10.1140/epjc/s10052-017-4852-3. arXiv: 1611.09661 [hep-ex]

Chapter 4
Object Reconstruction

Particles traversing the ATLAS detector are identified and reconstructed based on the combination of signatures they leave in each detector subsystem. Some example particle signatures are shown in Fig. 4.1.

The complex final state of $t\bar{t}H(\gamma\gamma)$ events can contain many types of physics objects: the decay of the Higgs boson produces a pair of photons, and the decay of the $t\bar{t}$ system can produce jets (including b-jets) and/or leptons.

The following Sections review the methods for constructing particle tracks and interaction vertices, as well as the reconstruction algorithms and requirements placed on photons, electrons, muons, hadronic jets, and missing transverse energy in ATLAS. Reconstructed tau leptons are not used in the analyses presented in this Thesis, and the description of tau reconstruction is left to [2].

The reconstruction of each type of physics object is subject to experimental uncertainties due to the detector geometry and resolution, as well as the object reconstruction algorithms. These uncertainties are treated as systematics, and they are fully accounted for in the statistical analyses performed in later Chapters. For a summary of experimental uncertainties, see Table 6.3.

4.1 Tracks and Vertices

Charged particle trajectories, or **tracks**, are reconstructed from the pattern of charge deposits ("hits") in the Inner Detector. The high granularity and resolution of the ID systems allows for accurate reconstruction of the position of each hit. The curvature of each track provides a measurement of the particle's charge and momentum with resolution $\Delta p_T / p_T$ that decreases with the p_T of the track [3]. The multi-stage algorithms used to construct tracks from large sets of ID hits are described in [4]. Tracks are required to have $p_T > 500$ MeV and $|\eta| < 2.5$.

J. E. Dickinson, *ATLAS Measurements of the Higgs Boson Coupling to the Top Quark in the Higgs to Diphoton Decay Channel*, Springer Theses,
https://doi.org/10.1007/978-3-030-86368-5_4

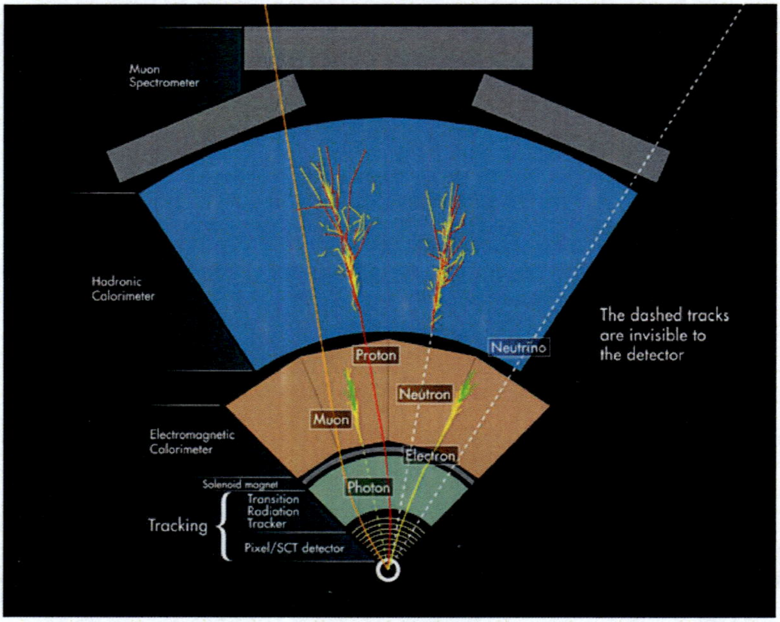

Fig. 4.1 Example signatures left in the ATLAS detector by different types of particles, shown in a cross sectional slice of the barrel [1]

Interaction vertices are are determined by an iterative fitting procedure [5]. A fit is performed over the selected tracks and a seed vertex to extract the best fit position of the vertex. Tracks that are incompatible with the calculated vertex are removed, and the fit is repeated. Tracks that are removed can be used to fit additional vertices. Vertex candidates with < 2 compatible tracks are discarded.

Each event can contain multiple vertices, and the number of vertices per event increases with pileup. It is common in ATLAS analyses to select the vertex with the maximum sum of track p_T^2 as the primary interaction vertex. However, $H \rightarrow \gamma\gamma$ events can have low charged particle multiplicity in the final state (e.g. gluon fusion), and in these cases it is insufficient to consider tracks alone. For $H \rightarrow \gamma\gamma$ measurements, the inputs to primary vertex selection must include not only track information from the ID, but also photon information from the ECAL. The primary vertex selection procedure in diphoton events is described in the following Section.

4.2 Photons

To reconstruct an electromagnetic shower, energy deposits in the calorimeters are collected into topological clusters, or topo-clusters [6]. These topo-clusters are constructed by grouping together calorimeter cells containing energy deposits that exceed the threshold for electronic noise and pileup. Topo-clusters that are associated with a secondary vertex in the ID are designated as converted photon candidates (which have undergone $\gamma \rightarrow e^+e^-$ in the ID volume). Topo-clusters that cannot be associated with any tracks or vertices are designated as unconverted photon candidates. Topo-clusters that can be associated with a single charged particle track are not designated as photon candidates: these are vetoed in order to resolve potential ambiguity between photons and electrons.

In order to separate **prompt** photons, which originate from the primary proton–proton interaction, from background photons, additional identification and isolation criteria are applied. Background photons can be secondary photons produced inside of jets, or hadrons with photon-like signatures (such as π^0 hadrons, which decay to a photon pair). In addition, for the analyses discussed in this Thesis, all photon candidates must have $p_T > 25$ GeV and $|\eta| < 2.37$, with the region $1.37 < |\eta| < 1.52$ (the "crack" containing dead material upstream of the ECAL) vetoed.

High efficiency identification and accurate reconstruction of photons in ATLAS are essential for measurements in the $H \rightarrow \gamma\gamma$ decay channel. The photon energy resolution is calculated using multivariate regression trained on simulated samples. The absolute energy scale is calibrated using $Z \rightarrow ee$ decays and validated on radiative Z decays [7].

4.2.1 Photon Identification

In order to reject background photons, cuts are placed on discriminating variables that are sensitive to differences between prompt and background photons [8]. These variables include:

- Hadronic leakage: energy deposited in the hadronic calorimeter near an ECAL cluster
- Shower shape variables based on the second ECAL layer
- Shower shape variables based on the ECAL strips layer

Because the ECAL strips layer has fine granularity in η, shower shape variables can be constructed that are sensitive to the two-pronged structure characteristic of π^0 hadron decay.

Cuts on the discriminating variables are optimized in bins of $|\eta|$ to create multiple photon identification working points. The loose photon identification criteria give higher photon acceptance but lower background rejection compared to the medium and tight working points. For the tight photon identification working point, the cut values are optimized in bins of E_T as well as $|\eta|$, and performed separately for converted and unconverted photon candidates. The analyses described in following Chapters will make use of photon candidates passing both loose and tight identification requirements.

4.2.2 Photon Isolation

Background photons are usually accompanied by nearby hadronic activity, whereas prompt photons are well isolated. The isolation variable is defined as the transverse energy in a cone of angular size $R = 0.2$ around the photon axis. The transverse energy can be calculated from calorimeter clusters (subtracting off the contributions from pileup and the photon candidate itself):

$$E_T^{\text{cone20}} = \sum_{c \in \text{cone}} E_T^c - E_T^{\gamma \text{ clusters}} - E_T^{\text{pileup}} \tag{4.1}$$

The transverse energy can also be calculated from tracks with $p_T > 1$ GeV that originate at z within 3 mm of the primary vertex (subtracting off the contribution of any tracks associated with a converted photon candidate):

$$p_T^{\text{cone20}} = \sum_{t \in \text{cone}} p_T^t - p_T^{\gamma \text{ tracks}} \tag{4.2}$$

Photon candidates are considered isolated if the isolation variables are small compared to the photon transverse energy. For the remainder of this work, photon candidates are considered isolated if they satisfy

$$E_T^{\text{cone20}} < 0.065 \times E_T^{\gamma} \text{ and } p_T^{\text{cone20}} < 0.05 \times E_T^{\gamma} \tag{4.3}$$

This corresponds to the loose photon isolation working point defined in [8].

4.2.3 Primary Vertex Selection in Diphoton Events

For studies of the $H \rightarrow \gamma\gamma$ decay, selection of the correct diphoton vertex is critical to achieve an accurate measurement of the diphoton invariant mass, which depends not only on the energy of each photon candidate, but also on the angle between them (Eq. 1.1).

For $H \rightarrow \gamma\gamma$ measurements, the inputs to primary vertex selection include both track information from the ID and photon information from the ECAL. The ECAL was designed with fine granularity in η for the purpose of **photon pointing**: the vertex of the diphoton system can be identified with high precision by the ECAL alone (15 mm resolution in z at η=0).

A multivariate algorithm (neural network) is trained to select the primary vertex most compatible with the diphoton vertex from all vertices reconstructed by the ID [9]. For each vertex, the following variables are used to train the neural network:

- the z-position where the photon trajectories intersect the beam axis
- the scalar sum over all tracks associated with the vertex of p_T and p_T^2

- $\Delta\phi$ between the vector sum of the track momenta and the momentum of the diphoton system
- the position of photon conversion vertices, if one or both photons convert to e^+e^- in the ID.

The training sample is composed of simulated ggF events. The neural network selects a primary vertex that is within 0.3 mm of the Higgs production vertex in 79% of simulated ggF events at $\sqrt{s} = 13$ TeV. For Higgs production modes with higher multiplicity final states (such as $t\bar{t}H$), this increases to 84–97% of events [10]. For $t\bar{t}H(\gamma\gamma)$, the vertex selected by the neural network and the vertex with the maximum sum of track p_T^2 are identical in > 96% of simulated events.

4.3 Electrons

Like photon candidates, electron candidates are constructed from a topo-cluster of energy deposits in the ECAL. However, the ID signature of an electron differs from that of a photon: electron candidates are associated with a single charged particle track. As for photons, the electron energy resolution is calculated using multivariate regression trained on simulated samples. The absolute energy scale is calibrated using $Z \to ee$ decays and validated on radiative Z decays [7, 8].

For the analyses following in later Chapters, electron candidates must have $p_T >$ 15 GeV and $|\eta| < 2.47$ with the region $1.37 < |\eta| < 1.52$ vetoed. In addition, electron candidates must satisfy the identification and isolation criteria detailed in this Section.

4.3.1 Electron Identification

In order to separate prompt electrons from secondary or fake electrons, a likelihood discriminant is constructed from sensitive variables [11]. These include

- Hadronic leakage: energy deposited in the hadronic calorimeter near an ECAL cluster
- Shower shape variables based on all three ECAL layers
- Agreement between the ID track and ECAL cluster in measured angle and energy of the electron candidate
- Parameters of the track in the ID, including the amount of transition radiation measured in the TRT.

Cuts on the likelihood discriminant are optimized in bins of E_T and $|\eta|$ to give about 88% electron efficiency for typical electroweak processes. Electron candidates in the remainder of this Thesis are required to satisfy these cuts, which correspond to the medium electron identification working point defined in [11].

4.3.2 Electron Isolation

As in the case of photons, background electrons are usually accompanied by nearby hadronic activity. The isolation variables used for electrons are analogous to those used for photons (Eqs. 4.1 and 4.2). However, because electrons from the decay of very high energy particles can be highly columnated, the track isolation variable $p_T^{\text{varcone20}}$ is defined in a cone of variable size: $R = \min(10\text{ GeV}/p_T^e, 0.2)$.

Electron candidates are considered isolated if the isolation variables are small compared to the electron transverse energy. For the remainder of this work, electron candidates are considered isolated if they satisfy

$$E_T^{\text{cone20}} < 0.20 \times p_T^e \text{ and } p_T^{\text{varcone20}} < 0.15 \times p_T^e \tag{4.4}$$

This corresponds to the loose electron isolation working point defined in [8].

4.4 Muons

Muon tracks are initially reconstructed separately in the ID (see Sect. 4.1) and the MS. In the MS, hit patterns in each layer are grouped into straight segments, which are further combined into tracks [12]. Muon tracks with $0.1 < |\eta| < 2.5$ are required to have at least three hits in at least two MS MDT layers. In the very central region of $|\eta| < 0.1$, hits are required in at least one MDT layer and holes larger than one MDT layer are not allowed. Forward muons with $2.5 < |\eta| < 2.7$ are required to leave hits in at least three MS MDT/CSC layers. A χ^2 fit is performed to the MS hits making up each track, and the track is accepted if the χ^2 passes the selection criteria. Muon tracks in the MS and ID are combined by performing a global fit to the muon track using hits in both the ID and MS.

The muon momentum scale and resolution are calculated from $J/\psi \rightarrow \mu\mu$ and $Z \rightarrow \mu\mu$ decays [12].

For the analyses following in later Chapters, muon candidates must have $p_T >$ 15 GeV and $|\eta| < 2.7$ in addition to satisfying the identification and isolation criteria detailed in this Section.

4.4.1 Muon Identification

For muon candidates in the central region $|\eta| < 2.5$, the compatibility of the charge-to-momentum ratio measured in the ID and MS is calculated:

$$\frac{|(\frac{q}{p})_{\mathrm{ID}} - (\frac{q}{p})_{\mathrm{MS}}|}{\sqrt{\sigma^2((\frac{q}{p})_{\mathrm{ID}}) + \sigma^2((\frac{q}{p})_{\mathrm{MS}})}} \tag{4.5}$$

where σ indicates the uncertainty on the measurement. Muon candidates are required to have ID/MS compatibility < 7. Because the region $2.5 < |\eta| < 2.7$ is outside of ID acceptance, muon candidates reconstructed in the MS only are accepted in this range.

Muon candidates in the remainder of this Thesis are required to satisfy these cuts, which correspond to the medium muon identification working point defined in [12].

4.4.2 Muon Isolation

Prompt muons and muons from the decay of a W, Z, or Higgs boson are well isolated, whereas background muons (e.g. from the decay of hadrons) are typically associated with nearby hadronic activity. The isolation variables used for muons are analogous to those used for electrons. For muon candidates, the variable-sized cone used to calculate the track isolation is allowed to reach a maximum of $R = 0.3$.

Muon candidates are considered isolated if the isolation variables are small compared to the muon transverse energy. For the remainder of this work, muon candidates must satisfy

$$E_T^{\mathrm{cone20}} < 0.30 \times p_T^{\mu} \text{ and } p_T^{\mathrm{varcone30}} < 0.15 \times p_T^{\mu} \tag{4.6}$$

This corresponds to the fixed-cut loose muon isolation working point defined in [12].

4.5 Hadronic Jets

As for photons and electrons, the reconstruction of hadronic jets begins with the formation of topo-clusters from energy deposits in the calorimeters [6, 13]. The topo-clusters are used as input to the **anti-**k_T jet-finding algorithm [14], which groups topo-clusters together based on their p_T and angular separation. Though many jet-finding algorithms have been proposed, the anti-k_T algorithm has the advantage of producing jets of a roughly conical shape, while also satisfying the criteria for infrared and collinear (IRC) safety:

1. Infrared safe: the jet finding is not impacted by the addition/removal of very low p_T constituents
2. Collinear safe: the jet finding is not impacted by splitting one high p_T constituent into two collinear constituents

For the remainder of this work, jets are constructed using the anti-k_T algorithm with $R = 0.4$ and are required to have $p_T > 25$ GeV and $|y| < 4.4$. Jets within $\Delta R < 0.4$ of a selected photon or electron are rejected. For jets with $p_T < 50$ GeV and $|\eta| < 2.4$, compatibility with the primary vertex is evaluated using the Jet Vertex Tagger (JVT) likelihood discriminant [15]. In order to suppress contribution from pileup, incompatible jets are rejected.

4.5.1 Flavor Tagging

Jets initiated by bottom quarks contain B-hadrons. Due to their comparatively long lifetimes ($c\tau \sim 450\ \mu$m), these B-hadrons travel a characteristically long distance in the detector before decaying. Thanks to high granularity and good resolution in the pixel detector, such secondary vertices and other track-based observables are measured with high precision. Using multivariate techniques, the track and vertex information is compiled into a single discriminant that separates b-initiated jets from jets initiated by light quarks [16, 17]. Because these discriminants rely on information collected by the Inner Detector, b-tagging is only possible for jets with $|\eta| < 2.4$.

Figure 4.2 shows the performance of the MV2 algorithm for identifying b-jets, with the suffix c20 (c10) indicating that 15% (7%) of the light jet sample is composed of charm jets [18]. The horizontal axis shows the percentage of b-jets that are correctly identified in a simulated sample of $t\bar{t}$ events, and the vertical axis shows the rejection of light jets.

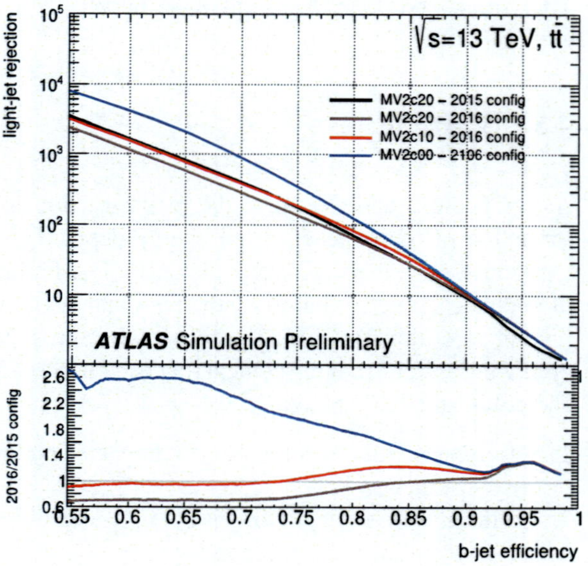

Fig. 4.2 Efficiency of b-jet identification versus rejection of light jets for ATLAS multivariate b-tagging discriminants in $t\bar{t}$ Monte Carlo [18]. The name MV2c20 (MV2c10) indicates that 15% (7%) of the light-jet sample is composed of charm jets

Multiple working points with different b-jet identification efficiencies are commonly used in ATLAS: these include 65, 70, 77, and 85%. For the remainder of this work, a **b-tagged jet** refers to a jet passing the MV2c10 77% working point: that is, 77% of b-jets in a simulation of $t\bar{t}$ events are correctly identified, which corresponds to $\sim$1 in 100 light jets being misclassified as a b-jet (see Fig. 4.2).

It is important to note that the additional tracking space point provided by the IBL led to substantial improvement in ATLAS b-tagging performance. With no changes in the b-tagging algorithm, rejection of light jets increased by a factor of up to four for a fixed b-jet efficiency, with the greatest improvement at working points between 70 and 80% [19].

4.6 Missing Transverse Energy

The incoming protons in ATLAS have approximately zero momentum in the plane transverse to the beam. High p_T neutrinos, or BSM particles with very low interaction cross sections (such as dark matter candidates), pass through the detector without interacting, leaving a large momentum imbalance in the transverse plane: this is referred to as **missing transverse energy (missing E_T)**.

Missing E_T is computed as the negative sum of the transverse momentum of all reconstructed photons, electrons, muons, and jets, plus a soft term reconstructed from tracks that are not associated with these objects [20, 21]. The z component of the missing energy cannot be calculated, since the parton–parton center of mass frame has unknown (and possibly substantial) longitudinal momentum. The missing E_T resolution $\Delta E_T^{\text{miss}} (\sim \sqrt{E_T^{\text{miss}}})$ ranges from about 10 to 25 GeV for $t\bar{t}$ events in ATLAS .

The **missing E_T significance** can be defined in numerous ways. In the following Chapters, an event's missing E_T significance defined as

$$\text{Missing } E_T \text{ significance} \equiv E_T^{\text{miss}} / \sqrt{H_T} \tag{4.7}$$

where H_T is the scalar sum of the p_T of all jets in the event.

References

1. Pequenao J, Schaffner P (2013) How ATLAS detects particles: diagram of particle paths in the detector. https://cds.cern.ch/record/1505342
2. Measurement of the tau lepton reconstruction and identification performance in the ATLAS experiment using pp collisions at $\sqrt{s} = 13$ TeV. Tech. rep. ATLAS-CONF- 2017-029. CERN, Geneva, 2017. https://cds.cern.ch/record/2261772
3. Early inner detector tracking performance in the 2015 data at $\sqrt{s} = 13$ TeV. Tech. rep. ATL-PHYS-PUB-2015-051. CERN, Geneva, 2015. http://cds.cern.ch/record/2110140

4. Cornelissen T et al (2007) Concepts, design and implementation of the ATLAS new tracking (NEWT). Tech. rep. ATL-SOFT-PUB-2007-007. ATL-COM-SOFT-2007-002. CERN, Geneva, 2007. https://cds.cern.ch/record/1020106
5. Aaboud M et al. (2017) Reconstruction of primary vertices at the ATLAS experiment in Run 1 proton–proton collisions at the LHC. Eur Phys J C 77.5 (2017):332. https://doi.org/10.1140/epjc/s10052-017-4887-5. arXiv: 1611.10235 [physics.ins-det]
6. Aad G et al (2017) Topological cell clustering in the ATLAS calorimeters and its performance in LHC Run 1. Eur Phys J C 77:490. https://doi.org/10.1140/epjc/s10052-017-5004-5. arXiv: 1603.02934 [hep-ex]
7. Aaboud M et al (2019) Electron and photon energy calibration with the ATLAS detector using 2015–2016 LHC proton-proton collision data. JINST 14(03):P03017. https://doi.org/10.1088/1748-0221/14/03/P03017. arXiv: 1812.03848 [hep-ex]
8. Aad G et al (2019) Electron and photon performance measurements with the ATLAS detector using the 2015–2017 LHC proton-proton collision data. JINST 14(12):P12006. https://doi.org/10.1088/1748-0221/14/12/P12006. arXiv:1908.00005 [hep-ex]
9. Aad G et al (2014) Measurement of Higgs boson production in the diphoton decay channel in pp collisions at center-of-mass energies of 7 and 8 TeV with the ATLAS detector. Phys Rev D 90(11):112015. https://doi.org/10.1103/PhysRevD.90.112015. arXiv: 1408.7084 [hep-ex]
10. The ATLAS Collaboration (2018) Measurements of Higgs boson properties in the diphoton decay channel with 36 fb^{-1} of pp collision data at $\sqrt{s} = 13$ TeV with the ATLAS detector. Measurements of Higgs boson properties in the diphoton decay channel with 36 fb^{-1} of pp collision data at $\sqrt{s} = 13$ TeV with the ATLAS detector. Phys Rev D 98(5):052005. arXiv:1802.04146. https://doi.org/10.1103/PhysRevD.98.052005. https://cds.cern.ch/record/2302492
11. Aaboud M et al (2019) Electron reconstruction and identification in the ATLAS experiment using the 2015 and 2016 LHC proton-proton collision data at $\sqrt{s} = 13$ TeV. Eur Phys J C 79(8):639. https://doi.org/10.1140/epjc/s10052-019-7140-6. arXiv: 1902.04655 [physics.ins-det]
12. Aad G et al (2016) Muon reconstruction performance of the ATLAS detector in proton–proton collision data at $\sqrt{s} = 13$ TeV. Eur Phys J C 76(5):292. https://doi.org/10.1140/epjc/s10052-016-4120-y. arXiv: 1603.05598 [hep-ex]
13. Aaboud M et al (2017) Jet energy scale measurements and their systematic uncertainties in proton-proton collisions at $\sqrt{s} = 13$ TeV with the ATLAS detector. Phys Rev D 96(7):072002. https://doi.org/10.1103/PhysRevD.96.072002. arXiv: 1703.09665 [hep-ex]
14. Cacciari M, Salam GP, Soyez G (2008) The anti-k_t jet clustering algorithm. JHEP 04:063. https://doi.org/10.1088/1126-6708/2008/04/063. arXiv: 0802.1189 [hep-ph]
15. Aad G et al (2016) Performance of pile-up mitigation techniques for jets in pp collisions at $\sqrt{s} = 8$ TeV using the ATLAS detector. Eur Phys J C 76(11):581. https://doi.org/10.1140/epjc/s10052-016-4395-z. arXiv: 1510.03823 [hep-ex]
16. Aad G et al (2016) Performance of b-Jet Identification in the ATLAS experiment. JINST 11(04):P04008. https://doi.org/10.1088/1748-0221/11/04/P04008. arXiv: 1512.01094 [hep-ex]
17. Aad G et al (2019) ATLAS b-jet identification performance and efficiency measurement with $t\bar{t}$ events in pp collisions at $\sqrt{s} = 13$ TeV. Eur Phys J C 79(11):970. https://doi.org/10.1140/epjc/s10052-019-7450-8. arXiv: 1907.05120 [hep-ex]
18. Optimisation of the ATLAS b-tagging performance for the 2016 LHC Run. Tech. rep. ATL-PHYS-PUB-2016-012. CERN, Geneva, 2016. https://cds.cern.ch/record/2160731
19. Plots of b-tagging performance before and after the installation of the Insertable B-layer, 2017. http://atlas.web.cern.ch/Atlas/GROUPS/PHYSICS/PLOTS/FTAG-2017-005/
20. Aaboud M et al (2018) Performance of missing transverse momentum reconstruction with the ATLAS detector using proton-proton collisions at $\sqrt{s} = 13$ TeV. Eur Phys J C 78(11):903, 66. arXiv:1802.08168. https://doi.org/10.1140/epjc/s10052-018-6288-9. https://cds.cern.ch/record/2305380
21. E_T^{miss} performance in the ATLAS detector using 2015-2016 LHC p-p collisions. Tech. rep. ATLAS-CONF-2018-023. CERN, Geneva, 2018. http://cds.cern.ch/record/2625233

Chapter 5
Data and Monte Carlo Samples

5.1 Data: LHC Run 2

The results reported in this Thesis are based on Run 2 proton–proton collision data with center of mass energy $\sqrt{s} = 13$ TeV recorded by the ATLAS detector between 2015 and 2018. The full dataset amounts to an integrated luminosity of 139.0 ± 2.4 fb^{-1} once data quality requirements are imposed (in order to ensure all detector components are operational) [1]. Some results are also reported using partial datasets recorded during 2015–2016 (36.1 ± 1.2 fb^{-1}) and 2015–2017 (79.8 ± 1.6 fb^{-1}).

The integrated luminosity recorded by ATLAS is shown as a function of time in Fig. 5.1a. The mean number of interactions per bunch crossing ($< \mu >$) is shown in Fig. 5.1b for the full and partial datasets. For the 2015–2016 data taking period, $< \mu >$ was 23, while it increased to 37 for the 2017–2018 data taking period.

Data events used in Run 2 $H \rightarrow \gamma\gamma$ analyses are initially selected by a trigger requiring two photon candidates with transverse momentum of at least 25 and 35 GeV for the sub-leading and leading photon, respectively. During 2015–2016, photon candidates were required to pass a loose online identification requirement at the trigger level. Due to the increase instantaneous luminosity in 2017–2018, the trigger requirement was tightened to a medium selection to keep the total trigger rate 20 Hz [2] (see Fig. 5.2).

The diphoton trigger efficiency is calculated from radiative Z decays. This efficiency is shown as a function of the E_T and η of the lower-p_T photon in Fig. 5.3. Once the full diphoton event selection is applied (described in later Chapters), the average trigger efficiency is found to be > 99 and $> 98\%$ for the 2015–2016 and 2017–2018 data taking periods, respectively.

J. E. Dickinson, *ATLAS Measurements of the Higgs Boson Coupling to the Top Quark in the Higgs to Diphoton Decay Channel*, Springer Theses,
https://doi.org/10.1007/978-3-030-86368-5_5

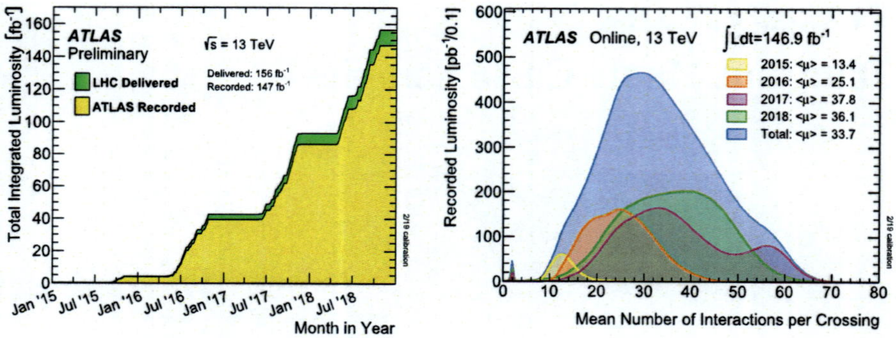

(a) Luminosity (b) Pileup

Fig. 5.1 Summary of ATLAS integrated luminosity (**a**) and pileup conditions (**b**) during Run 2 data taking

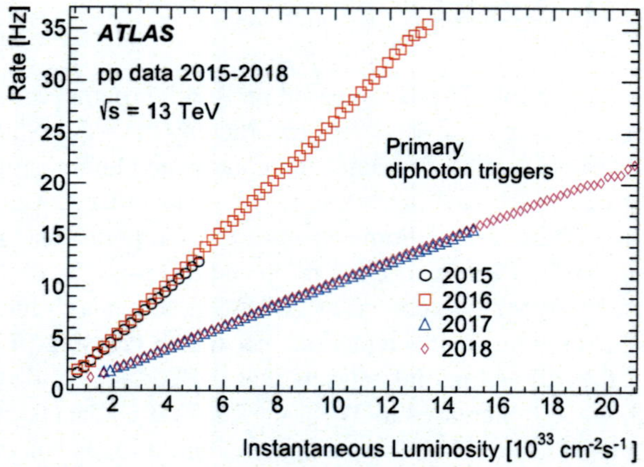

Fig. 5.2 The diphoton trigger rate is shown separately for each year of run 2 data taking [2]. The rate increases linearly with instantaneous luminosity. The difference in slope between the 2015–2016 and 2017–2018 periods is due to the change in photon identification requirement

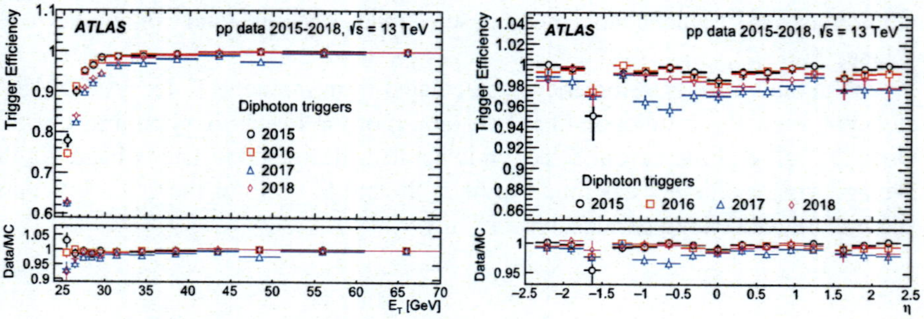

(a) Sub-leading photon E_T (b) Sub-leading photon η

Fig. 5.3 Efficiency of the diphoton trigger as a function of the E_T and η of the sub-leading (lower p_T) photon. The efficiency is calculated from radiative Z decays [2]

5.1.1 Data Control Samples

Events used in $H \to \gamma\gamma$ analysis are required to contain two photons passing the tight identification and isolation requirements described in Sect. 4.2. Data passing these requirements are referred to as **tight/isolated (TI) data**.

Events passing trigger requirements but failing at least one of the identification or isolation requirements are referred to as **non-tight/isolated (NTI) data**. The size of the NTI dataset (normalized to luminosity) is larger in 2015–2016 than in 2017–2018 due to the switch from looser to tighter photon identification requirement at trigger level.

It is assumed that data events with diphoton invariant mass far from the Higgs signal peak ($|m_{\gamma\gamma} - 125| > 2$) are comprised only of background processes. This region is referred to as the **data sidebands**.

The NTI data and NTI/TI data sidebands are used for multiple purposes in later Chapters, including

- to model the continuum background $m_{\gamma\gamma}$ distribution for the selection of a fit function (Sect. 6.2.2)
- to model the continuum background in the training of multivariate algorithms (Chap. 9)
- to estimate the continuum background yield under the Higgs signal peak ($m_{\gamma\gamma} = 125 \pm 2$ GeV), for calculation of number-counting significance (Sect. 9.1)
- to extract the relative contribution of different processes to the continuum background (Sect. 9.3)

The continuum background yield in the Higgs signal window (n_{TI}^{win}) is estimated by applying a scale factor to the event yield in one of the data sideband regions (either NTI or TI):

$$\begin{aligned} n_{TI}^{\text{win}} &= n_{NTI}^{\text{SB}} \times f_1 \times f_2 \\ n_{TI}^{\text{win}} &= n_{TI}^{\text{SB}} \times f_1 \end{aligned} \tag{5.1}$$

The scale factors f_1 and f_2 are calculated separately in the hadronic and leptonic channels according to the following formulae:

$$f_1 = \frac{n_{NTI}^{\text{win}}}{n_{NTI}^{\text{SB}}};\ f_2 = \frac{n_{TI}^{\text{SB}}}{n_{NTI}^{\text{SB}}} \tag{5.2}$$

The scale factor f_1 scales the number of NTI events in the data sidebands to the number of events in the 125 ± 2 GeV mass window. The second term f_2 scales the number of NTI to the number of TI events by using the ratio calculated in the sidebands. Only f_1 is applied to scale the number of TI sideband events to the number of TI events in the mass window. The scale factors evaluated on the full 139 fb^{-1} of Run 2 data are listed in Table 5.1.

Table 5.1 Scale factors for estimating continuum background under the Higgs boson mass peak ($m_{\gamma\gamma} = 125 \pm 2$ GeV) from data control regions

	Preselection	f_1	$f_1 \times f_2$
Hadronic	2γ, $0\ e/\mu$, ≥ 3 jets, ≥ 1 b-jet	0.103	0.013
Leptonic	2γ, $\geq 1\ e/\mu$, ≥ 1 b-jet	0.094	0.016

Table 5.2 Continuum background in the $m_{\gamma\gamma} = 125 \pm 2$ GeV window estimated by scaling the NTI and TI sidebands, compared to the fitted background in this region

Category	Scaled NTI SB	Scaled TI SB	Fitted	
$t\bar{t}H(\gamma\gamma)$ CP 1	0.86 ± 0.11	1.9 ± 0.44	1.3	$+0.41/-0.50$
$t\bar{t}H(\gamma\gamma)$ CP 2	1.3 ± 0.13	3.9 ± 0.63	3.0	$+0.55/-0.56$
$t\bar{t}H(\gamma\gamma)$ CP 3	0.90 ± 0.11	1.3 ± 0.37	0.87	$+0.34/-0.32$
$t\bar{t}H(\gamma\gamma)$ CP 4	0.90 ± 0.11	1.5 ± 0.40	1.2	$+0.38/-0.42$
$t\bar{t}H(\gamma\gamma)$ CP 5	1.7 ± 0.15	4.4 ± 0.68	3.9	$+0.63/-0.72$
$t\bar{t}H(\gamma\gamma)$ CP 6	2.0 ± 0.16	3.4 ± 0.59	2.8	$+0.51/-0.57$
$t\bar{t}H(\gamma\gamma)$ CP 7	1.5 ± 0.14	2.5 ± 0.50	2.2	$+0.57/-0.59$
$t\bar{t}H(\gamma\gamma)$ CP 8	6.2 ± 0.28	13.2 ± 1.2	10.4	$+1.1/-1.4$
$t\bar{t}H(\gamma\gamma)$ CP 9	6.3 ± 0.29	6.4 ± 0.81	5.4	$+0.59/-0.77$
$t\bar{t}H(\gamma\gamma)$ CP 10	0.88 ± 0.11	1.8 ± 0.42	1.4	$+0.85/-0.91$
$t\bar{t}H(\gamma\gamma)$ CP 11	12.0 ± 0.39	21.4 ± 1.5	17.6	$+2.6/-2.7$
$t\bar{t}H(\gamma\gamma)$ CP 12	58.5 ± 0.87	69.5 ± 2.7	57.9	$+2.9/-2.9$
$t\bar{t}H(\gamma\gamma)$ CP 13	0.91 ± 0.12	0.85 ± 0.28	0.63	$+0.28/-0.32$
$t\bar{t}H(\gamma\gamma)$ CP 14	1.7 ± 0.16	2.3 ± 0.46	1.9	$+0.41/-0.43$
$t\bar{t}H(\gamma\gamma)$ CP 15	0.83 ± 0.12	1.2 ± 0.34	0.96	$+0.34/-0.35$
$t\bar{t}H(\gamma\gamma)$ CP 16	2.1 ± 0.18	3.5 ± 0.57	2.95	$+0.52/-0.60$
$t\bar{t}H(\gamma\gamma)$ CP 17	2.3 ± 0.19	2.2 ± 0.45	1.75	$+0.45/-0.44$
$t\bar{t}H(\gamma\gamma)$ CP 18	0.85 ± 0.12	1.4 ± 0.36	1.25	$+0.54/-0.55$
$t\bar{t}H(\gamma\gamma)$ CP 19	2.4 ± 0.20	2.5 ± 0.49	2.23	$+0.45/-0.55$
$t\bar{t}H(\gamma\gamma)$ CP 20	3.6 ± 0.24	3.1 ± 0.54	2.92	$+0.42/-0.53$

Calculated as an example in the twenty $t\bar{t}H(\gamma\gamma)$ CP categories defined in Chap. 11

A comparison of yields calculated from the NTI data and the TI data sidebands with the fitted background yield is shown in Table 5.2, using the twenty categories of the $t\bar{t}H(\gamma\gamma)$ CP analysis (Chap. 11) as an example. Because the unscaled NTI sample is much larger than the TI sample, the statistical uncertainty on the scaled NTI prediction is smaller than the uncertainty on the TI. In most categories, the continuum background yield predicted by the scaled NTI data sidebands is smaller than that predicted by TI, and the TI prediction tends to agree better (within uncertainties) with the fitted background. The performance of these background estimates is further discussed in later Chapters.

Table 5.3 Summary of Monte Carlo generators used for matrix element calculation and parton showering

Generator	Purpose	Version	References
Pythia8	ME+shower, shower only	8.2.12	[3, 4]
Herwig7	ME+shower, shower only	7.03	[5, 6]
Powheg	ME only	2	[7–9]
MG5_aMC@NLO	ME only	2.6.2	[10, 11]
Sherpa	ME+shower	1.1	[12–15]
EvtGen	Decay of heavy hadrons	1.2.0	[16]

5.2 Simulated Samples

This Section provides an overview of the generation of the simulated Monte Carlo (MC) samples used in this work. The software used for matrix element calculation and parton showering is summarized in Table 5.3. In samples using Pythia8 or Herwig7 for parton showering, the EvtGen package is used to simulate the decays of heavy hadrons.

Once generated, all simulated Higgs boson events are passed through a Geant4 [17] simulation of the ATLAS detector and reconstructed with the same analysis software used for data [18]. Additional inelastic proton–proton interactions (pileup) are included in the simulation for all generated events such that the average number of interactions per bunch crossing reproduces that observed in data (Fig. 5.1b). The pileup collisions were produced using Pythia (8.1.86) with the A2 parameter tune [19] and the MSTW2008lo PDF set [20].

Simulated events are accompanied by a Monte Carlo weight: this weight indicates the relative probability of events in different regions of phase space, while allowing all regions of phase space to be populated somewhat uniformly. Statistical uncertainties on simulated distributions are given by the quadratic sum of the Monte Carlo weights. This corresponds to a variance σ^2 of

$$\sigma^2 = \sum_i w_i^2 \tag{5.3}$$

In order to keep these statistical uncertainties small, simulated samples typically contain $\geq 10\times$ the number of events expected in data.

Table 5.4 Summary of standard model Higgs boson Monte Carlo samples

Process	ME Generator	PDF set	Shower	Tune
ggF	Powheg	PDF4LHC15	Pythia8 (Herwig7)	AZNLO (H7UE)
VBF	Powheg	PDF4LHC15	Pythia8	AZNLO
WH	Powheg	PDF4LHC15	Pythia8	AZNLO
$q\bar{q} \rightarrow ZH$	Powheg	PDF4LHC15	Pythia8	AZNLO
$gg \rightarrow ZH$	Powheg	PDF4LHC15	Pythia8	AZNLO
$b\bar{b}H$	Powheg	PDF4LHC15	Pythia8	AZNLO
$t\bar{t}H$	Powheg	PDF4LHC15	Pythia8 (Herwig7)	A14 (H7UE)
$t\bar{t}H$	MG5_aMC@NLO	NNPDF30	Pythia8 (Herwig7)	A14 (H7UE)
$tHjb$	MG5_aMC@NLO	NNPDF30	Pythia8 (Herwig7)	A14 (H7UE)
tWH	MG5_aMC@NLO	NNPDF30	Pythia8 (Herwig7)	A14 (H7UE)

5.2.1 *SM Higgs Boson Samples*

Monte Carlo samples of Higgs boson events with the $H \rightarrow \gamma\gamma$ decay were generated for all production modes summarized in Table 5.4. For all samples, the simulation uses Higgs boson mass $m_H = 125$ GeV and width $\Gamma_H = 4.07$ MeV. A K**-factor** is applied to all SM samples that scales the prediction from the generator to the most accurate available theoretical calculation at $m_H = 125.09$ GeV (see Table 2.1). The rate is also scaled by the branching fraction of $H \rightarrow \gamma\gamma$ (0.277%).

Gluon fusion (ggF) is simulated at next-to-next-to-leading-order (NNLO) accuracy in QCD using the Powheg NNLOPS program [21], with the PDF4LHC15 PDF set [22]. The parton-level events are passed to Pythia for computation of parton showering, hadronization and underlying event effects using the AZNLO set of parameters that are tuned to data [23]. The sample is normalized such that it reproduces the total cross section predicted by the N^{3}LO QCD calculation with NLO electroweak corrections applied (Table 2.1). An alternative parton shower sample for ggF uses the same Powheg NNLOPS configuration for the matrix element calculation, but is interfaced instead to Herwig7 with the H7UE parameter tune [6].

Vector boson fusion (VBF), vector boson associated Higgs production (including $q\bar{q} \rightarrow WH$, $q\bar{q} \rightarrow ZH$, and $gg \rightarrow ZH$), and $b\bar{b}H$ production are all generated at next-to-leading-order (NLO) accuracy in QCD using Powheg with the PDF4LHC15 PDF set. The parton shower is generated by Pythia8 using the AZNLO parameter set.

Standard Model $t\bar{t}H$ production is simulated using multiple matrix element generators. All $t\bar{t}H$ samples are normalized to the cross section calculated at NLO in QCD and EW (see Table 2.1). The nominal $t\bar{t}H$ sample is generated at NLO with Powheg and PDF set PDF4LHC15 and interfaced to Pythia8 for parton showering using the A14 parameter tune [19]. An alternative parton shower sample for $t\bar{t}H$ uses the same Powheg configuration for the matrix element calculation, but is interfaced to Herwig7 for parton showering with the H7UE parameter tune.

A Standard Model $t\bar{t}H$ sample is also generated with MG5_aMC@NLO for use in conjunction with the BSM samples described in Sect. 5.2.2. The matrix element is calculated at NLO accuracy with the NNPDF30 PDF set [24], and the parton shower is generated using Pythia8 and the A14 parameter tune. The dynamical renormalization (μ_R) and factorization (μ_F) scales are defined as one half the scalar sum of the transverse masses of all final state particles ($\frac{1}{2}\sum m_T$). An alternative parton shower sample uses the same MG5_aMC@NLO configuration for matrix the element calculation, but is interfaced to Herwig7 for parton showering with the H7UE parameter tune.

Higgs production in association with a single top quark, a b quark and a light quark ($tHjb$) is produced at NLO accuracy in QCD (no electroweak correction) using MG5_aMC@NLO with the four-flavor scheme in the NNPDF30 PDF set. The parton-level events are showered using Pythia8 and the A14 parameter tune. The renormalization and factorization scales are defined as one half the scalar sum of the transverse masses of all final state particles ($\frac{1}{2}\sum m_T$). An alternative parton shower sample for $tHjb$ uses the same MG5_aMC@NLO configuration for the matrix element calculation, but is interfaced to Herwig7 for parton showering with the H7UE parameter tune.

Associated production of a Higgs boson with a single top quark and a W boson (tWH) is produced at NLO accuracy in QCD (no electroweak correction) using MG5_aMC@NLO with the NNPDF30 PDF set, and the parton showering is performed using Pythia8 and the A14 parameter tune. The renormalization and factorization scales are defined as one half the scalar sum of the transverse masses of all final-state particles ($\frac{1}{2}\sum m_T$). At NLO, there is interference between the tWH and $t\bar{t}H$ processes [25], as demonstrated in Fig. 2.12. The contribution to tWH from doubly resonant diagrams ($t\bar{t}H$ overlap) is removed, and the interference between the diagrams is maintained as a component of the tWH signal. An alternative parton shower sample for tWH uses the same MG5_aMC@NLO configuration for the matrix element calculation, but is interfaced to Herwig7 for parton showering with the H7UE parameter tune.

5.2.2 BSM Higgs Samples

The Higgs Characterization (HC) model [26] is implemented in the MG5_aMC@NLO generator, enabling the generation of Monte Carlo samples with different values of the CP mixing parameters defined in the effective Lagrangian in Eq. 2.28. The simulated samples and their parameter settings are listed in Table 5.5.

As in the SM case, the MG5_aMC@NLO samples are interfaced with Pythia8 for parton showering and EvtGen, and the NNPDF30 PDF set is used (five-flavor scheme for $t\bar{t}H$ and tWH, four-flavor scheme for $tHjb$). The renormalization and factorization scales are again defined as $\frac{1}{2}\sum m_T$. As a cross check, samples with SM settings ($\alpha = 0°$, $\kappa_t = 1$) are produced in the HC model and compared to the nominal SM samples: agreement is found to be good for all processes. Since the cross

Table 5.5 Parameters used in the HC model to produce samples with CP mixing in the Higgs-top coupling

κ_t	α	$\cos\alpha$	$\kappa_{SM} = 1/\cos\alpha$
1	0° (SM, CP even)	1	1
1	15°	0.965926	1.035276
1	30°	0.866025	1.154701
1	45°	0.707107	1.414214
1	60°	0.5	2
1	75°	0.258819	3.863703
1	90° (CP odd)	0.000001	10^6
−1	0	1	1
0.5	0	1	1
2	0	1	1
2	45	0.707107	1.414214

For $\alpha = 90°$: due to numerical precision, $\cos\alpha$ strictly equal to zero cannot be generated, and a value approaching it (10^{-6}) and a defined value for κ_{SM} (10^6) are used for the pure CP odd Monte Carlo samples

Table 5.6 NLO cross section times branching ratio for the $t\bar{t}H$, $tHjb$, tWH, and ggF processes for different CP scenarios, normalized by the K-factor

Parameters		Normalized $\sigma \times B_{\gamma\gamma}$ [fb]			
κ_t	α	$t\bar{t}H$	$tHjb$	tWH	ggF
1	0° (SM)	1.150	0.169	0.034	33.1
1	15°	1.113	0.177	0.038	–
1	30°	0.995	0.207	0.048	–
1	45°	0.827	0.266	0.064	53.7
1	60°	0.666	0.382	0.087	–
1	75°	0.545	0.548	0.116	–
1	90°	0.500	0.753	0.150	74.4
−1	0°	–	1.980	0.307	–
0.5	0°	–	0.264	0.030	–
2	0°	–	0.666	0.190	–
2	45°	–	0.570	0.226	–

Unfilled entries indicate that no sample with these parameters was generated

section of ggF production also varies with κ_t and α, MG5_aMC@NLO is also used to generate ggF+2 jets samples at three different values of the CP mixing angle α.

The same K-factors derived for the SM case are applied to all BSM samples, since it was found that the K-factors change little with α. For each sample, the cross section (with K-factor applied) times $H \to \gamma\gamma$ branching fraction is shown in Table 5.6.

5.2.3 Background Samples

Backgrounds for $H \to \gamma\gamma$ include non-Higgs processes with ISR/FSR photons and events containing one or more objects that are misidentified as photons. Such processes result in a background spectrum that is continuous and decreasing in the diphoton mass variable $m_{\gamma\gamma}$, and it is therefore referred to as the **continuum background**.

The exact rate and composition of the continuum background is difficult to model, and the background estimation in $H \to \gamma\gamma$ analyses is done using a data-driven method (see Sect. 6.2.2). However, simulated samples are needed to validate analysis methods and study the composition of the background in data. Large Monte Carlo samples are particularly critical for the determination of the functional form (not the parameters) of the continuum background $m_{\gamma\gamma}$ spectrum in specific regions of phase space, and for the estimation of the uncertainty on the continuum background shape (see description of the spurious signal test in Sect. 6.2.2).

In regions targeting $t\bar{t}H(\gamma\gamma)$, the dominant contributions to the continuum background arise from $t\bar{t}\gamma\gamma$ and $\gamma\gamma$+jets processes. The $t\bar{t}\gamma\gamma$ background is generated at leading order in QCD using MG5_aMC@NLO and the NNPDF PDF set, and interfaced to Pythia8 for parton showering (A14 tune). The renormalization and factorization scales are defined as $\frac{1}{2}\sum m_T$. Background events from continuum $\gamma\gamma$+jets production are simulated using the Sherpa event generator with the CT10 PDF set. The matrix element is calculated at leading order in QCD with the real emission of up to three additional partons, and interfaced with the Sherpa parton shower using the default parameter tune for the underlying-event activity.

Since the sample size required for good modeling of $\gamma\gamma$+jets and $t\bar{t}\gamma\gamma$ backgrounds is very large, a fast parametric simulation of the ATLAS detector response is used [18, 27].

References

1. Luminosity determination in pp collisions at $\sqrt{s} = 13$ TeV using the ATLAS detector at the LHC. Tech. rep. ATLAS-CONF-2019-021. CERN, Geneva, 2019. https://cds.cern.ch/record/2677054
2. Aad G et al (2020) Performance of electron and photon triggers in ATLAS during LHC Run 2. Eur Phys J C 80(1):47. https://doi.org/10.1140/epjc/s10052-019-7500-2. arXiv: 1909.00761 [hep-ex]
3. Sjöstrand T, Mrenna S, Skands PZ (2008) A brief introduction to PYTHIA 8.1. Comput Phys Commun 178:852. https://doi.org/10.1016/j.cpc.2008.01.036. arXiv: 0710.3820 [hep-ph]
4. Sjöstrand T et al (2015) An introduction to PYTHIA 8.2. Comput Phys Commun 191:159. https://doi.org/10.1016/j.cpc.2015.01.024. arXiv: 1410.3012 [bibhep-ph]
5. Bähr M et al (2008) Herwig++ physics and manual. Eur Phys J C 58:639. https://doi.org/10.1140/epjc/s10052-008-0798-9. arXiv: 0803.0883 [hep-ph]
6. Bellm J et al (2016) Herwig 7.0/Herwig++ 3.0 release note. Eur Phys J C 76(4):196. https://doi.org/10.1140/epjc/s10052-016-4018-8. arXiv: 1512.01178 [hep-ph]

7. Nason P (2004) A new method for combining NLO QCD with shower Monte Carlo algorithms. JHEP 11:040. https://doi.org/10.1088/1126-6708/2004/11/040 arXiv: hep-ph/0409146
8. Frixione S, Nason P, Oleari C (2007) Matching NLO QCD computations with Parton Shower simulations: the POWHEG method. JHEP 11:070. https://doi.org/10.1088/1126-6708/2007/11/070. arXiv: 0709.2092 [hep-ph]
9. Alioli S et al (2010) A general framework for implementing NLO calculations in shower Monte Carlo programs: the POWHEG BOX. JHEP 06:043. https://doi.org/10.1007/JHEP06(2010)043. arXiv: 1002.2581 [hep-ph]
10. Alwall J et al (2014) The automated computation of tree-level and next-to-leading order differential cross sections, and their matching to parton shower simulations. JHEP 07:079. https://doi.org/10.1007/JHEP07(2014)079. arXiv: 1405.0301 [hep-ph]
11. Frixione S, Webber BR (2002) Matching NLO QCD computations and parton shower simulations. JHEP 06:029 arXiv: hep-ph/0204244
12. Gleisberg T et al (2009) Event generation with SHERPA 1.1. JHEP 02:007. https://doi.org/10.1088/1126-6708/2009/02/007. arXiv: 0811.4622 [hep-ph]
13. Höche S et al (2009) QCD matrix elements and truncated showers. JHEP 05:053. https://doi.org/10.1088/1126-6708/2009/05/053. arXiv: 0903.1219 [hep-ph]
14. Gleisberg T, Höche S (2008) Comix, a new matrix element generator. JHEP 12:039. https://doi.org/10.1088/1126-6708/2008/12/039. arXiv: 0808.3674 [hep-ph]
15. Schumann S, Krauss F (2008) A Parton shower algorithm based on Catani-Seymour dipole factorisation. JHEP 03:038. https://doi.org/10.1088/1126-6708/2008/03/038. arXiv: 0709.1027 [hep-ph]
16. Lange DJ (2001) The EvtGen particle decay simulation package. Nucl Instrum Meth A 462:152. https://doi.org/10.1016/S0168-9002(01)00089-4
17. Agostinelli S et al (2003) Geant4–a simulation toolkit. Nucl Instrum Meth A 506:250. https://doi.org/10.1016/S0168-9002(03)01368-8
18. Aad G et al (2010) The ATLAS simulation infrastructure. Eur Phys J C 70:823–874. https://doi.org/10.1140/epjc/s10052-010-1429-9. arXiv: 1005.4568 [physics.ins-det]
19. Summary of ATLAS Pythia 8 tunes. Tech. rep. ATL-PHYS-PUB-2012-003. CERN, Geneva, 2012. https://cds.cern.ch/record/1474107
20. Martin AD et al (2009) Parton distributions for the LHC. Eur Phys J C63:189–285. https://doi.org/10.1140/epjc/s10052-009-1072-5. arXiv: 0901.0002 [hep-ph]
21. Hamilton K et al (2013) NNLOPS simulation of Higgs boson production. JHEP 10:222. https://doi.org/10.1007/JHEP10(2013)222. arXiv: 1309.0017 [hep-ph]
22. Butterworth J et al (2016) PDF4LHC recommendations for LHC Run II. J Phys G 43:023001. https://doi.org/10.1088/0954-3899/43/2/023001. arXiv: 1510.03865 [hep-ph]
23. Aad G et al (2014) Measurement of the $Z/\gamma*$ boson transverse momentum distribution in pp collisions at $\sqrt{s} = 7$ TeV with the ATLAS detector. JHEP 09:145. https://doi.org/10.1007/JHEP09(2014)145. arXiv: 1406.3660 [hep-ex]
24. Ball RD et al (2015) Parton distributions for the LHC Run II. JHEP 04:040. https://doi.org/10.1007/JHEP04(2015)040. arXiv: 1410.8849 [hep-ph]
25. Demartin F et al (2017) tWH associated production at the LHC. Eur Phys J C 77(1):34. https://doi.org/10.1140/epjc/s10052-017-4601-7. arXiv: 1607.05862 [hep-ph]
26. Demartin F et al (2014) Higgs characterisation at NLO in QCD: CP properties of the top-quark Yukawa interaction. Eur Phys J C 74(9):3065. https://doi.org/10.1140/epjc/s10052-014-3065-2. arXiv: 1407.5089 [hep-ph]
27. The new fast calorimeter simulation in ATLAS. Tech. rep. ATL-SOFT-PUB-2018-002. CERN, Geneva, 2018. http://cds.cern.ch/record/2630434

Chapter 6
Statistical Model

The detection of the Higgs to diphoton decay is possible due to the clean signature and excellent photon energy resolution of the ATLAS detector, as well as the high signal-to-background ratio of this channel compared to other Higgs boson decay modes. These factors make $H \to \gamma\gamma$ a powerful channel for Higgs boson measurements, despite the small branching fraction (0.277%). The $H \to \gamma\gamma$ channel is characterized by a peak in the diphoton invariant mass ($m_{\gamma\gamma}$) spectrum, which rises above a smoothly falling background. The size of the Higgs boson signal is extracted from a signal-plus-background maximum likelihood fit to the $m_{\gamma\gamma}$ spectrum in data.

The statistical tools required for a $H \to \gamma\gamma$ analysis are detailed in this Chapter. The first Section explains the statistical advantages of defining multiple analysis categories with different signal-to-background ratios, an approach that is used for the analyses detailed in later Chapters. Section 6.2 describes the parameterization of the shape of the signal and background $m_{\gamma\gamma}$ distributions. The likelihood model is introduced in Sect. 6.3, and sources of systematic uncertainty are described in Sect. 6.4. The final Section describes a common statistical procedure used to reweight data for presentation purposes.

6.1 Motivation for Categorization

In the analyses included in this Thesis, events passing a simple preselection are divided into multiple categories, and a maximum likelihood fit is performed simultaneously to the $m_{\gamma\gamma}$ spectra in all categories. The partition of selected events into categories helps to maximize the statistical significance Z: as long as the categories contain different ratios of signal to background events (S/B), the statistical significance of signal process S (e.g. $t\bar{t}H$) is higher in the multi-category case than in the single-category case.

J. E. Dickinson, *ATLAS Measurements of the Higgs Boson Coupling to the Top Quark in the Higgs to Diphoton Decay Channel*, Springer Theses,
https://doi.org/10.1007/978-3-030-86368-5_6

The short proof below demonstrates that this is true for categories containing large numbers of data events, where the statistical significance is well approximated by $Z = S/\sqrt{B}$. The conclusion holds for more complex formulations of the likelihood (including that outlined in Sect. 6.3), assuming that B is well estimated.

6.1.1 A Short Proof

Any region can be partitioned into two categories such that Category 1 contains S_1 signal and B_1 background events, and Category 2 contains S_2 signal and B_2 background events. The un-partitioned region contains $S_1 + S_2$ signal and $B_1 + B_2$ background events, and the corresponding signal significance is

$$Z = \frac{S_1 + S_2}{\sqrt{B_1 + B_2}} \tag{6.1}$$

Following the partition into two categories, the signal significance is the sum in quadrature of the single-category significances:

$$Z' = \sqrt{\frac{S_1^2}{B_1} + \frac{S_2^2}{B_2}} \tag{6.2}$$

The following demonstrates that $Z' > Z$ when $S_1/B_1 \neq S_2/B_2$. All event yields S_1, S_2, B_1, and B_2 must be ≥ 0.

$$Z'^2 - Z^2 = \frac{S_1^2}{B_1} + \frac{S_2^2}{B_2} - \frac{(S_1 + S_2)^2}{B_1 + B_2} \tag{6.3}$$

Applying a common denominator gives

$$= \frac{(S_1^2 B_2 + S_2^2 B_1)(B_1 + B_2) - B_1 B_2 (S_1 + S_2)^2}{B_1 B_2 (B_1 + B_2)} \tag{6.4}$$

Expanding and canceling like terms gives

$$= \frac{B_1^2 S_2^2 + B_2^2 S_1^2 - 2 B_1 B_2 S_1 S_2}{B_1 B_2 (B_1 + B_2)} \tag{6.5}$$

which simplifies to

$$= \frac{B_1 B_2}{B_1 + B_2} (\frac{S_1}{B_1} - \frac{S_2}{B_2})^2 > 0 \tag{6.6}$$

QED.

6.2 Signal and Background Shapes

The $H \rightarrow \gamma\gamma$ signal peak and the continuum diphoton background are modeled by analytic functions of $m_{\gamma\gamma}$. This Section outlines the criteria for selecting these functions, which are later fitted to data to extract cross section and signal strength results.

6.2.1 Signal Model

The **Double-Sided Crystal Ball function (DSCB)** is described by a Gaussian core and two asymmetric exponential tails:

$$f_{\mathrm{DSCB}}(m_{\gamma\gamma}) = N \times \begin{cases} e^{-t^2/2} & \text{if } -\alpha_{\mathrm{low}} \leq t \leq \alpha_{\mathrm{high}} \\ e^{-\frac{1}{2}\alpha_{\mathrm{low}}^2}\left[\frac{1}{R_{\mathrm{low}}}\left(R_{\mathrm{low}} - \alpha_{\mathrm{low}} - t\right)\right]^{-n_{\mathrm{low}}} & \text{if } t < -\alpha_{\mathrm{low}} \\ e^{-\frac{1}{2}\alpha_{\mathrm{high}}^2}\left[\frac{1}{R_{\mathrm{high}}}\left(R_{\mathrm{high}} - \alpha_{\mathrm{high}} + t\right)\right]^{-n_{\mathrm{high}}} & \text{if } t > \alpha_{\mathrm{high}} \end{cases} \tag{6.7}$$

where $t = (m_{\gamma\gamma} - \mu_{\mathrm{CB}})/\sigma_{\mathrm{CB}}$ and $R = \frac{n}{\alpha}$. The parameters μ_{CB} and σ_{CB} describe the mean and width of the Gaussian core, and α_{low}, α_{high}, n_{low}, and n_{high} describe the tails.

In each analysis category, the signal parameters are fit to the inclusive SM Higgs boson Monte Carlo sample (the sum of all SM Higgs boson production modes), which are generated with $m_H = 125$ GeV. The signal shape parameters in each category are taken from these fits. The Higgs boson mass has been measured by ATLAS and CMS [1] to be slightly higher than 125 GeV:

$$m_H = 125.09 \pm 0.21\ \text{(stat)}\ \pm 0.11\ \text{(syst) GeV} \tag{6.8}$$

The mean of the DSCB is therefore shifted by $m_H - 125 = 0.09$ GeV for consistency with the measured Higgs boson mass. This rigid translation is viable because the energy scale systematics are large compared to the size of this shift.

The Monte Carlo $m_{\gamma\gamma}$ distribution and fitted signal shape are shown in Fig. 6.1 for two example analysis categories targeting the $t\bar{t}H(\gamma\gamma)$ process (details in Chap. 9). In both categories, the DSCB captures the shape of the Monte Carlo signal. The fitted signal resolution (σ_{CB}) varies between categories depending on photon kinematics, and will be reported separately for each analysis in later Chapters.

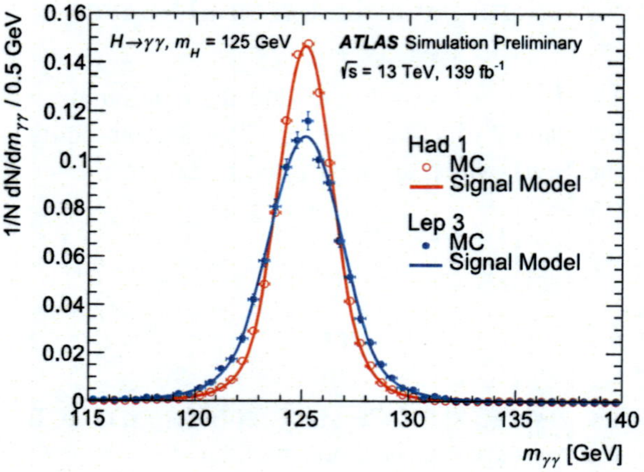

Fig. 6.1 Diphoton invariant mass spectrum in Higgs boson Monte Carlo events with the fitted DSCB function overlaid, shown for two categories targeting $t\bar{t}H(\gamma\gamma)$ events with high (low) signal-to-background ratio in red (blue). The $m_{\gamma\gamma}$ resolution (σ_{CB}) varies between the categories due to differences in photon kinematics

6.2.2 Background Model

The continuum background for $H \rightarrow \gamma\gamma$ decays is evaluated using a fully data-driven method. Simplified methods (including the estimation of continuum background yield from the data control regions described in Sect. 5.1.1) are used only for the optimization of analysis cuts and strategies.

The continuum background $m_{\gamma\gamma}$ distribution is modeled by a smoothly falling analytic function. The functional form is chosen using the **spurious signal test** to maximize sensitivity to signal and minimize the systematic uncertainty on the background shape. Full details of the spurious signal test can be found in [2].

The spurious signal test is used only to determine the form of the background function for the final signal-plus-background fit, not the values of the free parameters. The overall background normalization and the values of the parameters are determined in the final signal-plus-background fit.

To perform the spurious signal test, a template $m_{\gamma\gamma}$ distribution is constructed to model the continuum background in each analysis category. The templates are constructed from representative Monte Carlo or data control samples. Each template distribution is checked against the data sidebands to ensure that the chosen sample is a reasonable model of the background distribution in data. A χ^2 fit is used to determine this compatibility, which is reported alongside the spurious signal results in later Chapters.

The background template is fitted with a signal-plus-background model in the range 105 GeV $< m_{\gamma\gamma} <$ 160 GeV, where the background shape is the function under evaluation. The signal shape used in this fit is a DSCB function with parameters determined according to the procedure outlined in Sect. 6.2.1. The number of fitted signal events is computed for Higgs boson masses varying in intervals of 1 GeV from 121 to 129 GeV. Negative fitted signal is allowed: this indicates that the fit function over-predicts the continuum background. The spurious signal N_{sp} is taken to be the number of fitted signal events with largest magnitude in this 8 GeV window.

The spurious signal test is performed for a variety of candidate background functions:

- Exponential function: $f(m_{\gamma\gamma}) = e^{cm_{\gamma\gamma}}$
- Exponential function of a second order polynomial: $f(m_{\gamma\gamma}) = e^{c_1 m_{\gamma\gamma}^2 + c_2 m_{\gamma\gamma}}$
- Bernstein polynomial of order N=3,4,5: $B_N(m_{\gamma\gamma}) = \sum_{i=0}^{N} c_i \binom{N}{i} m_{\gamma\gamma}^i (1 - m_{\gamma\gamma})^{N-i}$
- Power law function: $f(m_{\gamma\gamma}) = m_{\gamma\gamma}^c$
- Dijet function: $f(m_{\gamma\gamma}) = m_{\gamma\gamma}^{c_1}(1 - m_{\gamma\gamma})^{c_2}$

The selected background function must satisfy at least one of the following criteria:

- $N_{sp} < 0.10 \times N_{s,\text{exp}}$, where $N_{s,\text{exp}}$ is the expected SM number of signal events in the category.
- $N_{sp} < 0.20 \times \sigma_{\text{lumi}}$, where σ_{lumi} is the statistical uncertainty on the fitted number of signal events when fitting the signal-plus-background model to the template. This ensures that the spurious signal is negligible compared with statistical uncertainty.

If more than one candidate function passes the test, the function with the fewest free parameters is selected. In cases where multiple functions with the same number of free parameters pass, the function with the smaller value of N_{sp} is chosen.

In the case of low statistics categories such as $t\bar{t}H$, a single-parameter function is greatly preferred in order to avoid fitting statistical fluctuations in the background templates. The spurious signal criteria are relaxed to accommodate 2σ local statistical fluctuations in the background template. This is done by defining a new variable:

$$\zeta_{sp} = \begin{cases} N_{sp} + 2\Delta_{MC}, & N_{sp} + 2\Delta_{MC} < 0 \\ N_{sp} - 2\Delta_{MC}, & N_{sp} - 2\Delta_{MC} > 0 \\ 0, & \text{otherwise} \end{cases} \tag{6.9}$$

where Δ_{MC} is the statistical uncertainty on the background template. The new variable ζ_{sp} should then satisfy the same criteria as N_{sp} before. Though the ζ_{sp} variable is used to choose the background functional form, N_{sp} (not ζ_{sp}) is used as the spurious signal uncertainty.

The number of spurious signal events in each category is treated as a systematic uncertainty on the signal yield. The selected functions and spurious signal systematics will be reported separately for each analysis in later Chapters.

6.3 Likelihood Model

All statistical tests in this work use the profile likelihood ratio as the test statistic:

$$\Lambda(x) = \frac{\mathcal{L}(x, \hat{\hat{\theta}}(x))}{\mathcal{L}(\hat{x}, \hat{\theta})} \tag{6.10}$$

where $\mathcal{L}$ is the likelihood function. In this equation, x represents the parameters of interest (POIs), and θ is the vector of all nuisance parameters (NPs), including systematic uncertainties and parameters describing the shape and normalization of non-Higgs background. The variables $\hat{x}$ and $\hat{\theta}$ denote the values of x and θ that maximize the unconditional likelihood estimate (all NPs are free), while $\hat{\hat{\theta}}(x)$ denotes the values of θ that maximize the conditional likelihood estimate at a fixed x.

The test statistic $-2\ln\Lambda$ is assumed to follow a χ^2 distribution with one degree of freedom [3]. Agreement of the measured x with a test hypothesis is quantified (in the asymptotic approximation) by a p-value from the observed value of $-2\ln\Lambda(x)$.

In the case of a single-category analysis, the likelihood function $\mathcal{L}$ is constructed from a signal-plus-background model. The parameterization of the signal model s and the choice of background functional form b for each category are described in detail in the previous Section. The resulting probability density functions are:

$$s(m_{\gamma\gamma};\mu_{\text{CB}},\sigma_{\text{CB}}) \tag{6.11}$$

$$b(m_{\gamma\gamma};\boldsymbol{\xi}) \tag{6.12}$$

The DSCB parameters in the signal function are fixed to the values found by following the procedure in Sect. 6.2.1. The parameters α^{low}, n^{low}, α^{high}, n^{high} are omitted in the argument of s for the sake of brevity. The background shape parameters $\boldsymbol{\xi}$ remain free in the fit.

The most common choice of POI is the **signal strength** μ, which gives the ratio of measured to SM cross sections:

$$\mu_i = \frac{\sigma_i^{\text{meas}}}{\sigma_i^{SM}} \tag{6.13}$$

In order to measure different POI, such as the κ parameters described in Sect. 2.3.3, the signal strengths can be re-paramtereized (see Table 2.3). Apart from the re-parameterization of the signal strength, the likelihood and test statistic remain unchanged. For the measurement of the CP properties of $t\bar{t}H$, the signal strengths are re-parameterized in terms of POI κ_t and mixing angle α (see Chap. 12).

The likelihood for a single category without systematics takes the form

$$\mathcal{L}(\mu, N_B, \boldsymbol{\xi}) = e^{-N_S-N_B}\prod_k\left[N_B b(m^k_{\gamma\gamma};\boldsymbol{\xi}) + N_S s(m^k_{\gamma\gamma};\mu_{\text{CB}},\sigma_{\text{CB}})\right] \tag{6.14}$$

where $m^k_{\gamma\gamma}$ are the measured data points, N_B the number of continuum background events, and N_S the number of measured Higgs boson events, which can be written

$$N_S = \sum_p \mu_p \times N_p \tag{6.15}$$

where N_p is the predicted yield of process p. For a multi-category analysis, the total likelihood function without systematics is the product of the likelihood in each category. All categories use the same parametrization for the likelihood (Eq. 6.14), and the POI are correlated across all categories. The background normalization and shape parameters (N_B^c and $\boldsymbol{\xi}^c$) are fitted independently in each category c. Measurements of the POI are extracted from an unbinned maximum likelihood fit over the total multi-category likelihood.

6.3.1 *Treatment of Systematics*

Systematic uncertainties are incorporated into the likelihood to account for known sources of imprecision in experimental measurements and theoretical calculations. Each systematic uncertainty has a value (sign and magnitude) σ and an associated **nuisance parameter (NP)** θ that is left free in the maximum likelihood fit. Systematic uncertainties in $H \rightarrow \gamma\gamma$ analyses can impact the measured Higgs signal yield from production process p (N_p), the mean of the Higgs signal shape (μ_{CB}), and the width of the Higgs signal shape (σ_{CB}).

For each systematic uncertainty on the Higgs signal yield from production process p, the parameter N_p in the likelihood is modified by a **response term** $(1 + \sigma_i \theta_i)$:

$$N_p \rightarrow N_p \prod_i (1 + \sigma_i \theta_i) \tag{6.16}$$

For each systematic uncertainty on the signal mean and resolution, response terms are added to μ_{CB} and σ_{CB} respectively:

$$\mu_{\text{CB}} \rightarrow \mu_{\text{CB}} \prod_i (1 + \sigma_i \theta_i) \tag{6.17}$$

$$\sigma_{\text{CB}} \rightarrow \sigma_{\text{CB}} \prod_i (1 + \sigma_i \theta_i) \tag{6.18}$$

A **constraint term** associated with each systematic uncertainty modifies the overall likelihood:

$$\mathcal{L}(\mu, \theta) \rightarrow \mathcal{L}(\mu, \theta) \prod_i F_i(r_i, \theta_i) \tag{6.19}$$

The means of the constraint terms r_i are **global observables**: these represent the expected values of the NPs θ_i. In the unconditional maximum likelihood fit, all global observables are zero. The constraint term $F(r, \theta)$ takes one of the following forms:

- Gaussian:

$$F(r,\theta) = \exp\left[-\frac{1}{2}\left(\frac{\theta - r}{\sigma}\right)^2\right] \tag{6.20}$$

- Log Normal:

$$F(r,\theta) = \frac{1}{(\theta - r)}\exp\left[-\frac{1}{2}\left(\frac{\ln(\theta - r)}{\sigma}\right)^2\right] \tag{6.21}$$

- Asymmetric: an asymmetric implementation of the log normal constraint with a smooth polynomial interpolation in between [4].

The forms of the constraints for each systematic uncertainty are summarized in Tables 6.2 and 6.3.

For the remainder of this work, the likelihood used in the maximum likelihood fit used includes the systematic uncertainties summarized in Sect. 6.4.

6.3.2 Generation of Asimov Datasets

For each Higgs boson measurement, expected results are evaluated on an **Asimov dataset** [3], which is used to represent the data that would be obtained for a specified value of the POI. Expected results (that is, the results that would be obtained if the data followed the Standard Model) can be derived from a fit to the signal-plus-background Asimov dataset. There are several possibilities for the evaluation of the signal-plus-background Asimov dataset.

Pre-fit Asimov does not use data in the region near the Higgs mass peak, but only data in the sideband region ($|m_{\gamma\gamma} - 125| > 5$ GeV). This Asimov prescription is particularly useful before an analysis is unblinded. A fit is performed to data in the sidebands only, and each global observable is set to the best fit value of the corresponding NP. The SM ($\mu = 1$) signal is super-imposed on the background obtained in this way.

Similarly, **post-fit Asimov** using the full diphoton mass range (105–160 GeV) can be used to compute the expected significance. In this case, a fit is performed to data in the full $m_{\gamma\gamma}$ range, and each global observable fixed to the best fit value of the corresponding NP. This fit can be performed in two ways:

1. Fix the signal strength μ fixed to 1, then perform the fit to data. Set each global observable to the best fit value of the corresponding NP. Generate the $m_{\gamma\gamma}$ distribution.
2. Perform the fit to data with μ profiled. Set each global observable to the best fit value of the corresponding NP. Set μ to 1 and generate the $m_{\gamma\gamma}$ distribution.

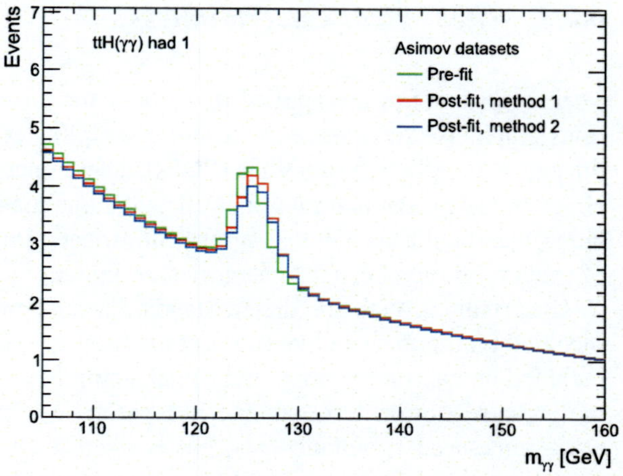

Fig. 6.2 Example signal-plus-background Asimov datasets in one $t\bar{t}H(\gamma\gamma)$ analysis category. The pre-fit Asimov dataset is shown in green, and the post-fit Asimov datasets generated with methods 1 and 2 are shown in red and blue respectively

The signal-plus-background Asimov datasets (and therefore the expected signal significances) obtained by these different methods are not necessarily the same. Figure 6.2 shows the pre-fit Asimov dataset (green) and the post-fit Asimov datasets obtained by methods 1 (red) and 2 (blue) in an example analysis category targeting the $t\bar{t}H(\gamma\gamma)$ process (details in Chap. 9). Differences are visible in the shape and normalization of the continuum background, as well as in the shape and normalization of the Higgs boson mass peak, due to the different treatment of the global observables in these three prescriptions.

In the analysis category shown in Fig. 6.2, an excess of events in the $m_{\gamma\gamma}$ region 125–130 GeV causes a shift in the position and width of the signal peak in the post-fit Asimov datasets relative to the pre-fit Asimov dataset (which is insensitive to data in the region 120–130 GeV). Because the fit used to generate the Asimov dataset in method 1 assumes a rigid $\mu = 1$ normalization of the signal peak, the observed excess in data results in a higher background normalization relative to method 2. In addition, profiling μ in method 2 can accommodate the excess in observed data without substantial pulls on signal systematic uncertainties. In method 1, these systematics are pulled to fit the excess, resulting in a higher signal normalization than method 2.

For the remainder of this work, expected results are derived on a post-fit Asimov dataset generated by profiling the POI (method 2, corresponding to the blue curve in Fig. 6.2).

6.4 Sources of Systematic Uncertainty

The systematic uncertainties on the $t\bar{t}H(\gamma\gamma)$ measurement can be divided into two primary categories: theoretical uncertainties and experimental uncertainties. The source, form, and impact of these uncertainties is summarized below.

6.4.1 Theoretical Uncertainties

Theory systematics are applied to account for missing higher order QCD corrections, imprecise knowledge of the strong coupling constant α_S, and uncertainties on the parton distribution functions (PDFs). These uncertainties can impact the overall Higgs boson production rates as well as the kinematics of Higgs boson events. The latter uncertainties (known as **migration uncertainties**) account for the possibility of event migration between categories, or in/out of selection entirely.

Uncertainties on the predicted Standard Model cross section of process p are not included on the measurement of cross section σ_p. However, these uncertainties are included on measurements of the signal strength μ_p, since this result parameterizes the cross section in terms of the SM prediction. In addition, if a process $q \neq p$ is not measured simultaneously, but is fixed to the SM expectation, then theory uncertainties on the inclusive SM σ_q are included on the measurement of both σ_p and μ_p. Migration uncertainties are included on measurements of both σ_p and μ_p.

The sign and magnitude of each theory uncertainty is calculated separately in each category and for each signal process: this includes the $t\bar{t}H$ process for the cross section measurement described in Chap. 10, and $t\bar{t}H$, $tHjb$, and tWH for the measurement of CP mixing angle described in Chap. 12. Theory uncertainties are evaluated by varying the QCD renormalization and factorization scales (μ_R, μ_F) up and down by a factor of two, and by varying the PDF parameters according to prescriptions laid out in [5] or [6].

For processes with rates much lower than the target process(es), inclusive uncertainties [5] can be applied in every category. The values of these can be found in Table 6.1. These are used for example for $tHjb$, tWH and $b\bar{b}H$ in Chap. 10.

For Higgs boson production modes with large rates (ggF, VBF, and VH), inclusive uncertainties are available. However, existing measurements of $V + b$-jets [7, 8] and $H + b$-jets in the $H \rightarrow ZZ \rightarrow 4\ell$ channel [9] indicate that there is substantial mis-modeling in regions where at least one b-jet is required. Figure 6.3 shows a

Table 6.1 QCD scale and PDF uncertainties on inclusive Higgs boson production cross sections at $\sqrt{s} = 13$ TeV with $m_H = 125.09$ GeV [5]

Process	Order (QCD)	QCD+ [%]	QCD− [%]	PDF [%]
ggF	N^3LO	4.6	6.7	3.2
VBF	NNLO	0.4	0.3	2.1
WH	NNLO	0.5	0.7	1.9
ZH	NNLO	3.8	3.0	1.6
$t\bar{t}H$	NLO	5.8	9.2	3.6
$b\bar{b}H$	5FS NNLO or 4FS NLO	20.1	23.9	–
$tHjb$	LO	6.5	14.7	3.7
tWH	NLO	4.9	6.7	6.3

The QCD uncertainties listed for the $b\bar{b}H$ process are combined QCD/PDF uncertainties

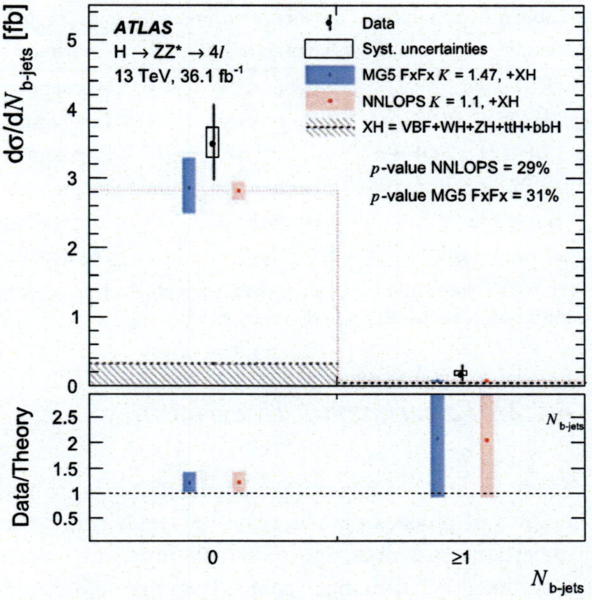

Fig. 6.3 Measured Higgs boson production cross section in bins of b-jet multiplicity in the $H \rightarrow ZZ \rightarrow 4\ell$ decay channel with 36.1 fb^{-1} of 13 TeV data [9]. The discrepancy between data and simulation in the ≥ 1 b-jets bin is $\mathcal{O}(100\%)$

discrepancy of $\mathcal{O}(100\%)$ between the predicted and measured rates of Higgs boson production with ≥ 1 b-jet, which is far too large to be covered by the inclusive uncertainties in Table 6.1. Since the preselection for the $t\bar{t}H(\gamma\gamma)$ analyses includes a b-jet requirement, a conservative 100% **heavy flavor uncertainty** is assigned to the ggF, VBF, and VH processes to account for this discrepancy.

Migration uncertainties due to Monte Carlo modeling can be calculated by comparing the predictions of different Monte Carlo generators. The **Underlying Event/Parton Showering (UEPS)** systematic accounts for uncertainty in the modeling of the underlying event, parton shower, and hadronization. This uncertainty is evaluated by comparing the same Monte Carlo generator interfaced with different parton showering algorithms (typically Pythia and Herwig). The relative UEPS uncertainty on the yield n_p of process p is

$$\frac{n_p^{\text{Herwig}} - n_p^{\text{Pythia}}}{n_p^{\text{Pythia}}} \tag{6.22}$$

The theory uncertainties and associated nuisance parameters (NPs) are summarized in Table 6.2. The correlation of NPs across categories and the form of the constraint terms are also shown. Details on the different types of constraints are included in Sect. 6.3.1.

Table 6.2 Summary of the theory uncertainties incorporated into the $H \to \gamma\gamma$ likelihood model

Source	Number of NPs	Constraint type	Correlated
QCD scale	1 per process	Asymmetric	Yes
LHAPDF: PDF + α_S	1 per process	Log normal	Yes
PDF4LHC: PDF + α_S	31	Log normal	Yes
UEPS	1 per process	Log normal	Yes
Heavy Flavor	1 per process	Log normal	Yes
$H \to \gamma\gamma$ BR	1	Asymmetric	Yes

The PDF uncertainty is evaluated either inclusively using the LHAPDF prescription [6] or using the 31 eigenvector PDF4LHC scheme [5]

6.4.2 Experimental Uncertainties

Many sources of experimental uncertainty arise from the reconstruction and calibration of physics objects from low-level detector responses. The impact of these uncertainties is obtained by calculating systematic variations provided by ATLAS Combined Performance groups. Experimental uncertainties include

- **Luminosity**, which contributes an uncertainty of 1.7% to the total event yield [10]
- **Trigger**, which contributes an uncertainty of 0.4% to the total event yield [11, 12]
- **Pileup reweighting**
- **Jet uncertainties**, including uncertainties on tracks, vertex identification, jet energy scale and resolution, and flavor composition/tagging [13–17]
- **Photon uncertainties**, including energy scale, resolution, identification, and isolation [18, 19]
- **Electron uncertainties**, including energy scale, resolution, identification, and isolation [18–20]
- **Muon uncertainties**, including energy scale, resolution, identification, and isolation [21]
- **Missing** E_T [22, 23].

In a multi-category analysis, the sign and magnitude of these experimental uncertainties are calculated separately in each category.

The choice of an analytic function to model the continuum background also introduces a potential bias to the signal plus background fit. The spurious signal test (Sect. 6.2.2) quantifies the impact of a given background model on the fitted number of signal events. The spurious signal is therefore included as an experimental systematic uncertainty on the signal yield in each category.

Finally, the uncertainty of the measured Higgs boson mass from the ATLAS and CMS combination [1] is also applied to the mean of the signal DSCB shape (μ_{CB}) by following Eq. 6.17.

The experimental uncertainties and associated nuisance parameters (NPs) are summarized in Table 6.3. The correlation of NPs across categories and the form of

Table 6.3 Summary of the experimental uncertainties incorporated into the $H \to \gamma\gamma$ likelihood model

Source	Number of NPs	Constraint type	Correlated
Yield			
Luminosity	1	Log normal	Yes
Trigger efficiency	1	Log normal	Yes
Photon efficiency	3 (ID, isolation, trigger)	Asymmetric	Yes
Yield and Migration			
Flavor tagging	12	Asymmetric	Yes
Jets	25	Asymmetric	Yes
Jet flavor	14	Asymmetric	Yes
Electrons	2	Asymmetric	Yes
Muons	7	Asymmetric	Yes
Missing E_T	3	Asymmetric	Yes
Pileup	1	Asymmetric	Yes
Shape			
Photon energy scale	69 (40 merged)	Gaussiasn	Yes
Photon energy resolution	9 (5 merged)	Log normal	Yes
Spurious signal	1 per category	Gaussian	No
Measured m_H	1	Gaussian	Yes

the constraint terms are also shown. Details on the different types of constraints are included in Sect. 6.3.1.

6.5 Weighting Data for Presentation

Every data event carries equal weight in the statistical analysis. However, it can be useful to look at weighted data in order to visualize the power of a multi-category analysis. A weight of $\ln(1 + S/B)$ can be applied to each data event, where S (B) is the expected signal (background) in the analysis category containing the event. This weighting procedure serves to magnify the contribution of the most sensitive categories, which have the highest S/B. Figure 6.4 shows an example of this weighting procedure applied to diphoton mass distribution in the $t\bar{t}H(\gamma\gamma)$ categories described in Chap. 9.

The central value of a single bin in the weighted data distribution is given by the weighted sum over categories:

$$\lambda = \sum_c w_c n_c \tag{6.23}$$

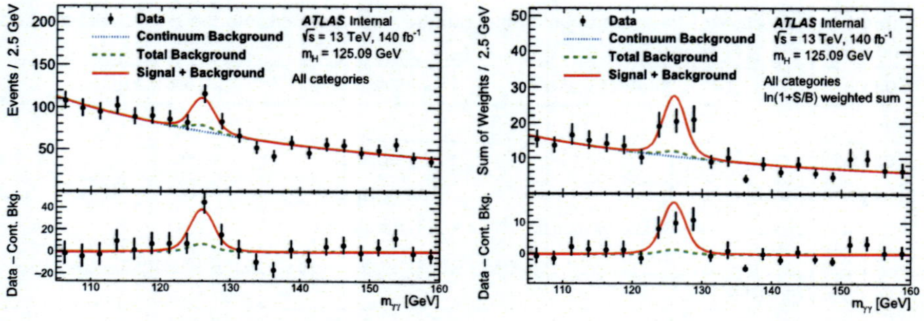

(a) Unweighted (b) Weighted

Fig. 6.4 Example diphoton invariant mass distribution in the unweighted (left) and $\ln(1+S/B)$ weighted (right) presentations. The discussion of the particulars of this data is deferred until Sect. 10.2

where n_c is the unweighted event yield and $w_c = \ln(1 + S_c/B_c)$ is the calculated weight in category c. The statistical uncertainty on each bin of the weighted data distribution is calculated by following the procedure outlined in [24]. A Poisson distribution $P(n_c/f, \lambda/f)$ is constructed for the scaled variable n_c/f with mean λ/f, where

$$f = \frac{\sum_c w_c^2 n_c}{\sum_c w_c n_c} \tag{6.24}$$

Statistical uncertainties are calculated on this distribution, and the scaling is undone by multiplying by f. The variance thus obtained is

$$f^2\sigma^2 = f\lambda = \sum_i w_i^2 n_i \tag{6.25}$$

The statistical error bars on Fig. 6.4b and similar Figures are derived according to this prescription.

References

1. Aad G et al (2015) Combined measurement of the Higgs Boson Mass in pp collisions at $\sqrt{s} = 7$ and 8 TeV with the ATLAS and CMS experiments. Phys Rev Lett 114:191803. https://doi.org/10.1103/PhysRevLett.114.191803. arXiv: 1503.07589 [hepspsex]
2. Hasib A et al (2016) Supporting document: probing Higgs production modes and couplings with the $H \rightarrow \gamma\gamma$ channel with the Run 2 of LHC in the ATLAS experiment. ATL-COM-PHYS-2016-1784. https://cds.cern.ch/record/2238687
3. Cowan G et al (2011) Asymptotic formulae for likelihood-based tests of new physics. Eur Phys J. C 71:1554. https://doi.org/10.1140/epjc/s10052-011-1554-0. arXiv: 1007.1727 [physics.data-an]

4. Cranmer K et al (2012) HistFactory: a tool for creating statistical models for use with RooFit and RooStats. Tech. rep. CERN-OPEN-2012-016. New York. https://cds.cern.ch/record/1456844
5. LHC Higgs Cross Section Working Group (2017) Handbook of LHC Higgs cross sections: 4. Deciphering the Nature of the Higgs Sector. Tech. rep. FERMILAB/CERN. https://cds.cern.ch/record/2227475
6. Buckley A et al (2015) LHAPDF6: parton density access in the LHC precision era. Eur Phys J C 75:132. https://doi.org/10.1140/epjc/s10052-015-3318-8. arXiv: 1412.7420 [hep-ph]
7. Aad G et al (2014) Measurement of differential production cross-sections for a Z boson in association with b-jets in 7 TeV proton-proton collisions with the ATLAS detector. JHEP 10:141. https://doi.org/10.1007/JHEP10(2014)141. arXiv: 1407.3643 [hepspsex]
8. Aad G et al (2013) Measurement of the cross-section for W boson production in association with b-jets in pp collisions at $\sqrt{s} = 7$ TeV with the ATLAS detector. JHEP 06:084. https://doi.org/10.1007/JHEP06(2013)084. arXiv: 1302.2929 [hepspsex]
9. Aaboud M et al (2017) Measurement of inclusive and differential cross sections in the $H \rightarrow ZZ* \rightarrow 4\ell$ decay channel in pp collisions at $\sqrt{s} = 13$ TeV with the ATLAS detector. JHEP 10:132. https://doi.org/10.1007/JHEP10(2017)132. arXiv: 1708.02810 [hepspsex]
10. Luminosity determination in pp collisions at $\sqrt{s} = 13$ *TeV using the ATLAS detector at the LHC*. Tech. rep. ATLAS-CONF-2019-021. CERN, Geneva, 2019. https://cds.cern.ch/record/2677054
11. Aaboud M et al (2017) Performance of the ATLAS trigger system in 2015. Eur Phys J C 77(5):317. https://doi.org/10.1140/epjc/s10052-017-4852-3. arXiv: 1611.09661 [hepspsex]
12. Aad G et al (2020) Performance of electron and photon triggers in ATLAS during LHC Run 2. Eur Phys J C 80(1):47. https://doi.org/10.1140/epjc/s10052-019-7500-2. arXiv: 1909.00761 [hepspsex]
13. Early inner detector tracking performance in the 2015 data at $\sqrt{s} = 13$ *TeV*. Tech. rep. ATL-PHYS-PUB-2015-051. CERN, Geneva, 2015. http://cds.cern.ch/record/2110140
14. Aaboud M et al (2017) Reconstruction of primary vertices at the ATLAS experiment in Run 1 proton–proton collisions at the LHC. Eur Phys J C 77(5):332. https://doi.org/10.1140/epjc/s10052-017-4887-5. arXiv: 1611.10235 [physics.ins-det]
15. Aaboud M et al (2017) Jet energy scale measurements and their systematic uncertainties in proton-proton collisions at $\sqrt{s} = 13$ TeV with the ATLAS detector. Phys Rev D 96(7):072002. https://doi.org/10.1103/PhysRevD.96.072002. arXiv: 1703.09665 [hepspsex]
16. Aad G et al (2016) Performance of b-jet identification in the ATLAS experiment. JINST 11(04):P04008. https://doi.org/10.1088/1748-0221/11/04/P04008. arXiv: 1512.01094 [hepspsex]
17. Aad G et al (2019) ATLAS b-jet identification performance and efficiency measure- ment with tt– events in pp collisions at $\sqrt{s} = 13$ TeV. Eur Phys J C 79(11):970. https://doi.org/10.1140/epjc/s10052-019-7450-8. arXiv: 1907.05120 [hepspsex]
18. Aaboud M et al (2019) Electron and photon energy calibration with the ATLAS de- tector using 2015–2016 LHC proton-proton collision data. JINST 14(03):P03017. https://doi.org/10.1088/1748-0221/14/03/P03017. arXiv: 1812.03848 [hepspsex]
19. Aad G et al (2019) Electron and photon performance measurements with the ATLAS detector using the 2015–2017 LHC proton-proton collision data. JINST 14(12):P12006. https://doi.org/10.1088/1748-0221/14/12/P12006. arXiv: 1908.00005 [hepspsex]
20. Aaboud M et al (2019) Electron reconstruction and identification in the ATLAS exper- iment using the 2015 and 2016 LHC proton-proton collision data at $\sqrt{s} = 13$ TeV. Eur Phys J C 79(8):639. https://doi.org/10.1140/epjc/s10052-019-7140-6. arXiv: 1902.04655 [physics.ins-det]
21. Aad G et al (2016) Muon reconstruction performance of the ATLAS detector in proton–proton collision data at $\sqrt{s} = 13$ TeV. Eur Phys J C 76(5):292. https://doi.org/10.1140/epjc/s10052-016-4120-y. arXiv: 1603.05598 [hepspsex]
22. Aaboud M et al (2018) Performance of missing transverse momentum reconstruction with the ATLAS detector using proton-proton collisions at $\sqrt{s} = 13$ TeV. Eur Phys J C 78(11):903, 66. arXiv:1802.08168. https://doi.org/10.1140/epjc/s10052-018-6288-9. https://cds.cern.ch/record/2305380

23. E_T^{miss} performance in the ATLAS detector using 2015-2016 LHC p-p collisions. Tech. rep. ATLAS-CONF-2018-023. CERN, 2018, Geneva. http://cds.cern.ch/record/2625233
24. Bohm G, Zech G (2010) Introduction to statistics and data analysis for physicists. DESY. ISBN: 9783935702416. http://www-library.desy.de/preparch/books/vstatmp_engl.pdf

Chapter 7
Machine Learning Techniques

The goal of Machine Learning (ML) is to make predictions about unknown data based on a large set of example data. As Machine Learning tools have developed and matured, they have been applied with great success to the proton–proton dataset provided by the LHC and the detailed Monte Carlo simulations produced by theorists.

This Chapter provides an overview of the use of the Machine Learning techniques used for the physics analyses included in this Thesis. Section 7.1 provides an overview of multivariate analysis (MVA) strategies, with a focus on classification problems. Section 7.2 describes a class of ML models known as **Boosted Decision Trees (BDTs)**, which are applied extensively throughout the following Chapters. The methods used to optimize ML models through hyper-parameter tuning are described in Sect. 7.3.

7.1 Multivariate Analysis

A physics analysis that makes use of Machine Learning techniques is called a **multivariate analysis (MVA)**. Given properties, or "features," $\vec{x}$ of a proton–proton event, an MVA aims predict another property y of the same event. Common features used in MVAs include the energy, flavor, and directional information of the physics objects (or combinations of objects) described in Chap. 4.

The target property y is predicted by finding a function F that approximates F_{true} such that

$$F_{\text{true}}(\vec{x}) = y \tag{7.1}$$

The allowed functional forms of the function F depend on the type of Machine Learning model used (Boosted Decision Tree, Neural Network, Support Vector Machine, etc.). The **complexity** of the model is related to the number of free parameters in the

J. E. Dickinson, *ATLAS Measurements of the Higgs Boson Coupling to the Top Quark in the Higgs to Diphoton Decay Channel*, Springer Theses,
https://doi.org/10.1007/978-3-030-86368-5_7

function F: a very complex model has many free parameters, while a simple model has few. The choice of **hyper-parameters** $\vec{h}$ determines the complexity of a model (see Sect. 7.3).

To begin MVA development, events with known $(\vec{x}, y)$ (e.g. from simulation) are divided into three statistically independent subsets:

1. A **training set**: example points to which the model's free parameters are fit
2. A **validation set**: used select the model's hyper-parameters (see Sect. 7.3)
3. A **testing set**: used to estimate the final performance of the model

These datasets need not be equal in size. In order to ensure that no bias is introduced, it is important that the model parameters and the hyper-parameters be optimized using independent datasets. The testing set is not used in optimization of the model, but kept as a statistically independent dataset to use in later analysis stages.

Using the training set, the free parameters of the model are chosen to minimize the **loss function**, which quantifies the accuracy of the estimated function F. A common choice of loss function is the mean squared error (MSE), or Euclidean distance between the predictions $F(\vec{x}_i)$ and observation y_i:

$$\text{MSE Loss } = L(F) = \sum_{i=1}^{N} (F(\vec{x}_i) - y_i)^2 \tag{7.2}$$

Another common choice for the loss function is the log loss, which takes the form

$$\text{Log Loss } = L(F) = \sum_{i=1}^{N} -\left[y_i \ln(F(\vec{x}_i)) + (1 - y_i)\ln(1 - F(\vec{x}_i))\right] \tag{7.3}$$

A model that is too complex will result in $F(\vec{x})$ that gives a low value of the loss function when evaluated on the training set, but not when evaluated on the validation set: this is known as **over-training**. A model that is not complex enough will result in $F(\vec{x})$ that performs poorly (high value of the loss function) on both the training and validation sets.

7.1.1 Classification

The MVAs used in the following Chapters address questions of **classification**, where the MVA aims to predict a label indicating what physics processes the event originated from.

A simple model will give a good approximation of F_{true} if the features $\vec{x}$ have very different shapes in the processes under consideration. For interactions that produce complex multi-particle final states, it is often the case that many features have modest shape differences, but no single feature offers powerful discrimination. In such cases, a classifier is a useful tool for combining several features into a single powerful discriminant.

7.1.1.1 Binary Classification

Binary classification is the simplest and most common type of classification algorithm. In binary classification, there are only two labels to choose from:

$$F_{\text{true}}(\vec{x}) = \begin{cases} 1 \text{ if signal (e.g. } t\bar{t}H) \\ 0 \text{ otherwise} \end{cases} \tag{7.4}$$

The output of a binary classifier evaluated on an event with features $\vec{x}$ is $F(\vec{x}) \in [0, 1]$, with values near 0 indicating more background-like and values near 1 indicating more signal-like. A threshold cut c_{th} is placed on the classifier output such that events with $F(\vec{x}) > c_{\text{th}}$ are labelled signal, and events with $F(\vec{x}) < c_{\text{th}}$ are labelled background.

The overall performance of a binary classifier model depends on:

- **Signal acceptance**, or true positive rate (a): the fraction of signal events that are labelled correctly as signal
- **Background rejection**, or true negative rate (r): the fraction of background events that are labelled correctly as background

An ideal classifier has 100% signal acceptance and 100% background rejection.

The **Receiver Operating Characteristic (ROC)** is commonly used to quantify the performance of a binary classifier. This curve is inscribed by scanning the threshold cut c_{th} and plotting

$$(r(c_{\text{th}}), a(c_{\text{th}})) \tag{7.5}$$

Some example ROC curves are shown in Fig. 7.1. The blue curve shows an ideal classifier, and the red curve shows the equivalent of random guessing. Two more realistic models, named Classifier 1 and Classifier 2, are shown in green and black respectively. The performance of Classifier 2 is better than Classifier 1, since the signal acceptance is higher for each value of the background rejection.

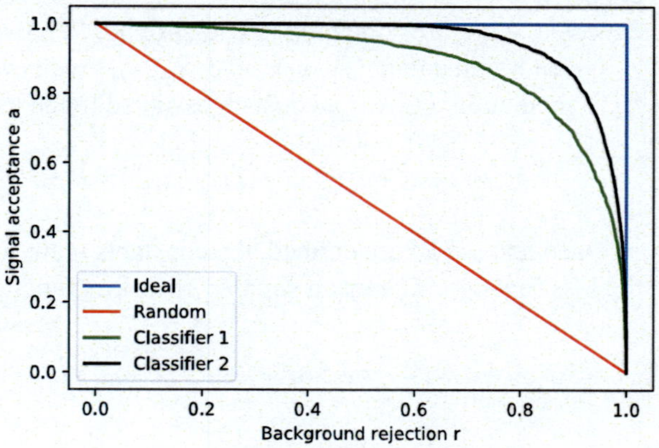

Fig. 7.1 Example ROC curves. The blue and red lines show an ideal and a random classifier, respectively. The green and black curves show examples of more realistic classifiers

The area under the curve (AUC) can also be computed for each ROC curve to measure the relative performance of different classifiers. A perfect classifier has ROC AUC = 1, and a random classifier has ROC AUC = 0.5. ROC curves and ROC AUC will be used to optimize the classifiers developed for the MVAs in later Chapters.

7.1.1.2 Multiclassification

Multiclassification problems require the function F to divide data in to $N_c > 2$ classes. In this case, the truth label $F_{\text{true}}(\vec{x})$ of an event in class c can be expressed a vector $\{\delta_{c1}, \ldots, \delta_{cN_c}\}$, where δ_{ij} is the Dirac delta function.

The output of a multiclassier takes the form $\{F_1(\vec{x}), \ldots, F_{N_c}(\vec{x})\}$ where $F_c(\vec{x}) \in [0, 1]$ for all classes c. This can be normalized to give a vector of the probabilities of the event falling into each class:

$$P_c(\vec{x}) = \frac{F_c(\vec{x})}{\sum_{i=1}^{N_c} F_i(\vec{x})} \tag{7.6}$$

The predicted class label assigned to each event corresponds to the class with highest probability in the output vector $\{P_1(\vec{x}), \ldots, P_{N_c}(\vec{x})\}$.

7.2 Boosted Decision Trees

Boosted Decision Tree (BDT) models have historically been a favorite MVA tool for high energy physicists. The basic building block of the BDT is the decision tree, which divides the feature space into rectangular regions (or "leaves") by a sequence of binary splits [1].

Before training a BDT, each event in the training set is assigned an event weight. For background events, the event weight is simply the Monte Carlo event weight (see Sect. 5.2). For signal events, the event weight is the product of the Monte Carlo event weight and a scale factor **RelativeNorm**. The RelativeNorm parameter specifies the expected relative normalization of the signal and background processes: if signal is expected to be rare compared to background, RelativeNorm is much less than 1.

Given training data $(\vec{x}_1, y_1), \ldots, (\vec{x}_N, y_N)$ with event weights $(w_1, \ldots, w_N)$, the BDT prediction $f(\vec{x}_i)$ is a constant assigned to the leaf R_j containing $\vec{x}_i$:

$$f(\vec{x}_i) = a_j \text{ for } R_j \ni \vec{x}_i \tag{7.7}$$

Once leaves $\{R\}$ are defined, the constants $\vec{a}$ are fully determined by minimizing the loss function. The mean squared error loss function in Eq. 7.2 is minimized by

$$a_j = \frac{\sum_{y_i \in R_j} w_i y_i}{\sum_{y_i \in R_j} w_i} \tag{7.8}$$

This is equivalent to the fraction of signal observations ($y_i = 1$) contained in the leaf R_j. For binary BDTs in the following Chapters, the inverse of the ROC AUC is used as the loss function that is minimized in the BDT training.

The leaf definitions are obtained by a sequence of binary partitions of the feature space. A greedy algorithm is used to determine the splitting variable x_n and cut value c such that the leaves $R_1 = \{\vec{x} : x_n < c\}$ and $R_2 = \{\vec{x} : x_n > c\}$ minimize the loss function. The same greedy algorithm is then applied separately to the leaves R_1 and R_2, then to their daughter leaves, etc. until some stopping criterion is reached. The constants $\vec{a}$, the splitting variables $\vec{x}_n$, and the cut values $\vec{c}$ are all free parameters of the BDT model.

Figure 7.2 shows a diagram of an example tree obtained in this way. In some events, a feature may be incalculable: for example, the leading lepton p_T is "missing" in an event containing no leptons. Features with missing values are deemed to pass all threshold cuts, as shown in Fig. 7.2.

If the splitting is allowed to proceed indefinitely, the final tree will have a single leaf for each event in the training set. Several hyper-parameters can be adjusted to define stopping conditions and prevent such egregious over-training:

- **MaxDepth**: the depth of the tree is limited to this maximum value. The number of leaves in the tree is at most 2^{MaxDepth}.
- **MinChildWeight**: the number of events in each leaf is required to be above this threshold.

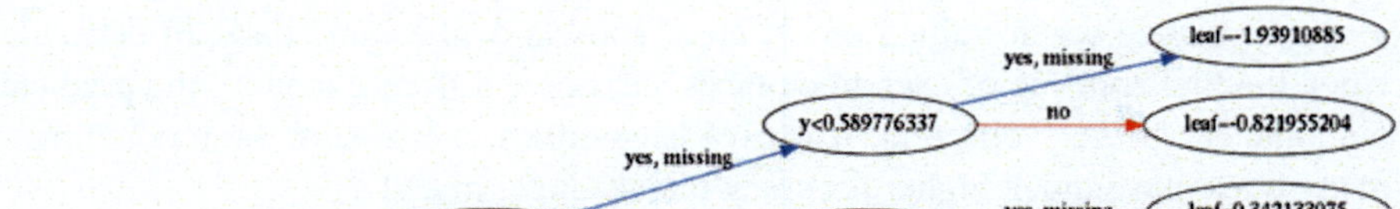

Fig. 7.2 Diagram of an example tree of depth 3, which partitions the feature space into eight rectangular regions. The sequence of cuts applied to features x and y is outlined, and the constant output for each leaf is listed

Additional conditions introducing penalties for large BDT weights (values of $\vec{a}$) and small improvements can be expressed as terms in a generalized loss function, called the **objective function**:

$$\text{Objective function } = C(f, \vec{a}) = L(f) + \Omega(\vec{a}) \tag{7.9}$$

where the "complexity function" Ω contains the following penalty terms:

- **L1 regularization**: controls the magnitude of the penalty terms linear in a_i.
- **L2 regularization**: controls the magnitude of the penalty terms quadratic in a_i.
- **MinLossReduction**: each splitting is required to improve the objective by at least this amount.

For a tree with weights $\vec{a} = \{a_1, \ldots, a_M\}$:

$$\Omega(\vec{a}) = \gamma M + \alpha \sum_{i=1}^{M} |a_i| + \lambda \sum_{i=1}^{M} a_i^2 \tag{7.10}$$

where α and λ are the L1 and L2 regularization hyper-parameters, and γ represents the minimum loss reduction. When these parameters are non-zero, the leaf definitions are chosen to minimize the full objective function.

7.2.1 Boosting

Decision trees are considered weak learners: that is, their predictions tend to have a large variance when trained on different subsets of the same data. In particular, when the first splitting of each tree places a cut on a different feature, the predicted outcome can be very different. **Additive Ensembles**, which sum the predictions of many trees, have much higher accuracy than a single tree [1].

The predicted outcome of an ensemble of T trees on features $\vec{x}$ is

$$F(\vec{x}) = \sum_{t=1}^{T} \beta_t f_t(\vec{x}) \tag{7.11}$$

The sequence of functions f_t and coefficients β_t are determined by a boosting algorithm. Popular boosting algorithms include Adaptive Boosting [2], Logit Boosting [3], and Gradient Boosting [4]. The Boosted Decision Trees described throughout this Thesis use **Gradient Boosting**, as implemented in the eXtreme Gradient Boosting (XGBoost) software package [5]. The algorithm for Gradient Boosting is outlined below.

An initial decision tree is trained according to the procedure already described, and β_0 is taken to be 1:

$$F^{(0)}(\vec{x}) = f_0(\vec{x}) = a_j^{(0)} \quad \text{for} \quad R_j^{(0)} \ni \vec{x} \tag{7.12}$$

Then, while $m < T$:

- For each event i, compute the pseudo-residuals for round m:

$$r_i^{(m)} = -\left.\frac{\partial L(f(\vec{x}_i))}{\partial f(\vec{x}_i)}\right|_{f=f_{m-1}} \tag{7.13}$$

- Re-weight the training data by taking $w_i' = w_i r_i^{(m)}$. This serves to enhance the importance of training instances that are poorly predicted by preceding trees. Each subsequent round therefore focuses on those events that are hardest to classify.
- Train a single tree f_m on the re-weighted training data.
- Compute α_m so that the loss function is minimized:

$$L(F^{(m-1)} + \alpha_m f_m) \tag{7.14}$$

- Set $\beta_m = \nu\alpha_m$, where ν is the **LearningRate** hyper-parameter. The introduction of the LearningRate parameter ($0 \leq \nu \leq 1$) slows the convergence of the model, ensuring that narrow minima are not overlooked.
- Update the model

$$F^{(m)}(\vec{x}) = F^{(m-1)}(\vec{x}) + \beta_m f_m(\vec{x}) = \sum_{t=1}^{m} \beta_t f_t(\vec{x}) \tag{7.15}$$

This procedure concludes after T rounds, or when early stopping conditions are met. The number of rounds T (**Rounds**) can be treated as a hyper-parameter that requires tuning. Alternatively, an early stopping condition can be implemented which terminates the boosting after k rounds pass without considerable improvement in the loss function.

In order to minimize over-training, a different random subsample of the training data can be used for each boosting round. This procedure is often referred to as bagging. The **Subsample** hyper-parameter can be tuned to specify what fraction of the total dataset is randomly selected at each boosting round.

Strategies for performing optimization of a multi-dimensional hyper-parameter space are discussed in Sect. 7.3.

7.2.2 *Multiclass BDT*

The binary BDT described above can be generalized to a multiclass BDT, which divides data into $N_c > 2$ classes. N_c binary BDTs are trained, where the signal of BDT i is taken to be events in class i, and the background is taken to be events from all other classes. The output of the multiclass BDT is the vector of the outputs of each binary BDT $\{F_1(\vec{x}), \ldots, F_{N_c}(\vec{x})\}$.

The multiclass loss function is minimized at each boosting step. To compute the multiclass loss function, the log loss function (Eq. 7.3) is computed for each binary tree and summed over all classes:

$$L_c(F_c) = \sum_{i=1}^{N} -\left[\delta_{ic} \ln(F_c(\vec{x}_i)) + (1 - \delta_{ic}) \ln(1 - F_c(\vec{x}_i))\right] \tag{7.16}$$

$$\text{Multiclass log loss} = \sum_{c=1}^{N_c} L_c(F_c) \tag{7.17}$$

7.3 Hyper-Parameter Optimization

The complexity of an ML model is determined by the choice of its hyper-parameters. Selection of appropriate values for all hyper-parameters is essential to avoid models that are highly sub-optimal (under-trained) or biased (over-trained). The optimal values of a model's hyper-parameters depend on the dataset, and are impossible to predict *a priori*.

In order to optimize hyper-parameters, the loss function $L(F_{\vec{h}})$ is re-parameterized as a function of n hyper-parameters $\vec{h} = (h_1, \ldots, h_n)$. The re-parametrized loss function $L(\vec{h})$ is then evaluated on the validation set, and the values of $\vec{h}$ that minimize L are chosen as the optimal hyper-parameters. The BDT hyper-parameters that are tuned in the following Chapters have been introduced above: MaxDepth, MinChildWeight, MinLossReduction, L1 Regularization, L2 Regularization, Rounds, LearningRate, Subsample, and RelativeNorm.

7.3.1 Grid Scan

The most straightforward approach to hyper-parameter optimization is to perform a grid scan over an n-dimensional rectangular phase space, and select from all points tested the $\vec{h}$ that gives minimal $L(\vec{h})$. This method is easily parallelized, but computationally expensive: scanning n hyper-parameters over s steps each requires training s^n BDTs. The interval of the scan and the size of each step must both be determined before running the scan, and poor choices of these values result in sub-optimal tuning.

The grid scan method of hyper-parameter optimization is applied in the training of the Top Reco BDT described in Chap. 8.

7.3.2 *Gaussian Processes Minimization*

A more complex but more efficient approach to hyper-parameter optimization is Bayesian minimization of $L(\vec{h})$ using Gaussian Processes [6], as implemented in the package scikit-optimize [7].

In this method, it is assumed that that $L(\vec{h})$ follows a multivariate Gaussian. Given the allowed range of each hyper-parameter, N_s starting points in the multi-dimensional hyper-parameter space are randomly selected. For each starting point $\vec{h}_i$, the BDT is trained using the corresponding hyper-parameter settings and $L(\vec{h}_i)$ is evaluated. The following procedure is then applied:

- Using the known points $(\vec{h}, L(\vec{h}))$, obtain posterior function L_m.
- Find the $\vec{h}_m$ that minimizes L_m.
- Train the BDT using hyper-parameters $\vec{h}_m$ to get $(\vec{h}_m, L(\vec{h}_m))$.
- Save the minimum of the known points $(\vec{h}, L(\vec{h}))$.
- Increment $m = m + 1$ and repeat until the function L_{N_p} has been obtained.

The hyper-parameters corresponding to the minimum L of the known points are taken to be optimal, and this training model is selected. Throughout this work, $N_s = 10$ and $N_p = 100$.

The Gaussian processes method of hyper-parameter optimization is applied in the training of the BDTs described in Chaps. 9 and 11.

References

1. Hastie T, Tibshirani R, Friedman J (2008) The elements of statistical learning, 2nd edn. Springer, Berlin
2. Freund Y, Schapire RE (1999) A short introduction to boosting. Proceedings of the sixteenth international joint conference on artificial intelligence. Morgan Kaufmann, pp 1401–1406
3. Friedman J, Hastie T, Tibshirani R (2000) Additive logistic regression: a statistical view of boosting. Ann Stat 38(2)
4. Friedman JH (2000) Greedy function approximation: a gradient boosting machine. Ann Stat 29:1189–1232
5. Chen T, Guestrin C (2016) XGBoost: a scalable tree boosting system. Proceedings of the 22nd ACM SIGKDD international conference on knowledge discovery and data Mining. KDD '16. ACM, San Francisco, California, USA, pp 785–794. ISBN: 978-1-4503-4232-2. https://doi.org/10.1145/2939672.2939785
6. Rasmussen CE, Williams CKI (2005) Gaussian processes for machine learning (adaptive computation and machine learning). The MIT Press. ISBN: 026218253X
7. Head T et al (2018) scikit-optimize/scikit-optimize: v0.5.2. Version v0.5.2. Mar 2018. https://doi.org/10.5281/zenodo.1207017

Chapter 8
Reconstruction of Top Decays

Three heavy particles are produced in $t\bar{t}H$ events (one Higgs boson and two top quarks), each of which decays to lighter particles immediately after they are produced. As a result, $t\bar{t}H$ events have final states with particularly high object multiplicity. In the case of $t\bar{t}H(\gamma\gamma)$, the Higgs boson is easy to identify and reconstruct with good mass resolution. However, the $t\bar{t}$-system is more complicated.

The decay of a top quark via $t \rightarrow bW$ is dominant, and all other modes are neglected. However, the subsequent decay of the W boson allows for many possibilities, which are enumerated in Table 8.1.

Events are grouped into different channels based on the number of leptons produced in the decay of the $t\bar{t}$ system. Because only electrons and muons are classified as leptons in the following Chapters, top decays to τ are only considered leptonic if the τ decay produces a prompt e or μ. The breakdown shown in Table 8.2 is obtained by folding the W decay rates in Table 8.1 with the branching fraction of $\tau \rightarrow e\bar{\nu}_e\nu_\tau$ or $\mu\bar{\nu}_\mu\nu_\tau$.

The all-hadronic (0ℓ) $t\bar{t}$ decay makes up more than half of $t\bar{t}H$ events. A Feynman diagram of this final state is shown in Fig. 8.1. Assuming that each final state parton corresponds approximately to one jet in $t\bar{t}H(\gamma\gamma)$ events (due to small rates of highly boosted objects), the hadronic decay of the top quark can be reconstructed by grouping together the correct set of three jets. For example, the jets corresponding to b_1, q_1, and q_1' in Fig. 8.1 can be combined into a reconstructed top candidate.

With many jets in each final state, finding the correct groupings of three jets (or **triplets**) is a tricky combinatorial problem. Additional complications include the presence of extra jets resulting from pileup or QCD radiation, or the failure of some jets from top quark decay to meet fiducial requirements.

This Chapter describes a novel method for reconstructing the decay of top quarks using a **Top Reconstruction BDT (Top Reco BDT)**. This algorithm is applied in later Chapters to distinguish between $t\bar{t}H$, tH, and backgrounds, as well as between SM and BSM $t\bar{t}H$ and tH scenarios. Section 8.1 gives an overview of the strategy

J. E. Dickinson, *ATLAS Measurements of the Higgs Boson Coupling to the Top Quark in the Higgs to Diphoton Decay Channel*, Springer Theses,
https://doi.org/10.1007/978-3-030-86368-5_8

Table 8.1 Branching fraction of W boson decay modes from the Particle Data Group [1]

Decay mode	Branching fraction (%)
$W \to e\nu_e$	10.75 ± 0.13
$W \to \mu\nu_\mu$	10.57 ± 0.15
$W \to \tau\nu_\tau$	11.25 ± 0.20
$W \to qq'$	67.60 ± 0.27

Table 8.2 Fraction of $t\bar{t}$ decays resulting in various numbers of leptons ($\ell = e$ or μ)

	0 τ_{had} (%)	1 τ_{had} (%)	2 τ_{had} (%)	Total (%)
0 ℓ	45.7	9.9	0.5	56.1
1 ℓ	34.1	3.7	0	37.8
2 ℓ	6.4	0	0	6.4

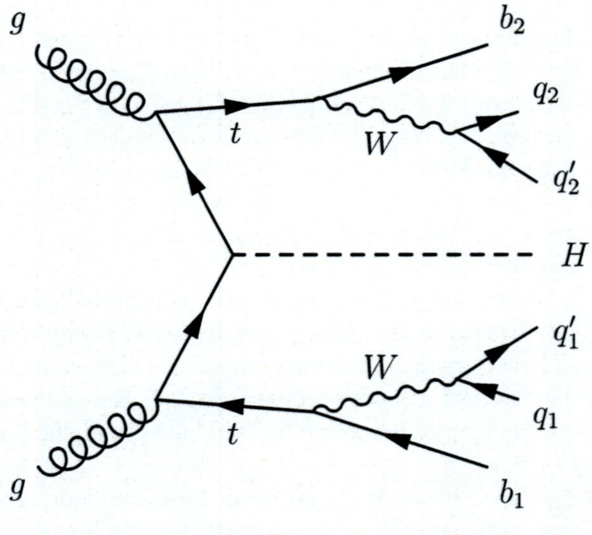

Fig. 8.1 Lowest order Feynman diagram of the $t\bar{t}H$ process with an all-hadronic decay of the $t\bar{t}$ system. Additional quarks and gluons, the result of pileup or QCD radiation, can further complicate this final state

for reconstructing the full $t\bar{t}$ system given the output of the Top Reco BDT, which is trained to identify the triplet most likely to be the result of a hadronic top quark decay. The training of the Top Reco BDT, including the construction of signal and background samples and the training variables tested, is described in Sect. 8.2.2. Section 8.3 reviews the performance of the Top Reconstruction BDT in events with different n_ℓ.

8.1 Reconstruction Strategy

8.1.1 Hadronic Channel

The hadronic channel preselection targets the all-hadronic decay of the $t\bar{t}$ system by requiring that each event contain 2 photons passing tight identification/isolation criteria, 0 leptons, ≥ 1 b-jet and ≥ 3 total jets. The same hadronic preselection is used throughout later Chapters. Since all preselected hadronic events contain at least three jets, there are always enough jets to construct at least one triplet. In events with more than three jets, all possible triplets are constructed. The Top Reco BDT score is evaluated for each triplet, and the triplet with highest score is selected as the primary top candidate.

In events with six or more jets, a second hadronic top can be fully reconstructed by following the same procedure and excluding jets already assigned to the first top. In events with four or five jets, the second top can only be partially reconstructed: it is taken to be the sum of the jets remaining after the primary top is reconstructed. In events containing exactly three jets, no second top is reconstructed.

Table 8.3 contains a summary of the reconstruction status of the primary and second tops in hadronic events with different jet multiplicity.

Because the average number of jets in $t\bar{t}H(\gamma\gamma)$ events passing hadronic selection is ~ 5.5, many events only contain a single reconstructible top. For this reason, the Top Reco BDT strategy laid out above prioritizes reconstruction of the primary top.

Alternative approaches are available that instead focus on the $t\bar{t}$ system. Rather than sequentially selecting the two top candidates with highest Top Reco BDT score, the average of the two Top Reco BDT scores can be maximized. Tests of this alternative method resulted in very similar performance to the sequential method. However, the computation time needed to calculate the average Top Reco BDT score of each triplet pair is too large for this method to be practically useful. Another alternative for reconstructing the $t\bar{t}$ system would be to train a BDT to identify the correct six jets corresponding to the all-hadronic top decays, as in [2].

Table 8.3 Summary of the reconstruction status of the primary and second tops in hadronic events with different jet multiplicity

n_{jets}	Top 1 (had)	Top 2 (had)
3	Full	None
4–5	Full	Partial
≥ 6	Full	Full

8.1.2 Leptonic Channel

The Top Reco BDT is trained to identify hadronic top decays, which are comprised of a b candidate and a hadronic W candidate (see the following Section). However, the same BDT can be applied to semi-leptonic top decays, which are instead comprised of a b candidate and a leptonic W candidate.

In events passing single-lepton preselection (2 photons passing tight identification/isolation criteria, 1 e or μ, and ≥ 1 b-jet), a leptonic W candidate is constructed from the lepton and missing transverse energy, which is attributed to a neutrino. The W candidate four-momentum is derived using a dedicated algorithm [3].

The lepton has well-measured four-momentum $(\mathbf{q}, E_\ell)$ and the neutrino is assigned four-momentum $(\mathbf{p}, E_\nu)$. The lepton-plus-neutrino system is then assigned the invariant mass of a W boson:

$$m_W^2 = (E_\ell + E_\nu)^2 - (\mathbf{q} + \mathbf{p})^2 \tag{8.1}$$

The neutrino is assumed to be massless, and the transverse components of the neutrino momentum are taken to be identical to the missing transverse energy ($p_x = E_x^{\text{miss}}$ and $p_y = E_y^{\text{miss}}$). This gives the following constraint on the z-component of the neutrino momentum, p_z:

$$m_W^2 = \left(E_\ell + \sqrt{(E_x^{\text{miss}})^2 + (E_y^{\text{miss}})^2 + p_z^2}\right)^2 - \left(\mathbf{q} + (E_x^{\text{miss}}, E_y^{\text{miss}}, p_z)\right)^2 \tag{8.2}$$

If the transverse mass $m_T < m_W$, then Eq. 8.2 has two real solutions for p_z. The solution with smaller $|p_z|$ is chosen, since this gives a momentum vector $\mathbf{p}$ that is closer in ΔR to the true neutrino momentum in more than half of events.

If $m_T > m_W$, the two solutions to Eq. 8.1 have a nonzero imaginary component. In this case, the difference between the transverse energy and neutrino transverse momentum (which is assumed to be zero in Eq. 8.1) is adjusted so that the imaginary component vanishes. This is equivalent to requiring $m_W = m_T$ in Eq. 8.2, or

$$(\mathbf{q}_T + \mathbf{p}_T)^2 = 0 \tag{8.3}$$

The resulting quadratic relationship between p_x and p_y gives two solutions for expressing p_y in terms of p_x: $p_y^{(1)}(p_x)$ and $p_y^{(2)}(p_x)$. For each solution, the distance between the neutrino transverse momentum and the missing transverse energy is calculated:

$$\begin{aligned} \delta^{(1)}(p_x) &= \sqrt{(p_x - E_x^{\text{miss}})^2 + (p_y^{(1)}(p_x) - E_y^{\text{miss}})^2} \\ \delta^{(2)}(p_x) &= \sqrt{(p_x - E_x^{\text{miss}})^2 + (p_y^{(2)}(p_x) - E_y^{\text{miss}})^2} \end{aligned} \tag{8.4}$$

Table 8.4 Summary of the reconstruction status of the primary and second tops in leptonic events with different lepton and jet multiplicities

n_ℓ	n_{jets}	Top 1 (semi-lep)	Top 2 (had)
1	1	Full	None
1	2–3	Full	Partial
1	≥ 4	Full	Full
2	Any	None	None

The distances $\delta^{(1)}$ and $\delta^{(2)}$ are minimized by $p_x^{(1)}$ and $p_x^{(2)}$, respectively. The corrected missing transverse energy is chosen to be

$$(p_x^{(1)}, p_y^{(1)}(p_x^{(1)})) \text{ or } (p_x^{(2)}, p_y^{(2)}(p_x^{(2)})) \tag{8.5}$$

The selected solution corresponds to the smaller of the minimized $\delta^{(1)}$ and $\delta^{(2)}$. Substituting the corrected missing transverse energy into Eq. 8.2 is guaranteed to yield exactly one real solution for p_z.

Once a W candidate has been constructed, the Top Reco BDT score is evaluated for every (leptonic W, jet) combination. The primary top is reconstructed from the leptonic W and the jet giving the highest Top Reco BDT score.

Since the single-lepton preselection requires one lepton and at least one b-jet in every event, there are always enough objects to reconstruct a semi-leptonic top. No top candidate can be reconstructed in dilepton events with this method, as this would require the missing E_T to be split between two semi-leptonic top decays.

In events with at least three jets remaining after semi-leptonic top reconstruction, a hadronic top can also be reconstructed from the jet triplet with highest BDT score. In events with two or three jets, the second top can only be partially reconstructed: it is taken to be the sum of the jets remaining after the primary top is reconstructed. In events containing only a single jet, no second top is reconstructed.

Table 8.4 contains a summary of the reconstruction status of the primary and second tops in leptonic events with different lepton and jet multiplicities.

8.2 Training of the Top Reco BDT

8.2.1 Constructing Signal and Background Samples

The goal of the Top Reco BDT is to identify jet triplets that correspond to the decay products of the same top quark: such triplets are designated as signal. The set of background triplets includes all possible wrong combinations. See Table 8.5 for a summary of the signal and background triplets in an event similar to that shown in Fig. 8.1. A triplet is correctly matched if its constituent W candidate (jet pair) and b candidate (single jet) are correctly matched to the corresponding truth particles, and these truth particles are daughters of the same top quark.

Table 8.5 Summary of signal and background triplets in an example hadronic $t\bar{t}H$ event

	b Candidate	W Candidate
Signal	b_1	$q_1 + q_1'$
Signal	b_2	$q_2 + q_2'$
Background	b_1	$q_2 + q_2'$
Background	b_1	$q_1 + q_2'$
Background	b_1	$q_2 + q_1'$
Background	b_1	b_2 + anything
Background	b_2	$q_1 + q_1'$
Background	b_2	$q_1 + q_2'$
Background	b_2	$q_2 + q_1'$
Background	b_2	b_1 + anything
Background	q_1, q_1', q_2 or q_2'	Anything
Background	QCD/pileup	Anything
Background	Anything	QCD/pileup + anything

b and q represent the jets that are truth matched to the corresponding particles in Fig. 8.1

Signal and background samples are constructed from simulated $t\bar{t}H(\gamma\gamma)$ events using the following procedure:

1. Find the W and b particle daughters of a truth top.
2. Follow the simulated decay chain of the W and b to identify the set of final state hadrons descending from each particle. Identify the two (one) truth jets containing the most hadrons descended from the W (b) (weighted by p_T).
3. Match each truth jet to the closest reconstructed jet in ΔR.
4. If this results in a set of three distinct jets (i.e. no jet is used twice in the triplet), this triplet is truth matched and marked as a signal triplet.
5. All other sets of three distinct jets are considered background triplets.

The training is performed on Powheg+Pythia8 $t\bar{t}H(\gamma\gamma)$ Monte Carlo events passing hadronic preselection. Details of this Monte Carlo sample are included in Chap. 5.

If all three truth jets selected at Step 2 are required to be within acceptance ($p_T > 25$ GeV, $|\eta| < 4.0$), only 59% of events passing hadronic preselection contain at least one truth matched triplet. In events with no truth matched triplet, either the top quark is matched to truth jets that fall out of acceptance, or multiple top decay products are merged into a single jet. Table 8.6 shows the possible failure modes of the truth matching algorithm and how frequently they occur. Truth jets falling below the p_T acceptance threshold are the main cause of failures in the truth matching.

If no cuts are placed on the truth jets matched to a top quark, 79% of events passing hadronic preselection contain at least one truth matched triplet. Relaxing the cuts on truth jets allows triplets to be matched through truth jets that re-enter acceptance after reconstruction. The remaining 21% of events with no truth matched triplet are dominated by events where two top decay products correspond to a single jet.

Table 8.6 Breakdown of the performance of the truth matching algorithm when acceptance cuts are applied to truth jets

	Truth jet has $p_T < 25$ GeV (%)	Truth jet has $\|\eta\| > 4.0$ (%)	Truth jets are merged (%)
Leading top	41	< 0.1	5
Sub-leading top	47	< 0.1	3

Jets falling below the p_T acceptance threshold are the main cause of failures in the truth matching. The leading top is defined as the one with higher p_T

The total number of possible jet triplets per event depends on the jet multiplicity:

$$n_{\text{triplet}} = \frac{1}{2} n_{\text{jet}} (n_{\text{jet}} - 1) \tag{8.6}$$

The factor of $\frac{1}{2}$ is due to the indistinguishability of the two jets comprising a W candidate. With an average of ~ 5.5 jets per event, the number of background triplets is far larger than the number of signal triplets (≤ 2 per event). Only 10% of background triplets are used in the training, and this subset still contains ten times more triplets than the signal training sample, which includes all truth matched triplets in 50% of Monte Carlo events. The other 50% of Monte Carlo events are reserved for validation and testing.

8.2.2 Training Variables

The discriminating variables used to separate the signal (truth matched) triplets from background triplets are:

- The four-vector information of the W boson (dijet) candidate
- The four-vector information of the b quark candidate
- The pseudo-continuous b-tag score of each of the three jets (that is, an integer indicating whether the jet is b-tagged at 85, 77, 70, or 65% working point)
- ΔR between the W and b candidates
- ΔR between the two jets comprising the W candidate
- The trijet mass, or mass of the top candidate.

The signal and background distributions for each of these variables are shown in Figs. 8.2, 8.3, 8.4 and 8.5. Compared to background triplets, truth matched triplets are composed of W and b candidates with high p_T and central η, and a peak is visible at the W mass in signal triplet distribution (Figs. 8.2 and 8.3). The b-tag score is high for truth matched b candidates and low for truth matched W candidates (Fig. 8.4). Figure 8.5 shows that the jets composing the W candidate are highly columnated in signal triplets, as are the W and b candidates. There is a clear peak in the trijet mass distribution of signal triplets at the top quark mass (Fig. 8.5).

The Top Reco BDT is trained using the XGBoost package [4] with these variables as input features. The Top Reco BDT discriminant (Fig. 8.6) shows that signal triplets

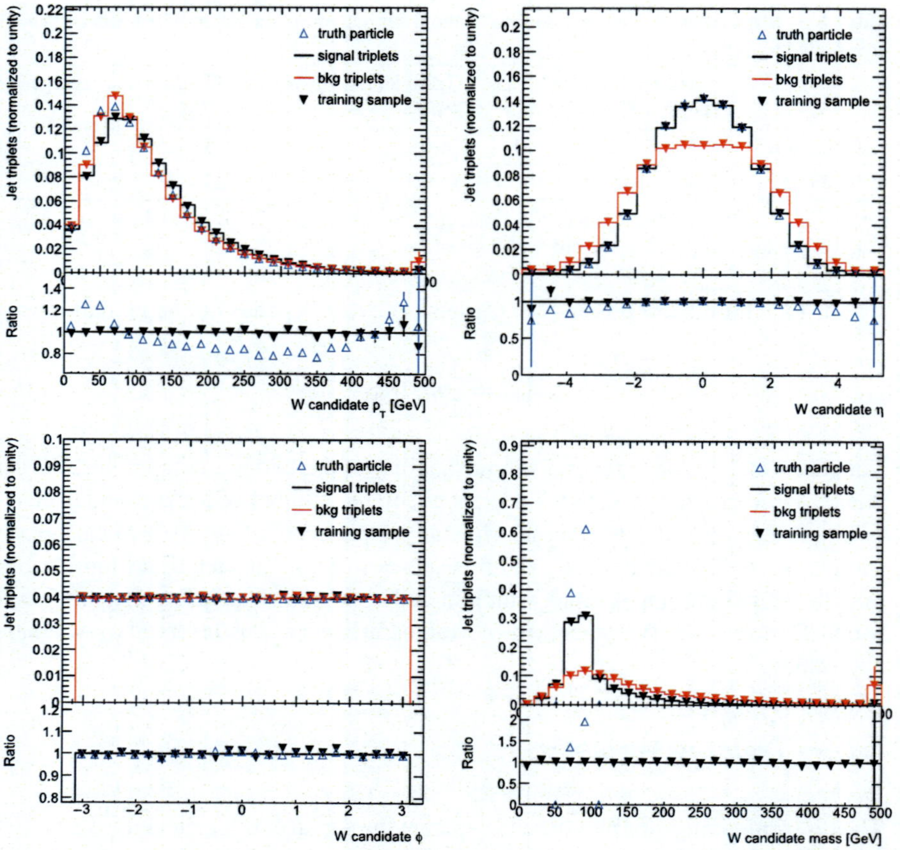

Fig. 8.2 Training variables for the Top Reco BDT: p_T, η, ϕ, and mass of the W candidate. Truth matched jet triplets (signal) and unmatched triplets (background) are shown in black and red, respectively. Due to high statistics, agreement is very good between the training (filled triangles) and testing (lines) samples. The open blue triangles show the distribution of truth particles

tend to have a very high BDT score, while the random background is spread evenly over the full BDT range. The area under the ROC curve is calculated to be 90.24%. The most important features for BDT training are calculated using the SHAP package [5]: the b-tag score of the b candidate is the dominant variable, followed by the triplet mass and the mass of the W candidate.

Additional training variables were tested, including

- The p_T, η, and ϕ of the top candidate (trijet system)
- The four-vector information of the W boson candidate *in the rest frame of the triplet*
- The four-vector information of the b candidate *in the rest frame of the triplet*
- ΔR between the W and b candidates *in the rest frame of the triplet*

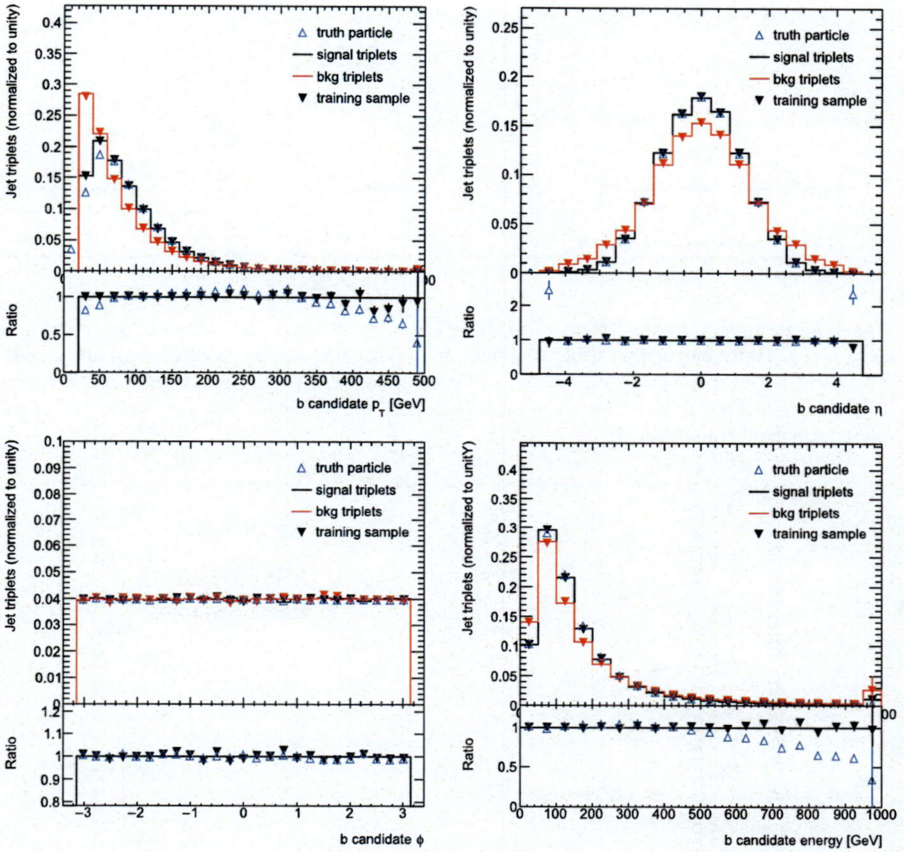

Fig. 8.3 Training variables for the Top Reco BDT: p_T, η, ϕ, and energy of the b candidate. Truth matched jet triplets (signal) and unmatched triplets (background) are shown in black and red, respectively. Due to high statistics, agreement is very good between the training (filled triangles) and testing (lines) samples. The open blue triangles show the distribution of truth particles, where applicable

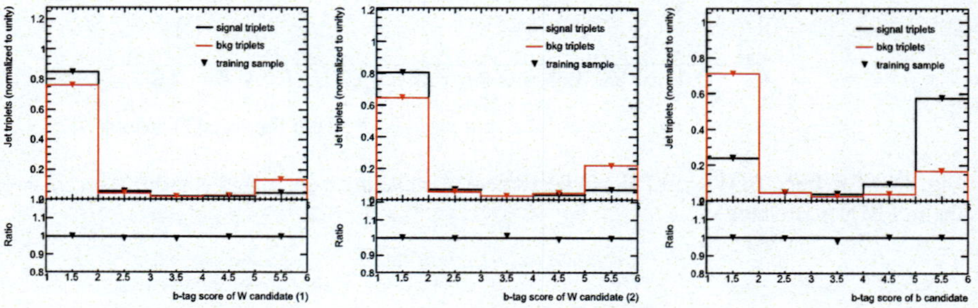

Fig. 8.4 Training variables for the Top Reco BDT: pseudo-continuous b-tag scores of each jet in the triplet. High values indicate that the b candidate is b-tagged at a tight working point. Truth matched jet triplets (signal) and unmatched triplets (background) are shown in black and red, respectively. Due to high statistics, agreement is very good between the training (filled triangles) and testing (lines) samples

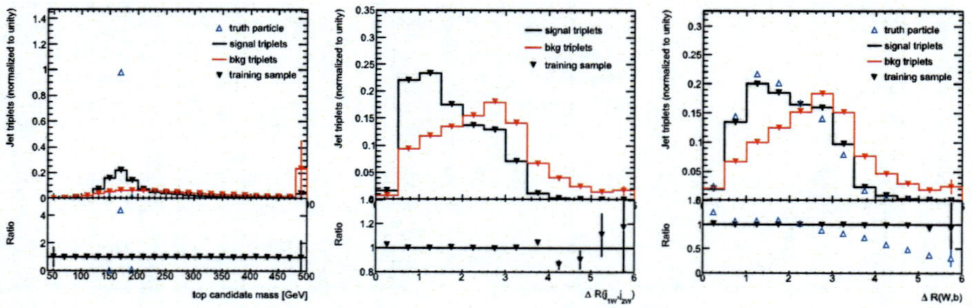

Fig. 8.5 Training variables for the Top Reco BDT: triplet mass, ΔR between the W jets, and $\Delta R(W, b)$. Truth matched jet triplets (signal) and unmatched triplets (background) are shown in black and red, respectively. Due to high statistics, agreement is very good between the training (filled triangles) and testing (lines) samples. The open blue triangles show the distribution of truth particles, where applicable

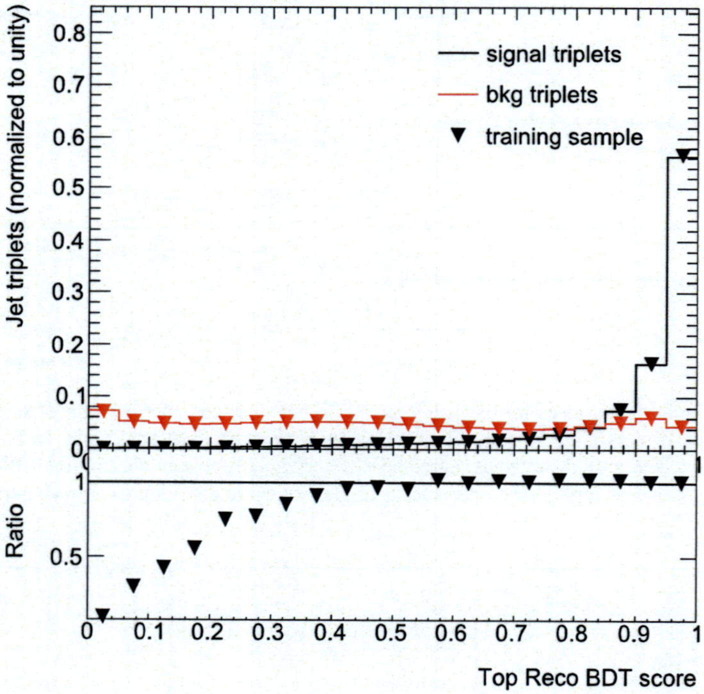

Fig. 8.6 Top Reco BDT score. Signal peaks near 1, while the background is evenly spread across the full BDT score range

- ΔR between the two jets comprising the W candidate *in the rest frame of the triplet*
- The number of jets in the event
- The p_T, η, ϕ, and mass of the system of jets recoiling from the triplet.

However, the inclusion of these variables resulted in poorer performance of the Top Reco BDT, as demonstrated in the next Section.

8.3 Performance

The performance of the Top Reco BDT is quantified by counting the number of events in which the algorithm identifies a top candidate (jet triplet) that is truth matched to a top quark. Similarly, the accuracy of W and b identification is quantified by counting number of events in which the selected W and b candidates are truth matched to the corresponding particle. Accuracies are calculated in Powheg+Pythia8 $t\bar{t}H(\gamma\gamma)$ Monte Carlo unless otherwise indicated. The subset of Monte Carlo events used to evaluate performance is statistically independent of that used in training.

The performance of several training setups is compared in this Section. The trainings are named as follows:

1. Nominal: the training variables listed at the beginning of Sect. 8.2.2
2. Nominal w/o m_t: same training variables used in Nominal, but without m_t
3. Nominal w/o m_W, m_t: same training variables used in Nominal, but without m_W and m_t
4. Boosted: same training variables used in Nominal, but with all kinematic variables calculated *in the rest frame of the triplet*
5. Boosted w/o m_t: same training variables used in Boosted, but without m_t
6. Boosted w/o m_W, m_t: same training variables used in Boosted, but without m_W and m_t
7. Boosted + p_T^t w/o m_W, m_t: same training variables used in Boosted w/o m_W, m_t, plus the p_T of the triplet.

Some comparisons are also made to the simple sum-p_T method, which takes the jet triplet with highest p_T as the top candidate.

The performance of the nominal Top Reco BDT is also compared on multiple Higgs Monte Carlo samples.

8.3.1 Hadronic Channel

The accuracy of each Top Reco BDT training setup is evaluated for all $t\bar{t}H(\gamma\gamma)$ events passing hadronic preselection. The results are shown in Table 8.7, where the accuracy is shown separately for top, W, and b identification. Note that in some events, the W and b are both correctly identified, but the top is not: in these cases, the

Table 8.7 Percentage of events passing hadronic preselection that contain a correctly reconstructed object (top, W, or b) in $t\bar{t}H(\gamma\gamma)$ Monte Carlo

Training	t-candidate (%)	W-candidate (%)	b-candidate (%)
Nominal	**37**	**46**	**57**
Nominal w/o m_t	35	45	57
Nominal w/o m_W, m_t	35	43	57
Boosted	25	33	53
Boosted w/o m_t	17	26	50
Boosted w/o m_W, m_t	17	25	51
Boosted + p_T^t w/o m_W, m_t	19	28	51

Shown for different BDT reconstruction algorithms

Table 8.8 Percentage of events passing hadronic preselection in which the Top Reco BDT correctly reconstructs one or two tops, compared across several $t\bar{t}H(\gamma\gamma)$ Monte Carlo samples

	Powheg+Pythia8 (%)	MG5_aMC@NLO+Pythia8 (%)	MG5_aMC@NLO+Herwig++ (%)
1 top correct	37	36	33
2 tops correct	5	4	4

W and b identified are not daughters of the same top quark. The Nominal training setup is selected for use in later Chapters based on its high accuracy in this channel.

The accuracy of top reconstruction for up to two tops is shown in different $t\bar{t}H(\gamma\gamma)$ Monte Carlo samples in Table 8.8. No strong dependence on generator or showering algorithm is observed.

The accuracy of the Top Reco BDT depends strongly on jet multiplicity. The fraction of events that contain no truth matched tops (irreducible background) is higher in events with fewer jets. At very high jet multiplicity, the total number of jet triplets increases rapidly (see Eq. 8.6), and the number of background triplets (combinatorial background) dominates the signal. These effects are demonstrated in Fig. 8.7 and Table 8.9, which show the dependence of the Top Reco BDT accuracy on the number of jets in the event. The Top Reco BDT performs best in events with five to seven jets.

The accuracy of each training setup in multiple $H \to \gamma\gamma$ Monte Carlo samples is shown in Table 8.10. The Nominal training has the strongest performance in all three samples.

The Top Reco BDT is also applied in two $H \to \gamma\gamma$ samples where no top quark is present at truth level: ggF and VBF. The kinematics of the selected top candidate in each sample are shown in Fig. 8.8. The tops reconstructed in $t\bar{t}H$ and tWH have high Top Reco BDT score and mass peak near 173 GeV. In the samples that do not contain true top quarks (ggF and VBF), only weak peaks are visible in these distributions.

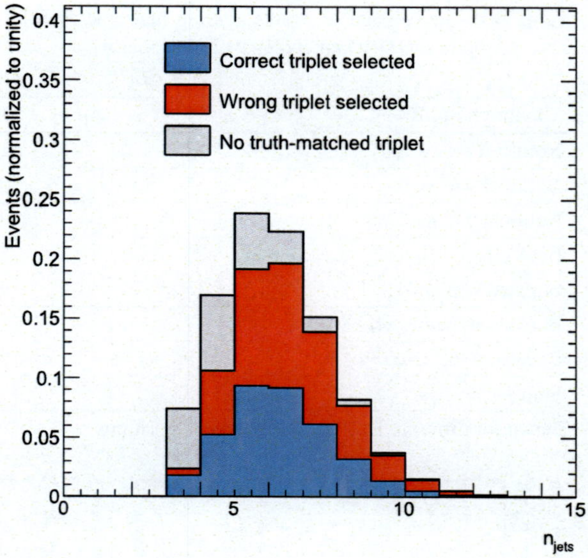

Fig. 8.7 Breakdown of hadronic $t\bar{t}H(\gamma\gamma)$ Monte Carlo events containing no top candidate (gray), a wrongly identified top candidate (red), and a correctly identified top candidate (blue). Shown in exclusive bins of jet multiplicity

Table 8.9 Percentage of hadronic $t\bar{t}H(\gamma\gamma)$ Monte Carlo events containing no correct top candidate, wrongly identified top candidate, and correctly identified top candidate in bins of jet multiplicity

n_{jets}	No truth-matched top (%)	Incorrectly identified top (%)	Correctly identified top (%)
3	58	18	24
4	37	11	30
5	19	44	37
6	12	50	38
7	8	54	38
8	6	58	36
9	5	61	34
10	4	64	32
11	5	64	31
12	4	67	29
13	4	69	27
14	1	72	27

8.3.2 Leptonic Channel

The accuracy of each Top Reco BDT training setup is evaluated for all $t\bar{t}H(\gamma\gamma)$ and $tWH(\gamma\gamma)$ events passing single-lepton preselection: the results are shown in Table 8.11.

In the leptonic channel, the Top Reco BDT is also applied to a $t\bar{t}\gamma\gamma$ Monte Carlo sample and a sample where no top quark is present at truth level ($ZH(\gamma\gamma)$). The

Table 8.10 Percentage of events passing hadronic preselection that contain a correctly reconstructed top in $t\bar{t}H(\gamma\gamma)$ and $tH(\gamma\gamma)$ Monte Carlo

	$t\bar{t}H$ (%)	$tHjb$ (%)	tWH (%)
Maximum in MC	79	51	52
Nominal	**37**	**31**	**21**
Nominal w/o m_t	35	30	20
Nominal w/o m_W, m_t	35	29	20
Boosted	25	25	15
Boosted w/o m_t	17	19	10
Boosted w/o m_W, m_t	17	19	10
Boosted + p_T^t w/o m_W, m_t	19	19	11
Sum p_T	4	4	3

Shown for different BDT reconstruction algorithms

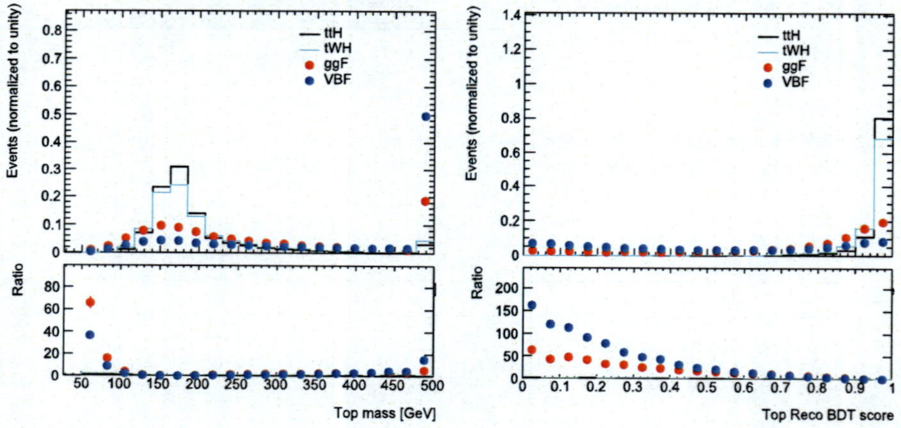

Fig. 8.8 Reconstructed top mass and Top Reco BDT score in events passing hadronic preselection. There are clear shape differences between samples with truth tops ($t\bar{t}H$, tWH) and without (ggF, VBF)

Table 8.11 Percentage of events passing leptonic preselection that contain a correctly reconstructed top in $t\bar{t}H(\gamma\gamma)$ and $tWH(\gamma\gamma)$ Monte Carlo

	$t\bar{t}H$ (%)	tWH (%)
Nominal	**76**	**47**
Nominal w/o m_t	77	48
Nominal w/o m_W, m_t	76	48
Boosted	71	44
Boosted w/o m_t	70	43
Boosted w/o m_W, m_t	71	44
Boosted + p_T^t w/o m_W, m_t	74	46
Sun p_T	19	9

Shown for different BDT reconstruction algorithms

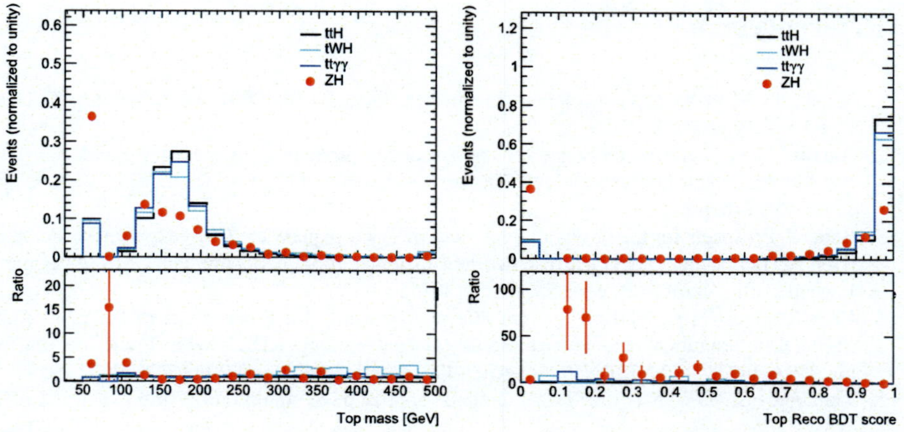

Fig. 8.9 Reconstructed top mass and Top Reco BDT score in events passing leptonic preselection. There are clear shape differences between samples with truth tops ($t\bar{t}H, tWH, t\bar{t}+\gamma\gamma$) and without ($ZH$). The underflow bin represents dilepton events, where no top candidate is reconstructed

Table 8.12 Percentage of $t\bar{t}H(\gamma\gamma)$ Monte Carlo events in which top candidates are accurately reconstructed

Accuracy	KLFitter (%)	Top Reco BDT (%)
Semi-leptonic top correct	25	23
Hadronic top correct	56	59
Both tops correct	19	20

The accuracy of KLFitter and the Top Reco BDT is compared

kinematics of the reconstructed semi-leptonic top candidate in each sample are shown in Fig. 8.9. The tops reconstructed in $t\bar{t}H$, tWH and $t\bar{t}\gamma\gamma$ have high Top Reco BDT score and mass peak near 173 GeV. In the ZH sample, weaker peaks are visible in these distributions.

In single lepton $t\bar{t}$ events, a log likelihood method is often used to assign final state objects to different parent top quarks. A comparison of one such algorithm (KLFitter [6]) is included in Table 8.12. Here the kinematic range is restricted to the range in which the KLFitter transfer functions are valid (that is, require all objects to have $|\eta| < 2.5$). Performance of the two algorithms is similar in this restricted η range. However, the Top Reco BDT has the advantage of being more widely applicable, both to the full η range and the case of all-hadronic $t\bar{t}$ decays.

References

1. Tanabashi M et al, Review of particle physics. Phys Rev 030001. https://doi.org/10.1103/PhysRevD.98.030001
2. Aaboud M et al (2018) Evidence for the associated production of the a top quark pair with the ATLAS detector. Phys Rev D 97(7). https://doi.org/10.1103/PhysRevD.97.072003. arXiv: 1712.08891 [hep-ex]
3. Bauer J, Prospects for the observation of electroweak top quark the CMS experiment. Perspektiven zur Beobachtung der elektroschwachen einzelner Top-Quarks mit dem CMS Experiment. Presented on 7. https://cds.cern.ch/record/1308713
4. Chen T, Guestrin C, XGBoost: a scalable tree boosting. In: Proceedings of the 22nd ACM SIGKDD international conference discovery and data mining, KDD'16, San Francisco, California, USA, pp 785–794. ISBN: 978-1-4503-4232-2. https://doi.org/10.1145/2939672.2939785
5. Lundberg S, Lee S-I (2017) A unified approach to interpreting model. In: CoRR. arXiv:1705.07874
6. Erdmann J et al (2014) A likelihood-based reconstruction algorithm pairs and the KLFitter framework. Nucl Instrum Meth A 748. https://doi.org/10.1016/j.nima.2014.02.029. arXiv: 1312.5595 [hep-ex]

Chapter 9
Selection of $t\bar{t}H(\gamma\gamma)$ Events

In the measurement of $t\bar{t}H(\gamma\gamma)$, events passing a simple preselection are divided into multiple categories, and a maximum likelihood fit is performed simultaneously to the $m_{\gamma\gamma}$ spectra in all categories. The preselection defines two channels:

1. Hadronic channel: 2 photons, 0 e/μ, ≥ 3 jets, ≥ 1 b-tagged jet (77% working point)
2. Leptonic channel: 2 photons, ≥ 1 e/μ, ≥ 1 b-tagged jet (77% working point).

The hadronic preselection targets the all-hadronic decay of the $t\bar{t}$ system, and the leptonic preselection targets events containing at least one semi-leptonic top decay.

In each channel, an XGBoost [1] Boosted Decision Tree is trained to separate simulated $t\bar{t}H(\gamma\gamma)$ events from continuum background, which is modeled by the NTI data control region (see Sect. 5.1.1). The figure of merit used to optimize these BDT trainings is described in Sect. 9.1. Section 9.2 describes the training and optimization of the BDT categories in both channels. Finally, Sect. 9.3 presents a study composition of the background in the $t\bar{t}H(\gamma\gamma)$ BDT categories.

9.1 Poisson Number-Counting Significance

Since the full likelihood model is quite complex, the multivariate $t\bar{t}H(\gamma\gamma)$ analysis categories are designed to maximize the **number-counting significance**. This number-counting significance is based on the Poisson distribution, which gives the probability of observing k events given an expectation of λ:

$$P(k, \lambda) = \frac{\lambda^k e^{-\lambda}}{k!} \tag{9.1}$$

J. E. Dickinson, *ATLAS Measurements of the Higgs Boson Coupling to the Top Quark in the Higgs to Diphoton Decay Channel*, Springer Theses,
https://doi.org/10.1007/978-3-030-86368-5_9

Assuming the background yield B is precisely known, the Poisson distribution can be used to approximate the likelihood. For an observation of $\mu S + B$ events, agreement of the observed μ with the null hypothesis ($\mu = 0$) is given by the p-value

$$-2\ln\frac{P(\mu S + B, B)}{P(\mu S + B, \mu S + B)} = 2(\mu S + B)\ln(1 + \frac{\mu S}{B}) - 2\mu S \tag{9.2}$$

In the case of $t\bar{t}H$, the background yield B includes an estimate of all non-$t\bar{t}H$ Higgs processes as well as the continuum diphoton background. Rather than exercising the full background estimation machinery, the number-counting significance relies on the estimation of continuum background yield from a data control region described in Sect. 5.1.1.

The number-counting significance gives a rough estimate of the absolute significance of a given set of categories. However, the *relative* performance of two sets of categories is well quantified by a comparison of their number-counting significances. This makes the number-counting significance a powerful figure of merit for the optimization of the $t\bar{t}H(\gamma\gamma)$ BDT categories.

In a single category, the expected significance of a Standard Model observation ($\mu = 1$) (compared to the null hypothesis) is:

$$Z = \sqrt{2(S + B)\ln(1 + \frac{S}{B}) - 2S} \tag{9.3}$$

If multiple categories are defined, the total expected significance Z_{tot} is the quadratic sum of the single-category significances.

The statistical uncertainty on Z_{tot} (denoted Y_{tot}) is dominated by the statistical uncertainty on the estimate of continuum background. All Higgs Monte Carlo samples have sufficiently high statistics that their contribution to this uncertainty is negligible. The total significance can be expressed as a function of integers n_i, which represent the number of raw events in the data control region in category i (i.e. the number of events in the NTI control region or TI sidebands). Standard propagation of uncertainties from n_i to Y_{tot} gives the following formula for the statistical uncertainty on Z_{tot}:

$$Y_{\text{tot}}^2 = \sum_i \left(\frac{\partial Z_{\text{tot}}}{\partial n_i}\right)^2 n_i \tag{9.4}$$

In cases where there is a large systematic uncertainty on one or more components of the background, it can also be useful to include this uncertainty in the number-counting significance. For example, it is possible to incorporate the heavy flavor uncertainty on gluon fusion (see Sect. 6.4) into the number-counting significance. The statistical and systematic uncertainties on the signal yield can be written

$$\sigma_{\text{stat}} = S/Z \tag{9.5}$$

$$\sigma_{\text{syst}} = \lambda n_{ggF} \tag{9.6}$$

where λ is the magnitude of the systematic uncertainty on ggF. The total significance then becomes

$$Z_{\text{total}} = \frac{S}{\sqrt{\sigma_{\text{stat}}^2 + \sigma_{\text{syst}}^2}} \tag{9.7}$$

Such systematic uncertainties are included in calculations where indicated.

9.2 Categorization of Events

Candidate $t\bar{t}H(\gamma\gamma)$ events must contain two photons with $p_T/m_{\gamma\gamma} > 0.35$ (0.25) for the leading (sub-leading) photon. The diphoton invariant mass must fall in the range 105–160 GeV, and both photons must pass the tight identification and isolation requirements described in Sect. 4.2.

Events are divided into two channels targeting different decays of the $t\bar{t}$ system using the hadronic and leptonic preselection above. In each channel, a Boosted Decision Tree is trained to separate simulated $t\bar{t}H(\gamma\gamma)$ events from continuum background. In both channels, the continuum background sample used in training corresponds to the NTI data control sample, where at least one photon fails the identification/isolation requirement (see Sect. 5.1.1). This sample has the high statistics necessary for BDT training, and it is statistically independent of the TI dataset on which the final maximum likelihood fit is performed.

The discriminating variables used in training are low-level inputs such as jet and lepton four-vectors. In most training variables, the true continuum background (the TI data sidebands) is well modeled by the NTI data control sample (see Figs. 9.1, 9.2, 9.3, 9.4, 9.7, 9.8, 9.9 and 9.10). Discrepancies between NTI and TI are small compared to the difference between $t\bar{t}H(\gamma\gamma)$ simulation and the NTI data control sample.

Once the BDTs are trained, four hadronic categories and three leptonic categories with different signal-to-background ratios are defined by placing cuts on the BDT discriminants. Events with low BDT score are rejected entirely. The remainder of this Section describes the BDT training and category definition procedure in detail.

The seven categories defined in this Section were first published by ATLAS in $t\bar{t}H(\gamma\gamma)$ measurements at 79.8 fb^{-1} ([2], Sect. 10.1), and the same categories are used in ATLAS $t\bar{t}H(\gamma\gamma)$ measurements at 139 fb^{-1} (Sect. 10.2). In addition, the same hadronic and leptonic BDT discriminants are used to define categories for the study of CP violation in $t\bar{t}H(\gamma\gamma)$ included in Chaps. 11 and 12.

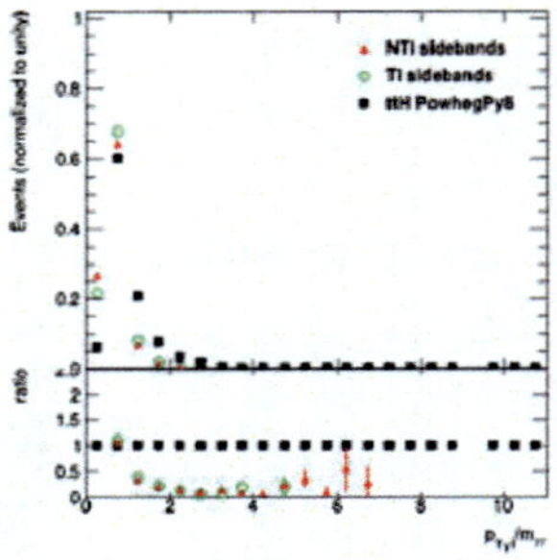

(a) Leading photon $p_T/m_{\gamma\gamma}$

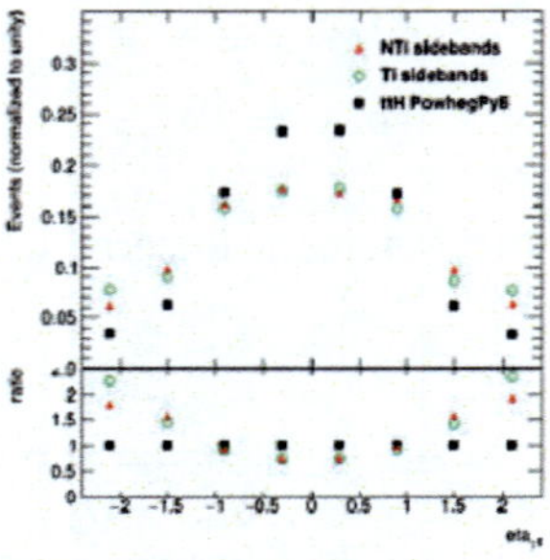

(b) Leading photon η

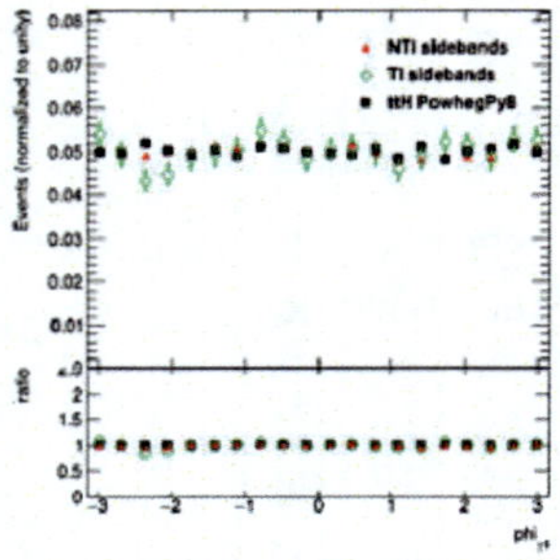

(c) Leading photon ϕ

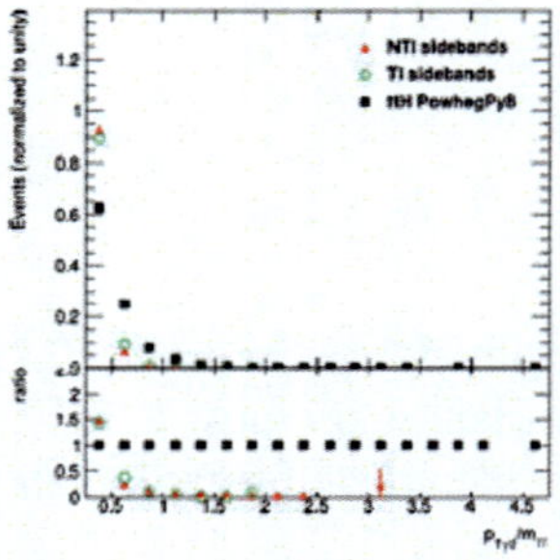

(d) Sub-leading photon $p_T/m_{\gamma\gamma}$

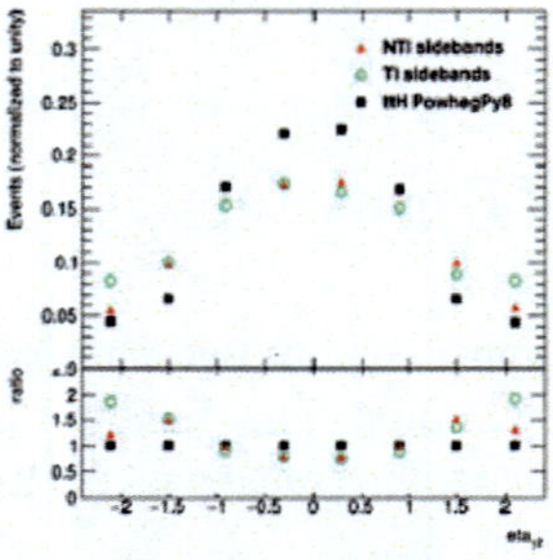

(e) Sub-leading photon η

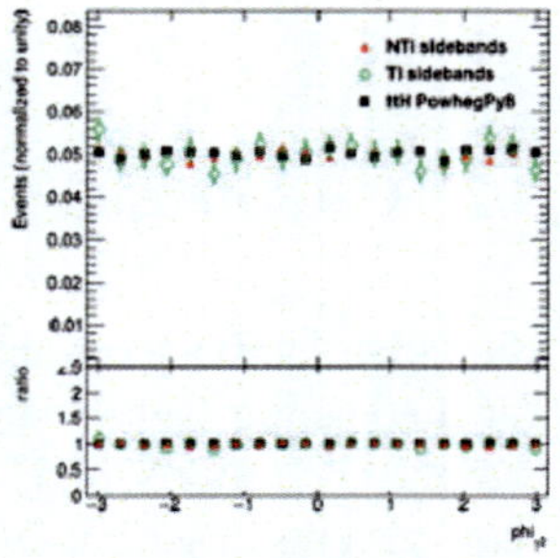

(f) Sub-leading photon ϕ

Fig. 9.1 Hadronic BDT training variables: $p_T/m_{\gamma\gamma}$, η, and ϕ of the leading and sub-leading photons. Black squares show $t\bar{t}H$ Monte Carlo signal, and red and green points represent the NTI and TI data sidebands, respectively

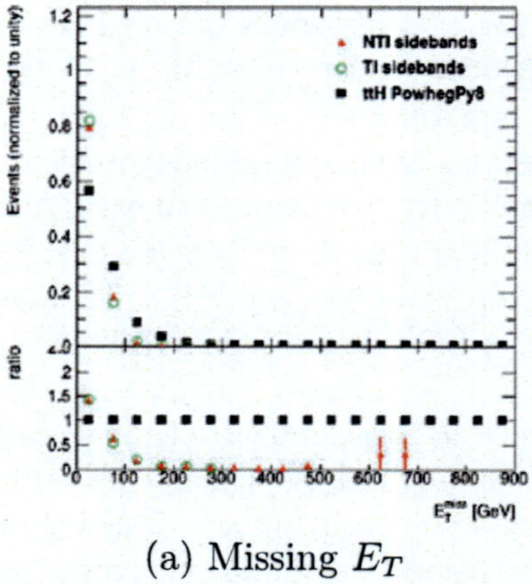

(a) Missing E_T

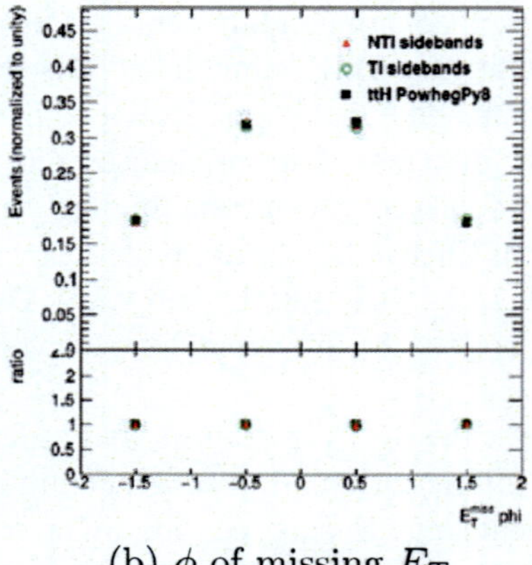

(b) ϕ of missing E_T

Fig. 9.2 Hadronic BDT training variables: missing E_T and ϕ direction of missing E_T. Black squares show $t\bar{t}H$ Monte Carlo signal, and red and green points represent the NTI and TI data sidebands, respectively

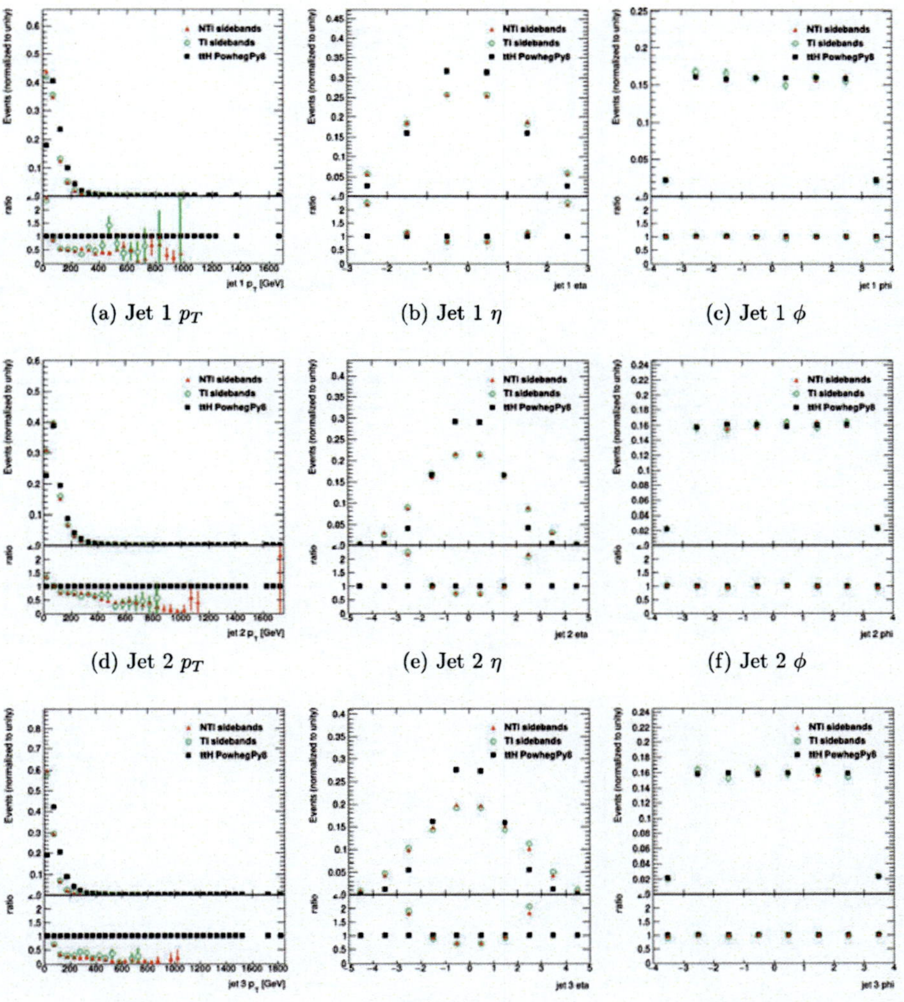

(a) Jet 1 p_T (b) Jet 1 η (c) Jet 1 ϕ

(d) Jet 2 p_T (e) Jet 2 η (f) Jet 2 ϕ

(g) Jet 3 p_T (h) Jet 3 η (i) Jet 3 ϕ

Fig. 9.3 Hadronic BDT training variables: p_T, η, and ϕ of the three jets with highest p_T. Black squares show $t\bar{t}H$ Monte Carlo signal, and red and green points represent the NTI and TI data sidebands, respectively

9.2.1 Hadronic Channel

The hadronic channel targets the all-hadronic decay of the $t\bar{t}$ system. In addition to the photon requirements detailed above, hadronic channel events must pass a preselection requiring at least three jets, with at least one b-tag at the 77% working point. Events with electrons or muons are vetoed.

A hadronic BDT is trained to separate $t\bar{t}H$ Monte Carlo (signal) from the continuum background, which is modeled by the NTI data control region. Both samples

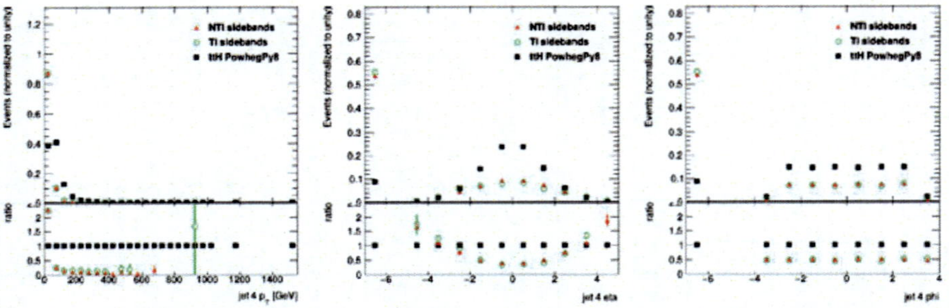

(a) Jet 4 p_T (b) Jet 4 η (c) Jet 4 ϕ

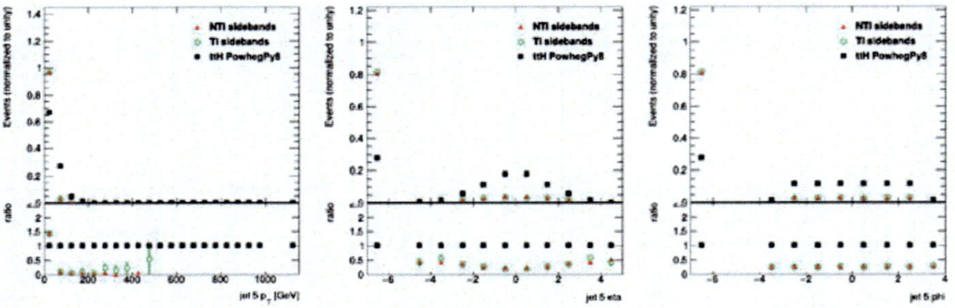

(d) Jet 5 p_T (e) Jet 5 η (f) Jet 5 ϕ

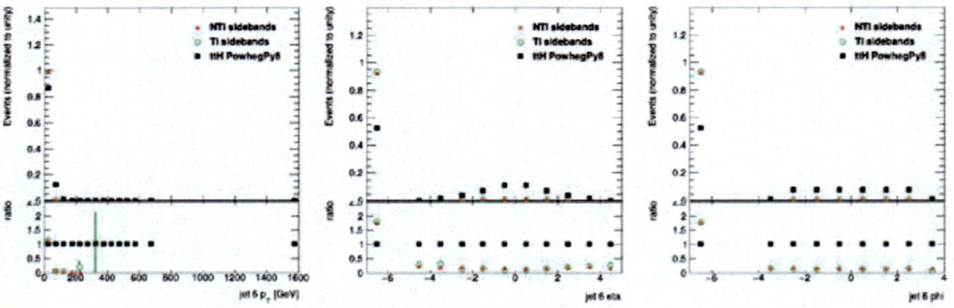

(g) Jet 6 p_T (h) Jet 6 η (i) Jet 6 ϕ

Fig. 9.4 Hadronic BDT training variables: p_T, η, and ϕ of the jets ranked fourth, fifth, and sixth in p_T. Black squares show $t\bar{t}H$ Monte Carlo signal, and red and green points represent the NTI and TI data sidebands, respectively. Events with no jet n are added to the underflow (left-most) bin for the kinematics of jet n

are divided into three subsets: 60% of each sample is used for training, another 20% for validation, and the remaining 20% for estimating the final sensitivity. In order to increase statistics in the training and validation sets, the cuts on photon $p_T/m_{\gamma\gamma}$ are replaced with flat cuts of $p_T > 35$ (25) GeV on the leading (sub-leading) photon.

The variables used to train the hadronic BDT are:

- $p_T/m_{\gamma\gamma}$, η, ϕ of the two photons. Photon p_T is scaled by $m_{\gamma\gamma}$ to reduce sculpting of the diphoton mass spectrum.

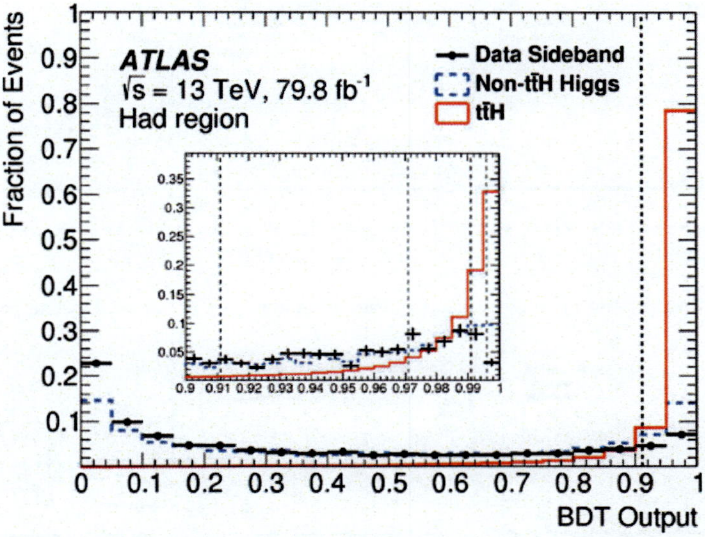

Fig. 9.5 Distribution of hadronic BDT score for $t\bar{t}H$ Monte Carlo signal (red), non-$t\bar{t}H$ Higgs processes (blue), and data sidebands (black) with 79.8 fb^{-1} of ATLAS data. Signal (red) and background (blue, black) processes peak at opposite ends of the BDT spectrum. Dashed vertical lines indicate the position of category boundaries, which are most visible in the inlaid plot that zooms in to the signal-like end of the spectrum

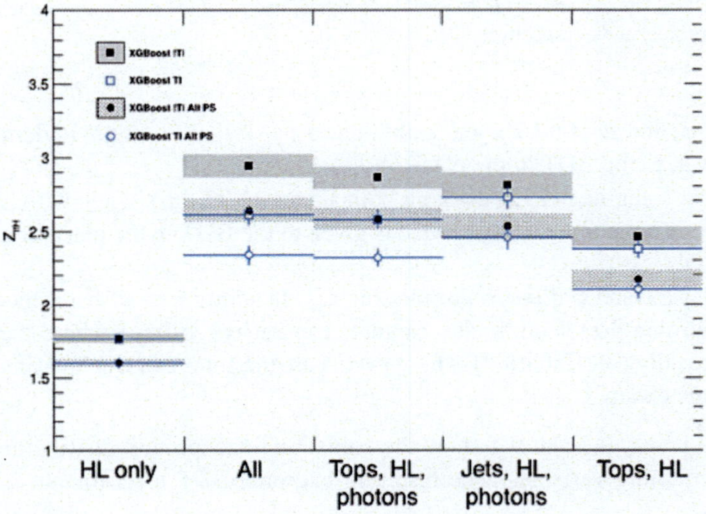

Fig. 9.6 Predicted number-counting significance of four hadronic categories, with different input variables used in BDT training. Black (blue) points represent significance calculations with continuum background estimated from the NTI (TI) data sidebands. Square points show significances that use the $t\bar{t}H$ yield from the nominal Monte Carlo sample, and circles show significances that use the yield from a sample using an alternative parton showering algorithm. The errors shown are statistical

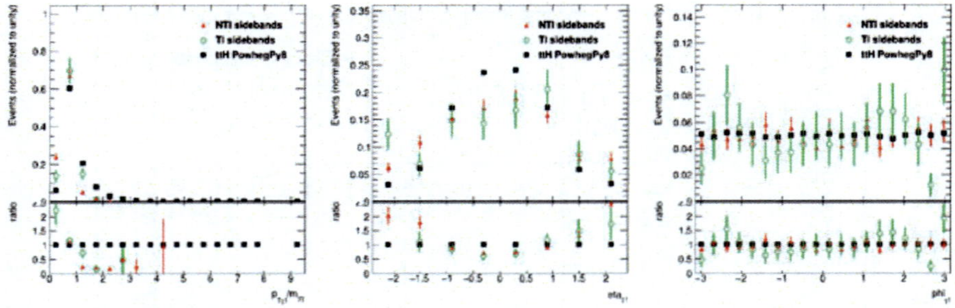

(a) Leading photon $p_T/m_{\gamma\gamma}$ (b) Leading photon η (c) Leading photon ϕ

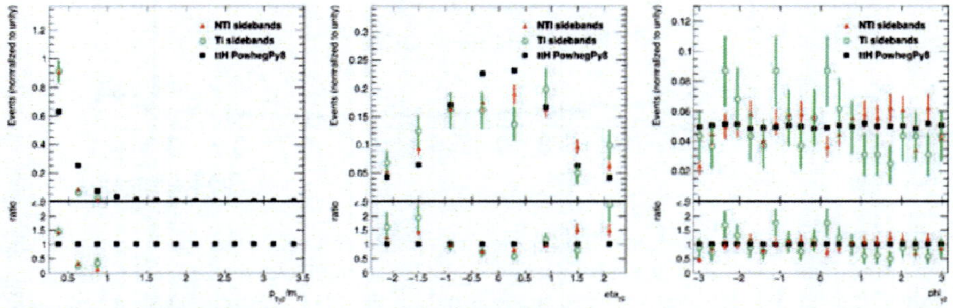

(d) Sub-leading photon $p_T/m_{\gamma\gamma}$ (e) Sub-leading photon η (f) Sub-leading photon ϕ

Fig. 9.7 Leptonic BDT training variables: $p_T/m_{\gamma\gamma}$, η, and ϕ of the leading and sub-leading photons. Black squares show $t\bar{t}H$ Monte Carlo signal, and red and green points represent the NTI and TI data sidebands, respectively

- p_T, η, ϕ, and E of the six jets with highest p_T. If there is no jet n, dummy values are given to the BDT in the place of jet n kinematics.
- Boolean b-tag flag (77% working point) for each of the six jets with highest p_T. If there is no jet n, a dummy value is given to the BDT in the place of a b-tag flag for jet n.
- Missing E_T and ϕ direction of missing E_T. In addition to all-hadronic $t\bar{t}$ decays, the hadronic preselection also captures top decays to $b\tau\nu_\tau$ where the τ decays hadronically (see Table 8.2). These events are the source of true missing E_T in the hadronic channel.

Figures 9.1, 9.2, 9.3 and 9.4 show the signal and background distributions in each of these training variables. The resulting hadronic BDT discriminant is shown in Fig. 9.5.

Four cuts are placed on the hadronic BDT discriminant to create five orthogonal regions: events in the region with lowest BDT score are rejected, and the remaining regions make up the four hadronic analysis categories. The cuts forming the category boundaries are treated as hyper-parameters of the analysis, and both the boundary values and BDT hyper-parameters are chosen to maximize the expected Standard Model $t\bar{t}H$ number-counting significance evaluated on the validation set. This sig-

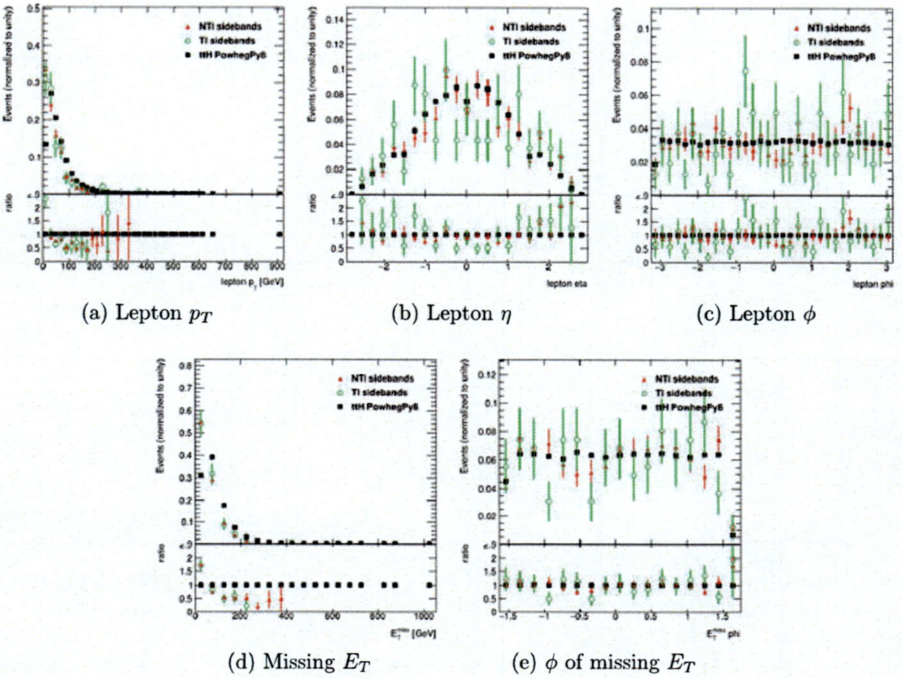

(a) Lepton p_T (b) Lepton η (c) Lepton ϕ

(d) Missing E_T (e) ϕ of missing E_T

Fig. 9.8 Leptonic BDT training variables: (top) p_T, η, and ϕ of the leading lepton, and (bottom) missing E_T and ϕ direction of missing E_T. Black squares show $t\bar{t}H$ Monte Carlo signal, and red and green points represent the NTI and TI data sidebands, respectively

nificance is given by Eq. 9.3 in each category, and the total significance is given by the sum in quadrature over all categories.

The category boundaries are determined from a brute force scan of all possible boundary combinations. In order to have sufficient data in each category to perform the maximum likelihood fit described in Chap. 6, a minimum of 0.8 continuum background events (predicted from NTI data) is required in each category in the range $123 < m_{\gamma\gamma} < 127$ GeV. This corresponds to $\sim$ 10 events in the full fit range of 105–160 GeV.

The optimal boundaries are shown in Table 9.1. 85% of the $t\bar{t}H$ signal events passing hadronic preselection are selected into these four categories, and 89% of the preselected continuum background events are rejected.

The predicted Higgs boson yield in each category is shown in Table 9.2, along with the estimated continuum background derived by scaling the NTI sidebands. These yields are calculated in the region $123 < m_{\gamma\gamma} < 127$ GeV. The expected number-counting significances of the total Higgs boson signal (Z_H), the $t\bar{t}H$ signal ($Z_{t\bar{t}H}$), and the single top plus Higgs signal (Z_{tH}) are shown in Table 9.3.

Many combinations of hadronic BDT training variables were tested before settling on the list above, including:

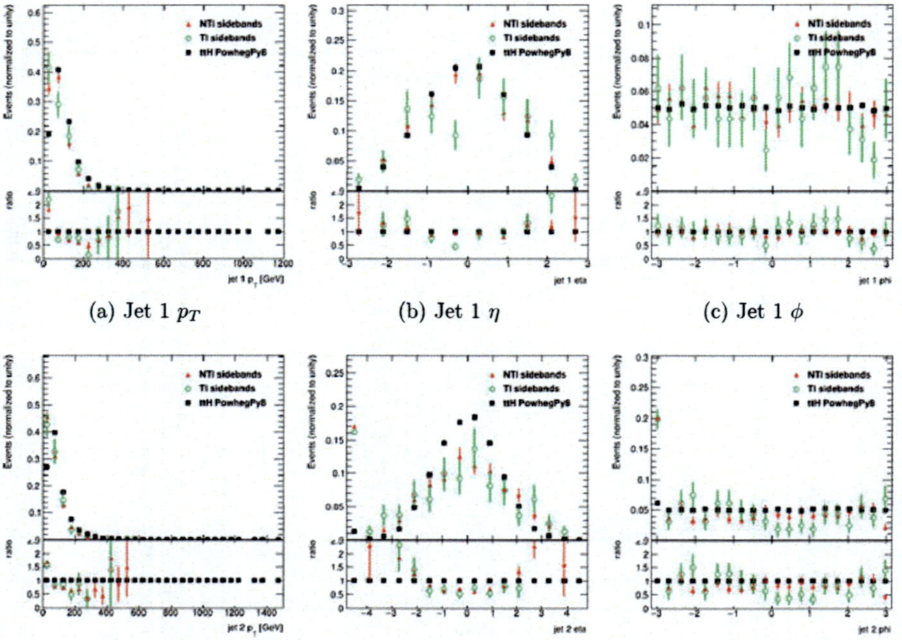

(a) Jet 1 p_T (b) Jet 1 η (c) Jet 1 ϕ

(d) Jet 2 p_T (e) Jet 2 η (f) Jet 2 ϕ

Fig. 9.9 Leptonic BDT training variables: p_T, η, and ϕ of the two jets with highest p_T. Black squares show $t\bar{t}H$ Monte Carlo signal, and red and green points represent the NTI and TI data sidebands, respectively. Events with no jet n are added to the underflow (left-most) bin for the kinematics of jet n

- High-level jet variables (HL): n_{jets}, $n_{\text{cen. jets}}$, $n_{b\text{-jets}}$, missing E_T, H_T
- Photons: $p_T/m_{\gamma\gamma}$ and η for the two photons
- Jets: p_T, η and pseudo-continuous b-tag score for up to six jets
- Tops: Top Reco BDT score (see Chap. 8), p_T, η, and m for the reconstructed top (t_1), p_T, η, and m of the system of jets recoiling against the top (t_2), $\Delta R(t, \gamma\gamma)$, $\Delta R(t_1, t_2)$, $\Delta R(t_2, \gamma\gamma)$.

Figure 9.6 compares the performance of several combinations of these variables in terms of number-counting significance. The set of high-level variables includes no photon-related variables, and there is near-perfect agreement between significance calculating using NTI sidebands and TI sidebands. The introduction of the photon variables increases the sensitivity significantly (compare, for example, "Tops, HL, photons" and "Tops, HL"). It is also interesting to note that replacing the low-level jet kinematics with reconstructed top variables gives very similar performance (compare "Tops, HL, photons" and "Jets, HL, photons").

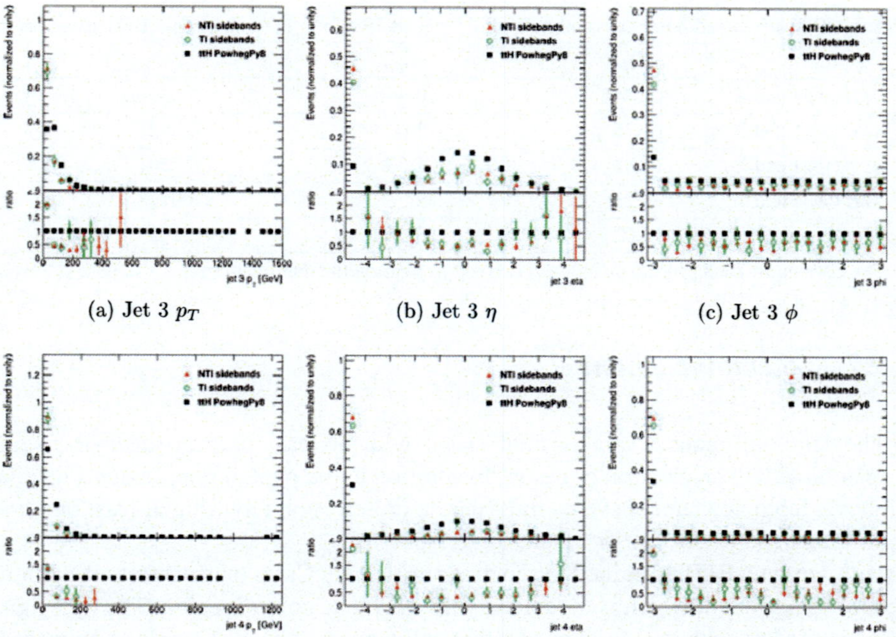

(a) Jet 3 p_T (b) Jet 3 η (c) Jet 3 ϕ

(d) Jet 4 p_T (e) Jet 4 η (f) Jet 4 ϕ

Fig. 9.10 Leptonic BDT training variables: p_T, η, and ϕ of the jets ranked third and fourth in p_T. Black squares show $t\bar{t}H$ Monte Carlo signal, and red and green points represent the NTI and TI data sidebands, respectively. Events with no jet n are added to the underflow (left-most) bin for the kinematics of jet n

Table 9.1 Ranges in hadronic BDT score defining the four hadronic $t\bar{t}H(\gamma\gamma)$ categories

Category	BDT range
$t\bar{t}H$ had 4	[0.911–0.971]
$t\bar{t}H$ had 3	[0.971–0.991]
$t\bar{t}H$ had 2	[0.991–0.996]
$t\bar{t}H$ had 1	[0.996–1]

Table 9.2 Predicted event yield in the region 123 GeV $< m_{\gamma\gamma} <$ 127 GeV at 79.8 fb^{-1} in the hadronic BDT categories

	$t\bar{t}H$	$tHjb$	tWH	ggF	VBF	WH	ZH	Cont. bkg.
$t\bar{t}H(\gamma\gamma)$ had 4	2.4	0.154	0.0657	0.657	0.19	0.122	0.338	26.5
$t\bar{t}H(\gamma\gamma)$ had 3	3.92	0.11	0.102	0.471	0.0911	0.0618	0.207	11.4
$t\bar{t}H(\gamma\gamma)$ had 2	2.95	0.0469	0.0835	0.146	0.0245	0.0178	0.0669	2.44
$t\bar{t}H(\gamma\gamma)$ had 1	3.93	0.0299	0.131	0.0897	0.0158	0.0159	0.0518	0.759

The continuum background estimate is derived by scaling the data yield in the NTI data sidebands

Table 9.3 Predicted number-counting significance at 79.8 fb^{-1} in the hadronic BDT categories

	Z_H	$Z_{t\bar{t}H}$	Z_{tH}
$t\bar{t}H(\gamma\gamma)$ had 4	0.746	0.448	0.0399
$t\bar{t}H(\gamma\gamma)$ had 3	1.38	1.06	0.0527
$t\bar{t}H(\gamma\gamma)$ had 2	1.81	1.53	0.0547
$t\bar{t}H(\gamma\gamma)$ had 1	3.23	2.73	0.0728
Combined	4.02	3.33	0.113

The continuum background estimate is derived by scaling the data yield in the NTI data sidebands

9.2.2 Leptonic Channel

The leptonic channel targets final states with at least one top decaying semi-leptonically to an electron or muon. In addition to the photon requirements detailed above, leptonic channel events must pass a preselection requiring at least one b-jet and at least one electron or muon.

A leptonic BDT is trained to separate $t\bar{t}H$ Monte Carlo (signal) from the continuum background, which is modeled by the NTI data control region. The $t\bar{t}H$ signal is required to contain at least one jet that is b-tagged at the 77% working point: 60% of this sample is used for training, another 20% for validation, and the remaining 20% for estimating the final sensitivity. The NTI data sample with ≥ 1 lepton contains limited statistics, and four statistically independent datasets are defined: one for training, one for BDT hyper-parameter optimization (validation 1), one for category boundary optimization (validation 2), and one for testing.

In the training and validation 1 samples, events are required to contain zero b-tagged jets and at least one un-tagged jet in the central η region, effectively modeling the background from non-top QCD processes. In addition, the cuts on photon $p_T/m_{\gamma\gamma}$ are replaced with flat cuts of $p_T > 35\,(25)$ GeV on the leading (sub-leading) photon, and the range of diphoton invariant masses is extended to $m_{\gamma\gamma} > 80$ GeV. 75% of this sample is used for training, and the other 25% for optimization of the BDT hyper-parameters.

In the validation 2 and testing samples, the same selection is applied that is used for the signal sample (≥ 1 b-jet, ≥ 1 lepton): 50% of this sample is used to optimize the BDT category boundaries, and the other 50% is reserved for estimating the final sensitivity.

The variables used to train the leptonic BDT are:

- $p_T/m_{\gamma\gamma}$, η, and ϕ of the two photons. Photon p_T is scaled by $m_{\gamma\gamma}$ to reduce sculpting of the diphoton mass spectrum.
- p_T, η, and ϕ of up to two leptons. If there is no second lepton, dummy values are given to the BDT in the place of second lepton kinematics.
- p_T, η, ϕ, and E of the four jets with highest p_T. If there is no jet n, dummy values are given to the BDT in the place of jet n kinematics.

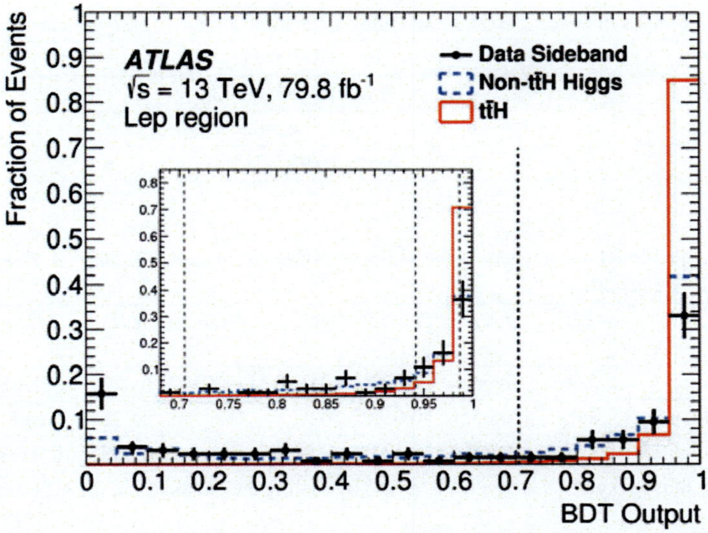

Fig. 9.11 Distribution of leptonic BDT score for $t\bar{t}H$ Monte Carlo signal (red), non-$t\bar{t}H$ Higgs processes (blue), and data sidebands (black) with 79.8 fb^{-1} of ATLAS data. Signal (red) and background (blue, black) processes peak at opposite ends of the BDT spectrum. Dashed vertical lines indicate the position of category boundaries, which are most visible in the inlaid plot that zooms in to the signal-like end of the spectrum

- Boolean b-tag flag (77% working point) for each of the four jets with highest p_T. If there is no jet n, a dummy value is given to the BDT in the place of a b-tag flag for jet n.
- Missing E_T and ϕ direction of missing E_T.

Figures 9.7, 9.8, 9.9 and 9.10 show the signal and background distributions in each of these training variables. The resulting leptonic BDT discriminant is shown in Fig. 9.11. The peak in the NTI background at high BDT score is composed primarily of $t\bar{t}\gamma\gamma$ events, which are not well rejected by the leptonic BDT (see Sect. 9.3.2).

Three cuts are placed on the leptonic BDT discriminant to create four orthogonal regions: events in the region with lowest BDT score are rejected, and the remaining regions make up the three leptonic analysis categories. The cuts forming the category boundaries are chosen to maximize the expected Standard Model $t\bar{t}H$ number-counting significance evaluated on the validation 2 set. This significance is given by Eq. 9.3 in each category, and the total significance is given by the sum in quadrature over all categories.

The category boundaries are determined from a brute force scan of all possible boundary combinations. In order to have sufficient data in each category to perform the maximum likelihood fit, a minimum of 0.8 continuum background events (predicted from NTI data) is required in each category in the range $123 < m_{\gamma\gamma} < 127$ GeV. This corresponds to ~ 10 events in the full fit range of 105–160 GeV.

Table 9.4 Ranges in leptonic BDT score defining the three leptonic $t\bar{t}H(\gamma\gamma)$ categories

Category	BDT range
$t\bar{t}H$ lep 3	[0.705–0.942]
$t\bar{t}H$ lep 2	[0.942–0.987]
$t\bar{t}H$ lep 1	[0.987–1]

Table 9.5 Predicted event yield in the region 123 GeV $< m_{\gamma\gamma} <$ 127 GeV at 79.8 fb^{-1} in the leptonic BDT categories

	$t\bar{t}H$	$tHjb$	tWH	ggF	VBF	WH	ZH	Cont. bkg.
$t\bar{t}H(\gamma\gamma)$ lep 3	0.66	0.0634	0.022	0.028	0.00102	0.0265	0.0189	2.84
$t\bar{t}H(\gamma\gamma)$ lep 2	1.85	0.0556	0.052	0	0	0.0226	0.013	2.35
$t\bar{t}H(\gamma\gamma)$ lep 1	3.98	0.0307	0.123	0.00587	0	0.00956	0.00994	1.67

The continuum background estimate is derived by scaling the data yield in the NTI data sidebands

Table 9.6 Predicted number-counting significance at 79.8 fb^{-1} in the leptonic BDT categories

	Z_H	$Z_{t\bar{t}H}$	Z_{tH}
$t\bar{t}H(\gamma\gamma)$ lep 3	0.465	0.368	0.045
$t\bar{t}H(\gamma\gamma)$ lep 2	1.16	1.06	0.0521
$t\bar{t}H(\gamma\gamma)$ lep 1	2.5	2.33	0.0643
Combined	2.8	2.58	0.0942

The continuum background estimate is derived by scaling the data yield in the NTI data sidebands

The optimal boundaries are shown in Table 9.4. 97% of the $t\bar{t}H$ signal events passing leptonic preselection (≥ 1 b-jet, ≥ 1 lepton) are selected into these four categories, and 43% of the preselected continuum background events are rejected.

The predicted Higgs boson yield in each category is shown in Table 9.5, along with the estimated continuum background derived by scaling the NTI sidebands. These yields are calculated in the region $123 < m_{\gamma\gamma} < 127$ GeV. The expected number-counting significances of the total Higgs boson signal (Z_H), the $t\bar{t}H$ signal ($Z_{t\bar{t}H}$), and the single top plus Higgs signal (Z_{tH}) are shown in Table 9.6.

Many combinations of leptonic BDT training variables were tested before settling on the list above, including:

- High-level jet variables (HL): $n_{\text{cen. jets}}$, missing E_T, H_T, transverse mass m_T, p_T (lepton + missing E_T)
- Photons: $p_T/m_{\gamma\gamma}$ and η for the two photons
- Jets: p_T, η, ϕ and b-tag flag for up to four jets
- Leps: p_T, η and ϕ of up to two leptons
- Tops: Top Reco BDT score, p_T, η, and m for the reconstructed top (t_1), p_T, η, and m of the system of jets recoiling against the top (t_2).

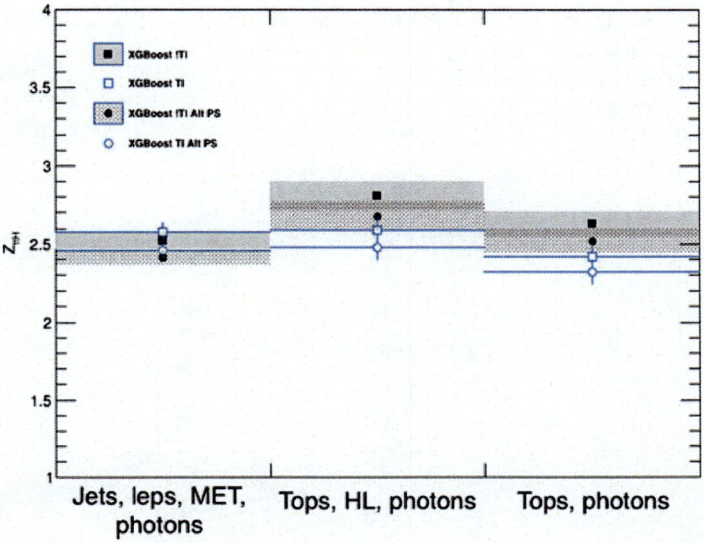

Fig. 9.12 Number-counting significance of three leptonic categories, with different input variables used in BDT training. Black (blue) points represent significance calculations with continuum background estimated from the NTI (TI) data sidebands. Square points show significances that use the $t\bar{t}H$ yield from the nominal Monte Carlo sample, and circles show significances that use the yield from a sample using an alternative parton showering algorithm. The errors shown are statistical

Figure 9.12 compares the performance of three combinations of these variables in terms of number-counting significance. Once again, replacing the low-level jet and lepton kinematics with reconstructed top variables gives very similar performance (compare "Tops, photons" and "Jets, leps, MET, photons").

9.3 Decomposition of Continuum Background

The continuum diphoton background included in the $t\bar{t}H$ analysis categories consists mainly of two processes: $\gamma\gamma$ + jets and $t\bar{t}\gamma\gamma$, with photons that can be either real or fake. The total continuum background is estimated using the data-driven method already described, but the relative contribution from each of these sources in the $t\bar{t}H$ categories has so far remained unexplored. This Section introduces several methods for decomposing the continuum background. Monte Carlo samples containing real photons are used to model the shapes of both real-photon and fake-photon background processes.

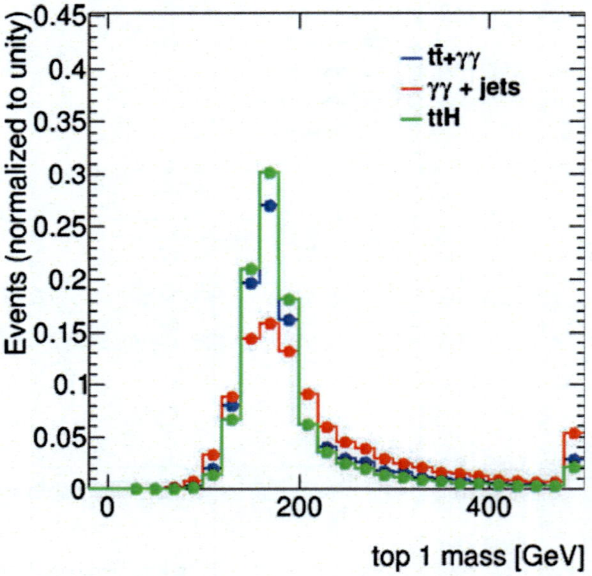

Fig. 9.13 The templates, normalized to unit area, of the top mass distribution in the $t\bar{t}H$, $t\bar{t}\gamma\gamma$ and $\gamma\gamma$ + jets Monte Carlo samples

9.3.1 Hadronic Channel

In the hadronic categories, the relative contribution of $t\bar{t}\gamma\gamma$ and $\gamma\gamma$ + jets is determined using a template fit method. A probability distribution function (PDF) f is constructed for each background process as a function of an event variable x. The PDFs are summed with a relative weight factor r, giving a combined PDF of

$$r \times f_{t\bar{t}\gamma\gamma}(x) + (1 - r) \times f_{\gamma\gamma}(x) + \frac{n_{t\bar{t}H}^{SM}}{n_{\text{data}}} \times f_{t\bar{t}H}(x) \tag{9.8}$$

The final term in this PDF represents the expected signal: $n_{t\bar{t}H}^{SM}$ is the number of SM signal events, n_{data} is the observed number of data events, and $f_{t\bar{t}H}(x)$ is the signal PDF. Equation 9.8 is fitted to data, and the best fit value of r gives the fraction of $t\bar{t}\gamma\gamma$ making up the total continuum background: this is referred to as the **top background fraction**.

The template fit method is most successful when the background PDFs $f_{t\bar{t}\gamma\gamma}$ and $f_{\gamma\gamma}$ are very different in shape. The variable x is chosen to be the reconstructed top candidate mass in order to exploit the shape difference between samples with and without true top quark (see Chap. 8 for details on the top reconstruction). Figure 9.13 shows the templates $f_{t\bar{t}\gamma\gamma}$, $f_{\gamma\gamma}$, and $f_{t\bar{t}H}$, which use a bin size of 20 GeV. The statistical uncertainty on the Monte Carlo templates is eclipsed by the size of the marker points.

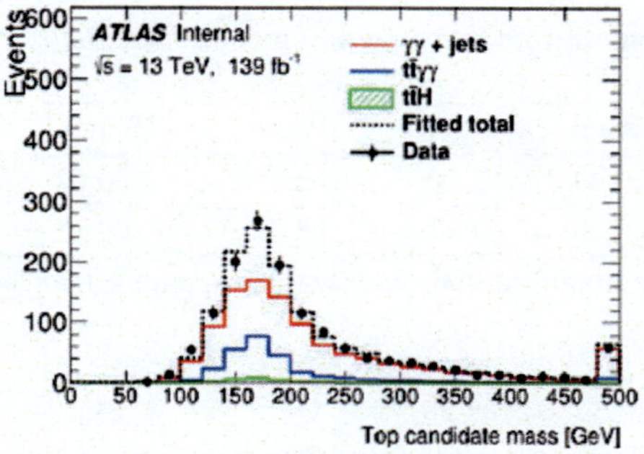

(a) $t\bar{t}H$ had 1-4

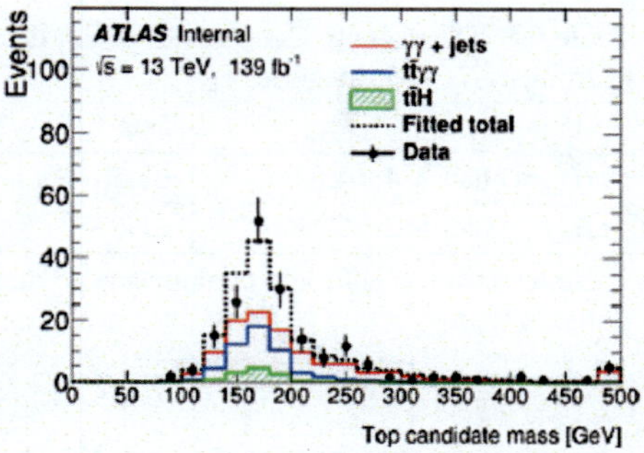

(b) $t\bar{t}H$ had 1-2

Fig. 9.14 Template fits to data in the hadronic $t\bar{t}H(\gamma\gamma)$ categories. The contributions from $t\bar{t}\gamma\gamma$ (blue) and $\gamma\gamma+$jets (red) are normalized to their fitted values of r and $(1-r)$. The $t\bar{t}H$ contribution (green) is fixed to the SM expectation. The fitted combined PDF is shown in the dashed black line. The left plot shows the fit for all four hadronic BDT categories, and the right shows only the two categories with highest S/B

Table 9.7 The top background fraction obtained in the hadronic $t\bar{t}H(\gamma\gamma)$ categories

	$n^{SM}_{t\bar{t}H}/n_{\text{data}}$ (%)	r (%)	Data (%)	Temp (%)
$t\bar{t}H$ had 1–4	2	21	±6	±2
$t\bar{t}H$ had 1–2	7	31	±16	±5

The uncertainty is split into components coming from data statistics (data) and from the template statistics (temp)

The template fits are shown in Fig. 9.14 in two regions: one region corresponding to all four hadronic $t\bar{t}H(\gamma\gamma)$ categories, and another region corresponding to the two hadronic categories with highest signal-to-background ratio (S/B).

The top background fractions obtained in each region are summarized in Table 9.7. The uncertainty on the top background fraction is divided into two components: a contribution from the statistics of the data, and a contribution from the statistics of the templates. The latter is determined by running 1000 toys per template fit and randomly varying the templates bin-by-bin. The variations are subject to a Gaussian constraint with a width corresponding to the 1σ statistical error on the bin. The standard deviation of the top background fractions from toys is taken as the error due to template statistics. The statistical uncertainty on the templates is small compared to the statistical uncertainty in data.

Using the top background fraction extracted from data, the number of $t\bar{t}\gamma\gamma$ and $\gamma\gamma+$ jets events in each region can be extracted:

$$n_{t\bar{t}\gamma\gamma} = r n_{\text{data}} \tag{9.9}$$

$$n_{\gamma\gamma} = (1-r) n_{\text{data}} \tag{9.10}$$

Table 9.8 Data/Monte Carlo scale factors determined from the top background fraction for the hadronic $t\bar{t}H(\gamma\gamma)$ categories

	$k_{t\bar{t}\gamma\gamma}$	Data	Temp	$k_{\gamma\gamma}$	Data	Temp
$t\bar{t}H(\gamma\gamma)$ had 1–4	4.87	±1.50	±0.44	1.00	±0.08	±0.02
$t\bar{t}H(\gamma\gamma)$ had 1–2	3.31	±1.67	±0.58	1.21	±0.31	±0.11

The uncertainty is split into components coming from data statistics (data) and from the template statistics (temp)

The ratio of this number of events to the prediction from Monte Carlo gives a data/Monte Carlo scale factor for each process. These scale factors are shown in Table 9.8.

The same procedure can be applied in the data sideband region only, with $n^{SM}_{t\bar{t}H} = 0$. The top background fractions and data/Monte Carlo scale factors obtained from the sidebands agree with those reported in Tables 9.7 and 9.8 within the statistical uncertainty.

9.3.2 Leptonic Channel

In the leptonic channel, the dilepton events provide a simple handle for background decomposition. Since it is highly unlikely that $\gamma\gamma$ +jets events contain more than one lepton, all dilepton events are taken to be the result of $t\bar{t}$ decays. In the data sidebands ($|m_{\gamma\gamma} - 125\,\text{GeV}| > 5\,\text{GeV}$), no contamination from $t\bar{t}H(\gamma\gamma)$ is expected, and all dilepton events are therefore attributed to $t\bar{t}\gamma\gamma$.

The number of dilepton events predicted by $t\bar{t}\gamma\gamma$ Monte Carlo is scaled to the dilepton event yield in the data sidebands (eight events at 139 fb^{-1}). This corresponds to a data/MC scale factor of 2.91 ± 1.3, with the uncertainty coming from the statistical error on the number dilepton events in data. This same scale factor is applied to the predicted yield of $t\bar{t}\gamma\gamma$ events containing ≥ 1 lepton, and the scaled prediction is subtracted from the number of ≥ 1 lepton events in data. The remaining events are attributed to $\gamma\gamma$ + jets. The data/MC scale factor for $\gamma\gamma$ + jets is extracted by solving:

$$n^{SB}_{\text{data}} = k_{t\bar{t}\gamma\gamma} n^{SB}_{t\bar{t}\gamma\gamma} + k_{\gamma\gamma} n^{SB}_{\gamma\gamma} \tag{9.11}$$

Tables 9.9 and 9.10 show the values of each parameter in this formula.

Applying the $t\bar{t}\gamma\gamma$ Monte Carlo scale factor to data in the full $m_{\gamma\gamma}$ range and including the Standard Model prediction for $t\bar{t}H$ gives the fractions listed in Table 9.11. As expected, the $t\bar{t}\gamma\gamma$ process is the dominant background in these categories, since $\gamma\gamma$+ jets is better rejected by a leptonic BDT that is trained on 0 b-jet background (refer to Sect. 9.2.2).

Table 9.9 The dilepton and inclusive ≥ 1 lepton event yields in the three leptonic $t\bar{t}H(\gamma\gamma)$ categories at $139\,\mathrm{fb}^{-1}$, counted in the $m_{\gamma\gamma}$ sidebands

	Data yield	Predicted $t\bar{t}\gamma\gamma$ yield	Predicted $\gamma\gamma$ + jets yield
Dilepton	8	2.75	0
≥ 1 leptons	118	29.42	3.92

Shown in data, $t\bar{t}\gamma\gamma$ Monte Carlo, and $\gamma\gamma$ + jets Monte Carlo

Table 9.10 Data/MC scale factors derived for each continuum background process in the leptonic channel

	$k_{t\bar{t}\gamma\gamma}$	$k_{\gamma\gamma}$
Dilepton	2.91 ± 1.3	n/a
≥ 1 leptons	2.91 ± 1.3	8.25 ± 7.22

The reported statistical error is due to the limited number of dilepton events

Table 9.11 The top background fraction obtained in the leptonic $t\bar{t}H(\gamma\gamma)$ categories

	$n_{t\bar{t}H}^{SM}/n_{\mathrm{data}}$	r
$t\bar{t}H(\gamma\gamma)$ lep 1–3	9%	$66\% \pm 23\%$

References

1. Chen T, Guestrin C, XGBoost: a scalable. In: Proceedings of the 22nd ACM SIGKDD international discovery and data mining, KDD'16, San Francisco, pp 785–794. ISBN: 978-1-4503-4232-2. https://doi.org/10.1145/2939672.2939785
2. The ATLAS Collaboration (2018) Observation of Higgs boson with a top quark pair at the LHC with the ATLAS 784:173–191. https://doi.org/10.1016/j.physletb.2018.07.035. https://cds.cern.ch/record/2621167. arXiv: 1806.00425

Chapter 10
First Observation of $t\bar{t}H$ Production

The observation of $t\bar{t}H$ production at the LHC represents the first direct evidence of Higgs boson coupling to quarks. The first observation of $t\bar{t}H$ above the 5σ threshold for discovery combined searches in multiple Higgs boson decay channels, including $H \rightarrow \gamma\gamma$ [1, 2]. The 2018 ATLAS $t\bar{t}H$ combination is discussed in Sect. 10.1. Section 10.2 describes the $t\bar{t}H(\gamma\gamma)$ results from ATLAS using the full Run 2 dataset, which gives a single-channel observed $t\bar{t}H$ significance of 4.9σ (4.5σ expected) [3].

10.1 Combination of Higgs Decay Channels

The full ATLAS $t\bar{t}H$ program includes searches in multiple Higgs boson decay channels. A statistical combination of all available $t\bar{t}H$ analyses was performed 2018 using a partial Run 2 dataset. The analyses included in this combination are summarized in Sect. 10.1.1. Section 10.1.2 includes a discussion of the treatment of systematic uncertainties, and the combined results are included in Sect. 10.1.3.

10.1.1 Input Analyses

10.1.1.1 Run 2 $t\bar{t}H(b\bar{b})$ with 36.1 fb^{-1}

The $t\bar{t}H(b\bar{b})$ analysis [4] targets the $H \rightarrow b\bar{b}$ decay in conjunction with at least one semi-leptonic top quark decay. The analysis included in the combination uses 36.1 fb^{-1} of ATLAS data collected in 2015–2016. Events are divided according to lepton multiplicity and b-jet multiplicity at different working points. Within each

J. E. Dickinson, *ATLAS Measurements of the Higgs Boson Coupling to the Top Quark in the Higgs to Diphoton Decay Channel*, Springer Theses,
https://doi.org/10.1007/978-3-030-86368-5_10

region, categories are defined using a variety of multivariate techniques. Categories containing high fractions of $t\bar{t}H$ are designated as signal regions. Control regions are used to constrain the rate of $t\bar{t}$ background processes, including $t\bar{t}+b\bar{b}$, $t\bar{t}+b$, $t\bar{t}$ + charm, and $t\bar{t}$ + light jets.

In the dilepton channel, four control regions and three signal regions are defined. In the single-lepton channel, six control regions and six signal regions are defined, including one signal region that targets boosted (high p_T) Higgs bosons and top quarks using large-radius ($R = 1.0$) jets. The $t\bar{t}H$ signal strength is determined by fitting all 19 categories. The rate of each background process is left free in the fit, with constraining power coming from the control regions.

10.1.1.2 Run 2 $t\bar{t}H$ (Multi-lepton) with 36.1 fb^{-1}

The $t\bar{t}H$ (multi-lepton) analysis [5] targets $H \to WW$, $H \to \tau\tau$, and non-resonant $H \to ZZ$ decays. The analysis included in the combination uses 36.1 fb^{-1} of ATLAS data collected in 2015–2016. Eight signal regions are defined based on lepton flavor and multiplicity, with two categories corresponding to the 4ℓ final state. In addition, four control regions are defined to target the $t\bar{t}W$, $t\bar{t}Z$, $t\bar{t}$, and diboson backgrounds.

The $t\bar{t}H$ signal strength is determined by fitting all twelve categories. The rate of each background process is left free in the fit, with constraining power coming from the control regions.

10.1.1.3 Run 2 $t\bar{t}H(\gamma\gamma)$ with 79.8 fb^{-1}

The $t\bar{t}H(\gamma\gamma)$ analysis targets the $H \to \gamma\gamma$ decay using the seven analysis categories described in Chap. 9. The analysis included in this $t\bar{t}H$ combination uses 79.8 fb^{-1} of 13 TeV ATLAS data collected between 2015 and 2017. The same analysis procedure is later applied to the full Run 2 dataset (139 fb^{-1}), and these results are reported in Sect. 10.2.

10.1.1.4 Run 2 $t\bar{t}H(ZZ \to 4\ell)$ with 79.8 fb^{-1}

The $t\bar{t}H(ZZ \to 4\ell)$ analysis [1] targets resonant $H \to ZZ$ decay with both Z bosons decaying leptonically to e or μ. This analysis uses 79.8 fb^{-1} of 13 TeV ATLAS data collected between 2015 and 2017. Events are selected that contain ≥ 1 b-jet and ≥ 4 leptons (two same-flavor/opposite-sign pairs) with 115 GeV $< m_{4\ell} <$ 130 GeV. Events are then sorted into two groups: had-enriched (≥ 4 jets and exactly 4 leptons), and lep-enriched (≥ 2 jets and ≥ 5 leptons). The had-enriched sample is further divided into one "Tight" category and one "Loose" category based on a BDT discriminant. This BDT is trained to separate $t\bar{t}H(ZZ \to 4\ell)$ Monte Carlo events from simulated background using the following training variables: p_T^{jj}, m^{jj},$\Delta\eta^{jj}$, $\Delta R_{\text{min}}(j, Z)$, $\Delta R(j_1, 4\ell)$, η_{Zepp}^{ZZ}, E_T^{miss}, n_{jet}, $n_{b\text{-jet}}$, H_T, and $H \to 4\ell$ matrix element.

10.1.1.5 Run 1 $t\bar{t}H$

The Run 1 ATLAS search for $t\bar{t}H$ production [6] combined three Higgs boson decay channels: $H(b\bar{b})$ [7], H (multi-lepton) [8], and $H(\gamma\gamma)$ [6]. With $4.5\,\text{fb}^{-1}$ at $\sqrt{s} = 7\,\text{TeV}$ and $20.3\,\text{fb}^{-1}$ at $\sqrt{s} = 8\,\text{TeV}$, the Run 1 observed $t\bar{t}H$ significance is 2.5σ, compared to 1.5σ expected. This corresponds to an observed signal strength of $\mu_{t\bar{t}H} = 1.82^{+0.83}_{-0.75}$.

10.1.2 Systematic Uncertainties

The Run 2 $t\bar{t}H(b\bar{b})$ and $t\bar{t}H$ (multi-lepton) analyses use data from 2015 to 2016. The experimental uncertainties on physics objects (jets, leptons, photons, etc.) are mostly correlated between these two analyses. The modeling uncertainty on $t\bar{t}V$, tV, 4-top, and diboson backgrounds are treated as fully correlated. The modeling uncertainty on $t\bar{t}$ in the $t\bar{t}H(b\bar{b})$ analysis is not correlated with $t\bar{t}H$ (multi-lepton) because the two analyses probe different regions of phase-space. See [5] for full discussion of correlations.

The Run 2 $t\bar{t}H(ZZ \to 4\ell)$ and $t\bar{t}H(\gamma\gamma)$ analyses use data from 2015 to 2017 and a later version of ATLAS analysis software than the 2015–2016 analyses. The experimental uncertainties on physics objects are treated as fully correlated between these two analyses, but mostly uncorrelated with $t\bar{t}H(b\bar{b})$ and $t\bar{t}H$ (multi-lepton).

The main experimental systematics that are correlated in the combination of these four analyses are uncertainties on the electron/photon energy scale and resolution, and on the electron efficiencies. The uncertainty on overall luminosity is treated as partially correlated between the 2015–2016 and 2015–2017 datasets: a 1.51% common uncertainty is assigned to both periods, and an additional 1.61% (1.27%) is applied for the 2015–2016 (2015–2017) period.

Theory uncertainties, including uncertainties on the $t\bar{t}H$ signal prediction, are fully correlated across all four 13 TeV analyses. The theory systematics are dominated by the Monte Carlo modeling of $t\bar{t}$ background processes and uncertainties on the $t\bar{t}H$ signal (including QCD scale and UEPS uncertainties).

The dominant sources of systematic uncertainty on the 13 TeV cross section measurement are summarized in Table 10.1. To determine the contribution from each group of systematics, the corresponding nuisance parameters are fixed to their best fit values from the unconditional maximum likelihood fit. The resulting error is then subtracted (in quadrature) from the total error to obtain the error due only to the systematics in question.

In the combination of Run 1 and Run 2 $t\bar{t}H$ results, almost all systematics are treated as uncorrelated. The only exceptions are the QCD scale uncertainty on $t\bar{t}H$, uncertainties on the Higgs boson branching fractions, and uncertainties on the electron/photon energy scale and resolution.

Table 10.1 Breakdown of the systematic uncertainties for the combined $t\bar{t}H$ cross section measurement at $\sqrt{s} = 13\,\text{TeV}$

Uncertainty source	$\Delta\sigma/\sigma$ (%)
Theory uncertainties	11.9
$t\bar{t}$ + heavy flavor	9.9
$t\bar{t}H$	6.0
Non-$t\bar{t}H$ Higgs production	1.5
Other backgrounds	2.2
Experimental uncertainties	9.3
Fake leptons	5.2
Jets and missing E_T	4.9
Electrons and photons	3.2
Luminosity	3.0
Tau leptons	2.5
Flavor tagging	1.8
Monte Carlo statistics	4.4

This is shown as the uncertainty due to each group of systematics ($\Delta\sigma$), as a fraction of the total cross section (σ). The numbers are obtained from quadratic subtraction of the total uncertainty and the uncertainty of the fit with the corresponding group fixed

10.1.3 Results of Combined Analysis

10.1.3.1 Run 2

A summary of all $\sqrt{s} = 13\,\text{TeV}$ analysis categories (including control regions) is shown in Fig. 10.1. The $t\bar{t}H$ signal is clearly visible above the background in the categories with high signal-to-background ratio (on the right-hand side of the plot). The combined $t\bar{t}H$ significance is observed to be 5.8σ in the 13 TeV analyses, corresponding to a signal strength of $\mu_{t\bar{t}H} = 1.32\,^{+0.32}_{-0.28}$. The expected significance is 4.9σ.

The combined $t\bar{t}H$ cross section at $\sqrt{s} = 13\,\text{TeV}$ is measured to be

$$\sigma_{t\bar{t}H} = 670 \pm 90\ (\text{stat.})\ ^{+110}_{-100}\ (\text{syst.})\ \text{fb} \qquad (10.1)$$

This is in agreement with the Standard Model prediction of $507\,^{+35}_{-50}$ fb. The measured cross section from each constituent analysis is shown in Fig. 10.2a. No events are observed in the $t\bar{t}H \to ZZ \to 4\ell$ categories, so an upper limit is placed on the cross section in this channel. The measured $t\bar{t}H$ cross section in each Higgs boson decay channel is shown in Fig. 10.2b. The $H \to b\bar{b}$ and $H \to \gamma\gamma$ decay channel cross sections differ slightly from the analysis-level cross sections due to some small contamination of other Higgs boson decays in these categories.

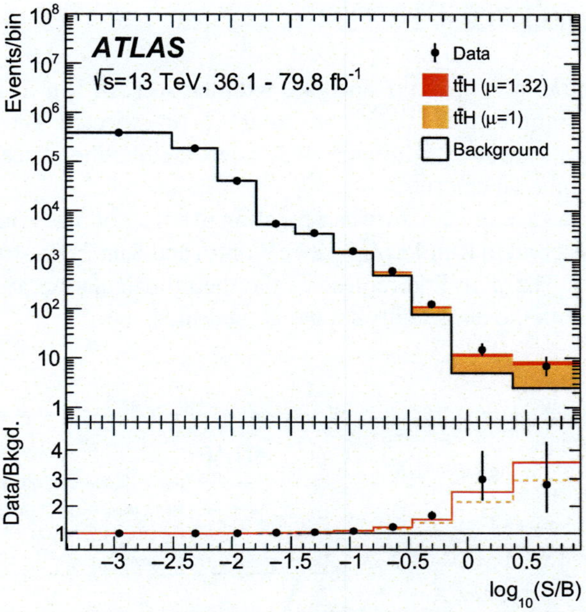

Fig. 10.1 Summary of all $t\bar{t}H$ analysis categories, binned in the logarithm of the signal-to-background ratio (S/B) of the category. The $t\bar{t}H$ signal is clearly visible above the background in the most powerful categories (those with highest S/B)

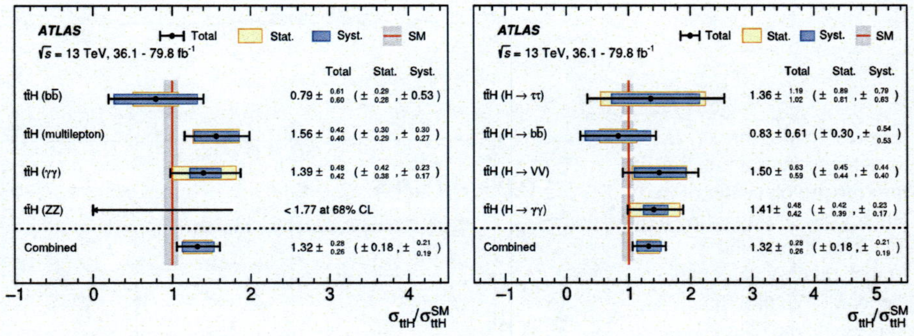

(a) Per analysis (b) Per decay mode

Fig. 10.2 Measured $t\bar{t}H$ cross section broken down by **a** ATLAS analysis and **b** Higgs boson decay mode, normalized to the Standard Model prediction. The yellow (blue) bars represent the statistical (systematic) component of the uncertainty on the measured cross section. The Standard Model prediction is shown in red, with theory errors in gray

10.1.3.2 Run 1 and Run 2

Combining the four Run 2 $t\bar{t}H$ analyses with the ATLAS Run 1 analysis gives an observed $t\bar{t}H$ significance of 6.3σ, compared to 5.1σ expected. The combined signal strength is $\mu_{t\bar{t}H} = 1.38\ ^{+0.31}_{-0.27}$. Figure 10.3 shows the likelihood scans of the Run 1, Run 2, and combined datasets.

The $t\bar{t}H$ cross section is shown as a function of center of mass energy in Fig. 10.4. The data correspond to Run 1 ($\sqrt{s} = 8$ TeV only) and Run 2 (Eq. 10.1) cross section measurements. The pink line represents the theory calculation at NLO in QCD + NLO in EW, with the uncertainty shown as a purple band.

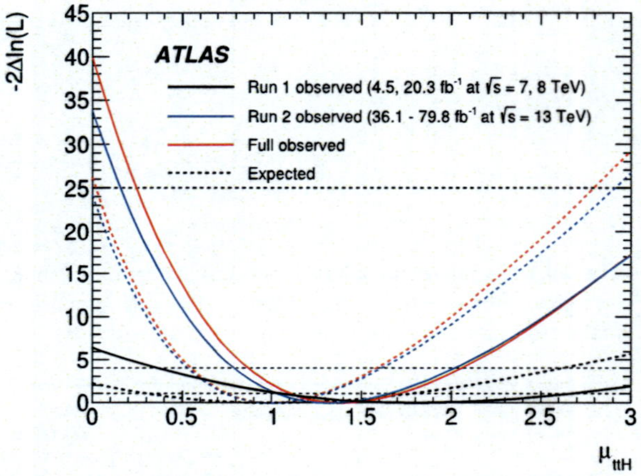

Fig. 10.3 Expected (dashed) and observed (solid) likelihood scans of the signal strength $\mu_{t\bar{t}H}$ for the Run 1 (black), Run 2 (blue), and combined datasets (red)

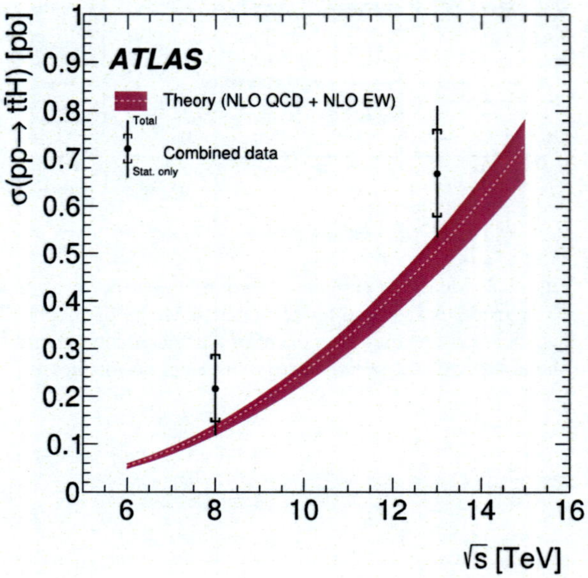

Fig. 10.4 Measured $t\bar{t}H$ cross section as a function of proton–proton center of mass energy $\sqrt{s}$. Data is shown in black with total and statistical only uncertainties denoted. The pink line represents the theory calculation at NLO in QCD + NLO in EW, with the uncertainty shown as a purple band

10.2 $t\bar{t}H$ in the Diphoton Decay Channel

The $t\bar{t}H(\gamma\gamma)$ categorization described in Chap. 9 was developed and optimized on the partial LHC Run 2 dataset of 79.8 fb^{-1}, collected during 2015–2017. This same categorization is applied to the full Run 2 dataset of 139 fb^{-1}, and no re-optimization of the categories is performed. This Section reports the full Run 2 results, which are included in [3].

10.2.1 Signal and Background Shapes

In each analysis category, the signal shape is determined by fitting a Double-Sided Crystal Ball (DSCB) function to a sample of inclusive Higgs boson Monte Carlo (see Sect. 6.2.1). The best fit DSCB parameters for each $t\bar{t}H(\gamma\gamma)$ category are shown in Table 10.2, along with S_{90}, the smallest $m_{\gamma\gamma}$ range containing 90% of Higgs signal.

For the hadronic $t\bar{t}H(\gamma\gamma)$ categories, the functional form of the continuum background is extracted from templates created from NTI data. To enhance statistics in these templates, events with no b-tagged jets are also included, and the b-tag status of all central jets are set to "true" when computing BDT score. For the leptonic $t\bar{t}H(\gamma\gamma)$ categories, the background templates are constructed from $t\bar{t}\gamma\gamma$ Monte Carlo, since this process dominates the continuum background once the BDT selection has been applied. In order to enhance statistics, no photon identification or isolation cuts are applied to the leptonic background templates.

The functional form of the continuum background and the spurious signal are listed in Table 10.3.

Table 10.2 Best fit parameter values for the signal $m_{\gamma\gamma}$ shape in each of the seven $t\bar{t}H(\gamma\gamma)$ categories at 139 fb^{-1}

Category	μ_{CB} (GeV)	σ_{CB} (GeV)	α_{low}	α_{high}	n_{low}	n_{high}	S_{90} (GeV)
$t\bar{t}H(\gamma\gamma)$ had 4	125.15	1.55	1.7	1.5	5.0	13.3	3.00
$t\bar{t}H(\gamma\gamma)$ had 3	125.16	1.52	1.7	1.5	5.0	12.8	2.96
$t\bar{t}H(\gamma\gamma)$ had 2	125.14	1.46	1.7	1.4	4.6	16.6	2.84
$t\bar{t}H(\gamma\gamma)$ had 1	125.14	1.29	1.7	1.5	4.7	16.6	2.48
$t\bar{t}H(\gamma\gamma)$ lep 3	125.16	1.74	1.6	1.7	6.1	5.8	3.30
$t\bar{t}H(\gamma\gamma)$ lep 2	125.13	1.62	1.7	1.4	4.5	21.6	3.13
$t\bar{t}H(\gamma\gamma)$ lep 1	125.12	1.44	1.7	1.4	4.9	13.5	2.80

Note that these are the fitted values for the MC samples at $m_H = 125$ GeV

Table 10.3 Results of the spurious signal test for the $t\bar{t}H(\gamma\gamma)$ categories at 139 fb^{-1}

Category	Function	N_{sp}	χ^2/ndof	Prob(χ^2) (%)
$t\bar{t}H(\gamma\gamma)$ had 4	Power law	+1.07	0.82	82.2
$t\bar{t}H(\gamma\gamma)$ had 3	Power law	+0.658	1.14	22.2
$t\bar{t}H(\gamma\gamma)$ had 2	Power law	+0.756	1.31	6.47
$t\bar{t}H(\gamma\gamma)$ had 1	Exponential	+0.316	1.33	5.45
$t\bar{t}H(\gamma\gamma)$ lep 3	Exponential	+0.165	0.87	73.8
$t\bar{t}H(\gamma\gamma)$ lep 2	Power law	+0.430	1.33	5.45
$t\bar{t}H(\gamma\gamma)$ lep 1	Exponential	+0.241	1.51	0.981

The χ^2/ndof indicates the performance of a background only fit to the template in the $m_{\gamma\gamma}$ range 105–160 GeV, but is in general not a part of the spurious signal criteria

10.2.2 Systematic Uncertainties

The experimental uncertainties included in the $t\bar{t}H(\gamma\gamma)$ measurement are those summarized in Sect. 6.4.2. A conservative 100% heavy flavor uncertainty is applied to the predicted ggF, VBF, and VH yields in each category, and inclusive uncertainties from [9] are applied to the $b\bar{b}H$, $tHjb$, and tWH yields.

For the $t\bar{t}H$ signal yields, QCD scale variations and PDF uncertainties (31 eigenvector scheme) are evaluated separately in each category. In addition, an Underlying Event/Parton Showering (UEPS) uncertainty is calculated on the $t\bar{t}H$ yield in each category by comparing predictions from Monte Carlo samples using different showering algorithms: MG5_aMC@NLO+Pythia8 and MG5_aMC@NLO+Herwig++. The UEPS uncertainties are listed in Table 10.4.

The main sources of systematic uncertainty on the measurement of $t\bar{t}H$ cross section times $H \rightarrow \gamma\gamma$ branching fraction are summarized in Table 10.5. To determine the contribution from each group of systematics, the corresponding nuisance parameters are fixed to their best fit values from the unconditional maximum likeli-

Table 10.4 Underlying Event/Parton Showering (UEPS) uncertainty on $t\bar{t}H$ signal yields in each analysis category

Category	UEPS (%)
$t\bar{t}H$ had 4	+2.2
$t\bar{t}H$ had 3	−3.2
$t\bar{t}H$ had 2	−6.7
$t\bar{t}H$ had 1	−11.4
$t\bar{t}H$ lep 3	+5.1
$t\bar{t}H$ lep 2	−2.9
$t\bar{t}H$ lep 1	−9.0

The uncertainty is obtained by comparing event yields from $t\bar{t}H$ Monte Carlo samples with two different showering algorithms: MG5_aMC@NLO+Pythia8 and MG5_aMC@NLO+Herwig++

Table 10.5 Contribution of groups of systematic uncertainties to the total error on the measured $t\bar{t}H$ cross section times $H \to \gamma\gamma$ branching ratio with 139 fb^{-1}

Uncertainty source	$\Delta\sigma_{\text{low}}/\sigma$ (%)	$\Delta\sigma_{\text{high}}/\sigma$ (%)
Theory uncertainties	6.6	9.7
Underlying event/Parton shower	5.0	7.2
Heavy flavor	4.0	3.4
QCD scale	3.3	4.7
PDF+α_S	0.3	0.5
Experimental uncertainties	7.8	9.1
Photon energy resolution	5.5	6.2
Photon energy scale	2.8	2.7
Jet/missing E_T	2.3	2.7
Spurious signal	2.1	2.0
Photon efficiency	1.9	2.7
Luminosity and trigger	1.6	2.3
Higgs mass	1.6	1.5
Pileup	1.0	1.5
Flavor tagging	0.9	1.1
Leptons	0.4	0.6

This is shown as the uncertainty due to each group of systematics ($\Delta\sigma$), as a fraction of the total cross section (σ). The numbers are obtained from quadratic subtraction of the total uncertainty and the uncertainty of the fit with the corresponding group fixed

hood fit. The resulting error is then subtracted (in quadrature) from the total error to obtain the error due only to the systematics in question.

The dominant theory uncertainty is the UEPS uncertainty: since up to six jets in a single event are used in the BDT training (see Chap. 9), dependence on the parton showering algorithm is expected. The photon energy scale and resolution are the dominant experimental systematics, and their role will be discussed in the next Section.

10.2.3 Results

The diphoton mass spectrum in each hadronic (leptonic) category is shown in Fig. 10.5 (Fig. 10.6), with the signal-plus-background fits overlaid. The observed significance of $t\bar{t}H(\gamma\gamma)$ is 4.9σ, which corresponds to a signal strength of

$$\mu_{t\bar{t}H} = 1.38\ ^{+0.41}_{-0.36} = 1.38\ ^{+0.33}_{-0.31}\ (\text{stat.})\ ^{+0.13}_{-0.11}\ (\text{exp.})\ ^{+0.22}_{-0.14}\ (\text{th.}) \tag{10.2}$$

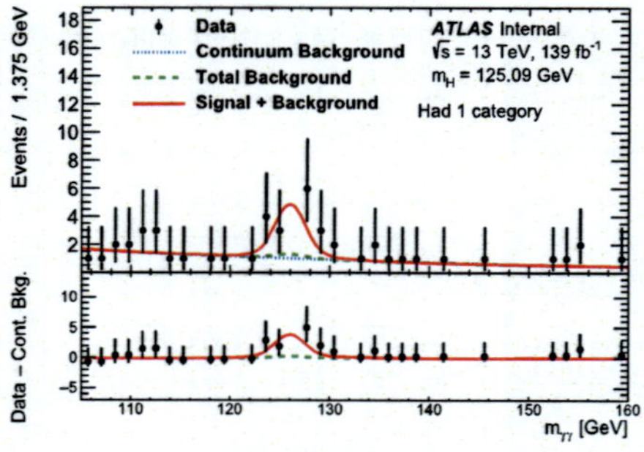

(a) $t\bar{t}H(\gamma\gamma)$ had 1

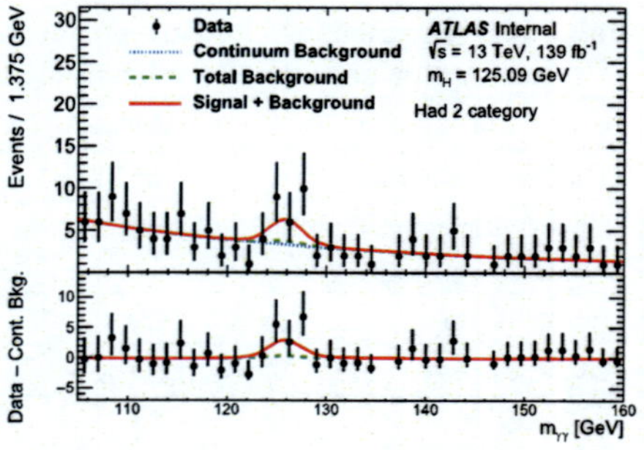

(b) $t\bar{t}H(\gamma\gamma)$ had 2

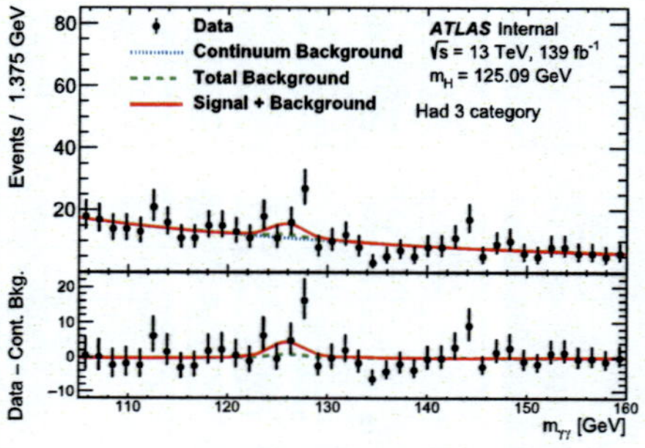

(c) $t\bar{t}H(\gamma\gamma)$ had 3

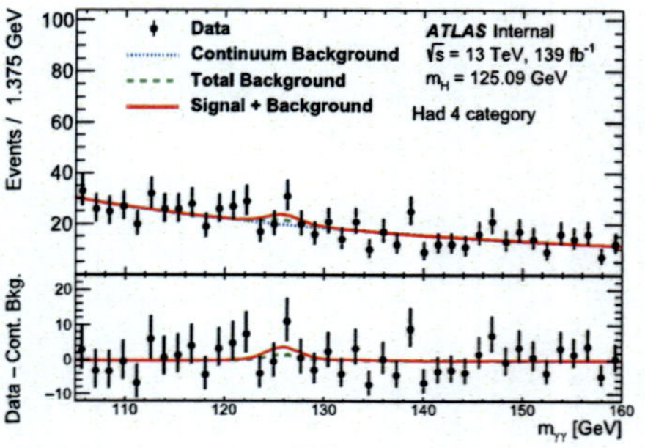

(d) $t\bar{t}H(\gamma\gamma)$ had 4

Fig. 10.5 Diphoton invariant mass spectrum in the four hadronic $t\bar{t}H(\gamma\gamma)$ categories with 139 fb^{-1} of ATLAS data. The tightest cut on BDT score is applied in **a** $t\bar{t}H(\gamma\gamma)$ had 1, and the loosest cut is applied in **d** $t\bar{t}H(\gamma\gamma)$ had 4. The data is shown in black, with the fitted continuum background distribution in blue, the total background (including non-$t\bar{t}H$ Higgs processes) in green, and the $t\bar{t}H$ signal plus background in red

The expected $t\bar{t}H(\gamma\gamma)$ significance on the full Run 2 dataset (assuming the Standard Model) is 4.5σ (pre-fit) and 4.2σ (post-fit).

Figure 10.7 shows the fitted signal and background yields in each category, calculated in the $m_{\gamma\gamma}$ window containing 90% of the Higgs boson signal (S_{90} in Table 10.2). These yields are also listed in Table 10.6.

In order to visually enhance the contribution of the most powerful analysis categories, the data and fit in each category are weighted by $\ln(1 + S/B)$ where S (B) is the expected signal (background) in the smallest $m_{\gamma\gamma}$ window containing 90% of the Higgs boson signal. This weighting procedure serves to magnify the contribution to the $m_{\gamma\gamma}$ spectrum of categories with highest S/B. The weighted diphoton mass spectrum (summed over all categories) is shown in Fig. 10.8. A description of the calculation of errors on this plot is included in Sect. 6.5.

The $t\bar{t}H$ cross section times the $H \rightarrow \gamma\gamma$ branching fraction is measured to be

$$\sigma_{t\bar{t}H} \times B_{\gamma\gamma} = 1.59\ ^{+0.43}_{-0.39}\ \text{fb}\ = 1.59\ ^{+0.38}_{-0.36}\ (\text{stat.})\ ^{+0.15}_{-0.12}\ (\text{exp.})\ ^{+0.15}_{-0.11}\ (\text{th.})\ \text{fb} \quad (10.3)$$

This is compared to a Standard Model prediction of $1.15\ ^{+0.09}_{-0.12}$ fb.

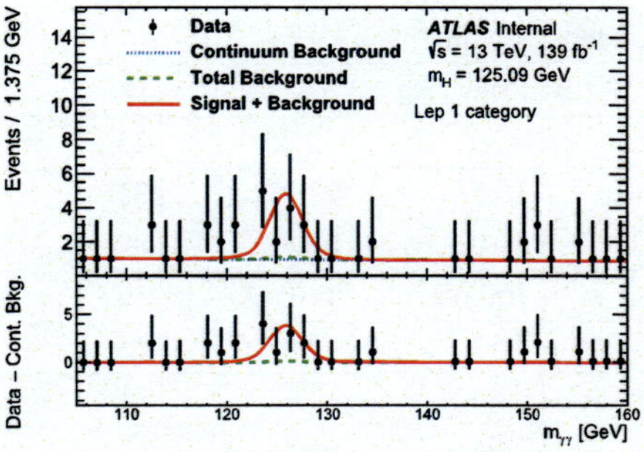

(a) $t\bar{t}H(\gamma\gamma)$ lep 1

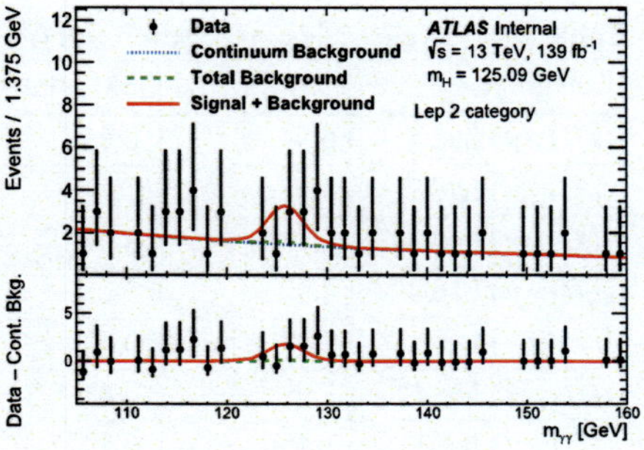

(b) $t\bar{t}H(\gamma\gamma)$ lep 2

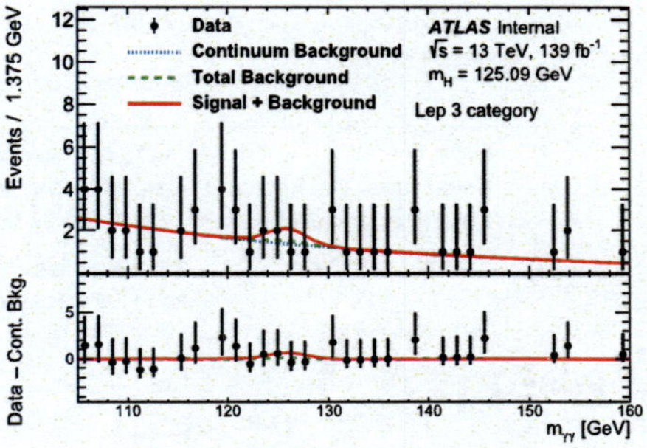

(c) $t\bar{t}H(\gamma\gamma)$ lep 3

Fig. 10.6 Diphoton invariant mass spectrum in the three leptonic $t\bar{t}H(\gamma\gamma)$ categories with 139 fb^{-1} of ATLAS data. The tightest cut on BDT score is applied in **a** $t\bar{t}H(\gamma\gamma)$ lep 1, and the loosest cut is applied in **c** $t\bar{t}H(\gamma\gamma)$ lep 3. The data is shown in black, with the fitted continuum background distribution in blue, the total background (including non-$t\bar{t}H$ Higgs processes) in green, and the $t\bar{t}H$ signal plus background in red

Fig. 10.7 Fitted number of $t\bar{t}H(\gamma\gamma)$ and background events in each analysis category with 139 fb^{-1} of ATLAS data. The fitted continuum background yield is shown in blue, the non-$t\bar{t}H$ Higgs background in green, and the $t\bar{t}H$ signal (normalized to the fitted $\mu = 1.38$) in red. The total number of data events per category is overlaid in black

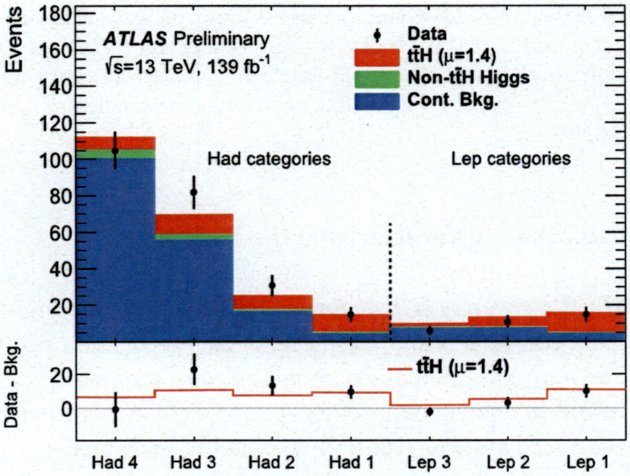

Table 10.6 Event yields in the seven $t\bar{t}H(\gamma\gamma)$ categories with 139 fb^{-1}

Category	SM $t\bar{t}H$	SM non-$t\bar{t}H$ Higgs	Cont. bkg.	SM Total	Data
$t\bar{t}H(\gamma\gamma)$ had 1	6.94	0.81	4.50	12.25	15
$t\bar{t}H(\gamma\gamma)$ had 2	5.58	1.11	16.53	23.22	31
$t\bar{t}H(\gamma\gamma)$ had 3	7.73	3.07	56.02	66.83	82
$t\bar{t}H(\gamma\gamma)$ had 4	4.92	5.03	100.65	110.59	105
$t\bar{t}H(\gamma\gamma)$ lep 1	7.89	0.42	4.57	12.88	15
$t\bar{t}H(\gamma\gamma)$ lep 2	3.92	0.43	7.46	11.82	11
$t\bar{t}H(\gamma\gamma)$ lep 3	1.45	0.49	7.54	9.48	6

The observed yield (data) is compared with the sum of expected $t\bar{t}H$ signal, background from non-$t\bar{t}H$ Higgs production, and continuum background. The continuum background yield is extracted from the unconditional maximum likelihood fit. Event yields are counted in the smallest $m_{\gamma\gamma}$ window containing 90% of the expected signal

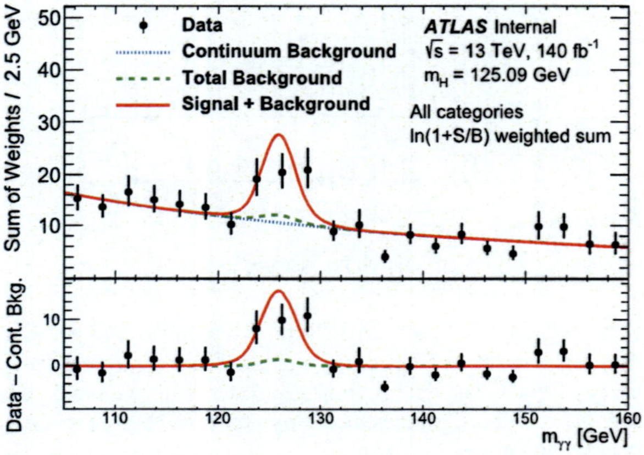

Fig. 10.8 Weighted diphoton invariant mass spectrum of all seven $t\bar{t}H(\gamma\gamma)$ categories with 139 fb^{-1} of ATLAS data. The data is shown in black, with the fitted continuum background distribution in blue, the total background (including non-$t\bar{t}H$ Higgs processes) in green, and the $t\bar{t}H$ signal plus background in red. The data and fit in each category are weighted by $\ln(1 + S/B)$, where S (B) is the expected signal (background) in the smallest $m_{\gamma\gamma}$ window containing 90% of the Higgs signal

10.2.3.1 Characterizing the Higgs Peak

In the seven $t\bar{t}H$ categories, the shape of the Higgs mass peak in data is broader and centered at higher m_H than expected. As a result, the photon energy scale and resolution systematics are pulled to fit the peak in data, making these the dominant experimental uncertainties. A number of tests are performed to quantify the difference between the observed and expected Higgs signal peak, and no significant tension with the expected Higgs boson signal shape is observed.

If the Higgs boson mass is allowed to float along with the $t\bar{t}H$ signal strength, the best fit occurs at $m_H = 126.45\ ^{+0.65}_{-0.63}$ GeV and $\mu_{t\bar{t}H} = 1.40\ ^{+0.42}_{-0.36}$. The tension between this measurement and the expectation ($m_H = 125.09$ GeV) is at the level of 2.15σ.

In order to fit the width of the Higgs mass peak, a scale factor is assigned to σ_{CB}, the resolution of the DSCB function. A scale factor of 1.0 corresponds to the expected width, while a scale factor larger than 1.0 corresponds to a broadened peak. This scale factor is allowed to vary along with $\mu_{t\bar{t}H}$ (for fixed m_H). The best fit is at a scale factor of $1.45\ ^{+0.45}_{-0.32}$ and $\mu_{t\bar{t}H} = 1.52\ ^{+0.46}_{-0.39}$. The tension between this measurement and the expected width (scale factor = 1.0) is 1.53σ.

Allowing all three parameters ($\mu_{t\bar{t}H}$, m_H, and width scale factor) to vary simultaneously, the best fit value lies at $m_H = 126.33\ ^{+0.64}_{-0.63}$ GeV, width scale factor of $1.35\ ^{+0.38}_{-0.29}$, and $\mu_{t\bar{t}H} = 1.53\ ^{+0.45}_{-0.38}$. The tension between this measurement and the expectation (scale factor = 1.0 and $m_H = 125.09$ GeV) is 1.68σ.

A rigid signal shape is obtained by fixing the photon energy scale and resolution systematics to their expected values. The observed significance in this model is reduced to 4.5σ, corresponding to a signal strength of $\mu_{t\bar{t}H} = 1.14\ ^{+0.38}_{-0.32}$. Within uncertainties, these results with fixed resolution are consistent with results where the mean and width of the signal peak are allowed to vary.

References

1. The ATLAS Collaboration (2018) Observation of Higgs boson production in association with a top quark pair at the LHC with the ATLAS detector. Phys Lett B 784:173–191. https://doi.org/10.1016/j.physletb.2018.07.035. https://cds.cern.ch/record/2621167. arXiv: 1806.00425
2. Sirunyan AM et al (2018) Observation of TTH production. Phys Rev Lett 120(23):231801. https://doi.org/10.1103/PhysRevLett.120.231801. arXiv: 1804.02610 [hep-ex]
3. Measurement of Higgs Boson production in association with a $t\bar{t}$ pair in the diphoton decay channel using 139 fb^{-1} of LHC data collected at $\sqrt{s} = 13$ TeV by the ATLAS experiment. Tech. rep. ATLAS-CONF-2019-004. CERN, Geneva. http://cds.cern.ch/record/2668103
4. Aaboud M et al (2018) Search for the standard model Higgs boson produced in association with top quarks and decaying into a $b\bar{b}$ pair in pp collisions at $\sqrt{s} = 13$ TeV with the ATLAS detector. Phys Rev D 97(7):072016. https://doi.org/10.1103/PhysRevD.97.072016. arXiv: 1712.08895 [hep-ex]
5. Aaboud M et al (2018) Evidence for the associated production of the Higgs boson and a top quark pair with the ATLAS detector. Phys Rev D 97(7):072003. https://doi.org/10.1103/PhysRevD.97.072003. arXiv: 1712.08891 [hep-ex]
6. Aad G et al (2016) Measurements of the Higgs boson production and decay rates and coupling strengths using pp collision data at $\sqrt{s} = 7$ and 8 TeV in the ATLAS experiment. Eur Phys J C 76(1):6. https://doi.org/10.1140/epjc/s10052-015-3769-y. arXiv: 1507.04548 [hep-ex]
7. Aad G et al (2015) Search for the standard model Higgs boson produced in association with top quarks and decaying into $b\bar{b}$ in pp collisions at $\sqrt{s} = 8$ TeV with the ATLAS detector. Eur Phys J C 75(7):349. https://doi.org/10.1140/epjc/s10052-015-3543-1. arXiv: 1503.05066 [hep-ex]

8. Aad G et al (2015) Search for the associated production of the Higgs boson with a top quark pair in multilepton final states with the ATLAS detector. Phys Lett B 749:519–541. https://doi.org/10.1016/j.physletb.2015.07.079. arXiv: 1506.05988 [hep-ex]
9. LHC Higgs Cross Section Working Group (2017) Handbook of LHC Higgs cross sections: 4. Deciphering the nature of the Higgs sector. Tech. rep. FERMILAB/CERN. https://cds.cern.ch/record/2227475

Chapter 11
CP-Sensitive Categorization of $t\bar{t}H(\gamma\gamma)$ Events

With the observation of the $t\bar{t}H$ process, the coupling between the Higgs boson and top quark has been directly established. This Chapter and the following focus on the direct measurement of the CP structure of the top Yukawa coupling using $t\bar{t}H$ and tH production in the $H \rightarrow \gamma\gamma$ decay channel.

As discussed in Chap. 2, the CP nature of the coupling (scalar, pseudoscalar or mixed) impacts the overall rates of $t\bar{t}H$ and tH production (Table 5.6) as well as sensitive kinematic distributions [1–3]. Because of the large enhancement in the tH cross section in BSM CP scenarios, the $tHjb$ and tWH processes are considered as signal in addition to $t\bar{t}H$. This Chapter presents the development of multivariate $t\bar{t}H(\gamma\gamma)$ CP categories that are sensitive to changes in $t\bar{t}H$ and tH kinematics with different mixing angles α. Results derived from this categorization using the full $H \rightarrow \gamma\gamma$ statistical machinery are presented in Chap. 12.

Section 11.1 reviews the figures of merit used to optimize the $t\bar{t}H(\gamma\gamma)$ CP category definitions. The most sensitive kinematic variables are reviewed in Sect. 11.2. Section 11.3 describes the training of CP BDTs to separate CP even and CP odd hypotheses, and Sect. 11.4 describes the procedure used to optimize the cut values defining category boundaries.

11.1 Number-Counting Limit on CP Mixing

Following a procedure analogous to that presented in Sect. 9.1, the Poisson distribution $P(k, \lambda)$ (Eq. 9.1) can be used to approximate the likelihood and a p-value can be calculated for various CP scenarios. Assuming the CP even hypothesis (SM, $\alpha = 0°$ and $\kappa_t = 1$), the expected single-category exclusion of a CP hypothesis with mixing angle α and coupling strength κ_t is given by the p-value

J. E. Dickinson, *ATLAS Measurements of the Higgs Boson Coupling to the Top Quark in the Higgs to Diphoton Decay Channel*, Springer Theses,
https://doi.org/10.1007/978-3-030-86368-5_11

$$-2\ln\frac{P(S_{0,1}+B,\, S_{\alpha,\kappa_t}+B)}{P(S_{0,1}+B,\, S_{0,1}+B)} = \tag{11.1}$$

$$2(S_{0,1}+B)\ln\left(\frac{S_{0,1}+B}{S_{\alpha,\kappa_t}+B}\right) - 2(S_{0,1}+B) + 2(S_{\alpha,\kappa_t}+B)$$

where S_{α,κ_t} is the number of signal events predicted at mixing angle α and coupling strength κ_t. In the measurement of CP properties of the Higgs-top coupling, the signal yield S_{α,κ_t} includes $t\bar{t}H$, $tHjb$, and tWH processes. The background yield B includes an estimate of other Higgs boson processes as well as the continuum diphoton background. Rather than exercising the full background estimation machinery, the number-counting significance relies on the estimation of continuum background yield from a data control region described in Sect. 5.1.1.

One figure of merit used to optimize the $t\bar{t}H(\gamma\gamma)$ CP categories is the expected exclusion of the CP odd hypothesis ($\alpha = 90°$ and $\kappa_t = 1$):

$$Z_{CP}(90) = \sqrt{2(S_{0,1}+B)\ln\left(\frac{S_{0,1}+B}{S_{90,1}+B}\right) - 2(S_{0,1}+B) + 2(S_{90,1}+B)} \tag{11.2}$$

The other figure of merit is the expected significance of SM $t\bar{t}H + tH$ (Eq. 9.3, with S = SM $t\bar{t}H + tHjb + tWH$): this is referred to as $Z_{t\bar{t}H+tH}$. In order to obtain the total $Z_{t\bar{t}H+tH}$ and $Z_{CP}(90)$, the single-category significances are added in quadrature.

The number-counting significance can be used to estimate the impact of systematic uncertainties on the signal and background. Since the measurement of CP in $t\bar{t}H(\gamma\gamma)$ is dominated by the statistical uncertainty on the number of background events, it is enough to compare the magnitude of systematic uncertainties to the background-only statistical uncertainty:

$$\delta = \frac{S_{0,1}}{Z_{t\bar{t}H+tH}} \tag{11.3}$$

For $\sigma_{\text{syst}} << \delta$, the impact of systematics is expected to be nearly negligible compared to the statistical uncertainty on the measurement. It is verified in Sect. 11.4 that the main theory uncertainties are small compared to δ in the optimized BDT categories.

Assuming $\kappa_t = 1$, a number-counting estimate of the 95% confidence limit on the mixing angle α can also be derived from the exclusion $Z_{CP}(\alpha)$. A curve can be calculated by interpolating $Z_{CP}(\alpha)$ calculated at all available Monte Carlo points (see Sect. 5.2.2). The value of α where this curve intersects $Z = 1.96$ corresponds to the 95% confidence limit.

11.2 Sensitive Observables

Using $t\bar{t}H$ and tH Monte Carlo samples generated with $\kappa_t = 1$ and different values of the CP mixing angle α (details in Sect. 5.2.2), the following distributions are obtained at particle level (i.e. before simulating the parton shower, hadronization, and detector response).

The most powerful shape discrimination between CP even ($\alpha = 0°$, $\kappa_t = 1$) and CP odd ($\alpha = 90°$, $\kappa_t = 1$) comes from the p_T of the Higgs boson. In $t\bar{t}H$, $tHjb$, and tWH, the Higgs boson tends to have higher p_T and more central in η in CP hypotheses with higher values of the mixing angle α. Figure 11.1 shows the Higgs boson p_T and η for different values of α.

The angular separation between top quarks also provides discrimination between CP hypotheses in $t\bar{t}H$. Figure 11.2 shows the angular distances $\Delta\eta$ and $\Delta\phi$ between the top and antitop, which are larger for higher values of α.

The p_T and η of the top quarks are shown in Fig. 11.3. These variables give some separation between CP even and CP odd in $tHjb$, but their discrimination power is limited in $t\bar{t}H$ and tWH. Another powerful variable is the invariant mass of the top quark plus Higgs boson system, shown in Fig. 11.4. In the case of $t\bar{t}H$ and tWH, the m_{tH} is generally larger in CP odd than in CP even, while the opposite trend is visible in $tHjb$.

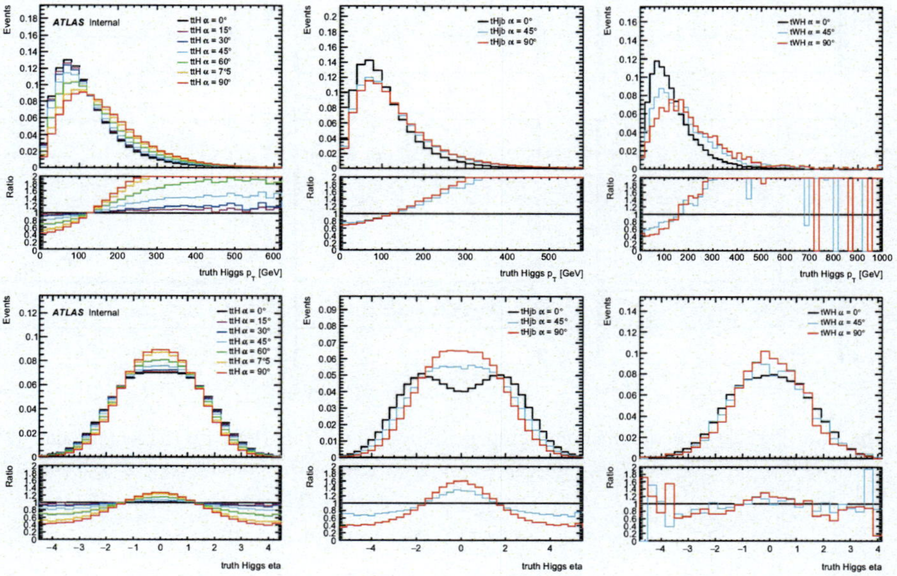

Fig. 11.1 Particle-level distributions of Higgs boson p_T and η in $t\bar{t}H$ (left), $tHjb$ (center), and tWH (right) Monte Carlo for different values of the CP mixing angle α. The coupling strength κ_t is fixed to 1

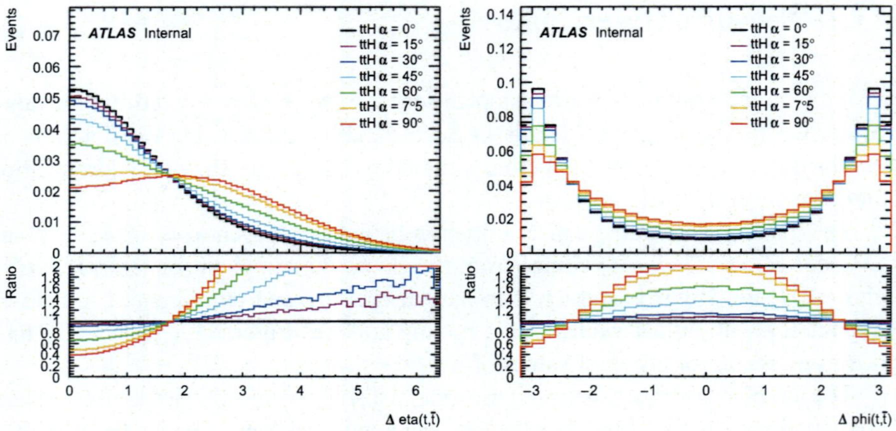

Fig. 11.2 Particle-level distributions of the angular separation between top and antitop quarks in $t\bar{t}H$ Monte Carlo for different values of the CP mixing angle α. The coupling strength κ_t is fixed to 1

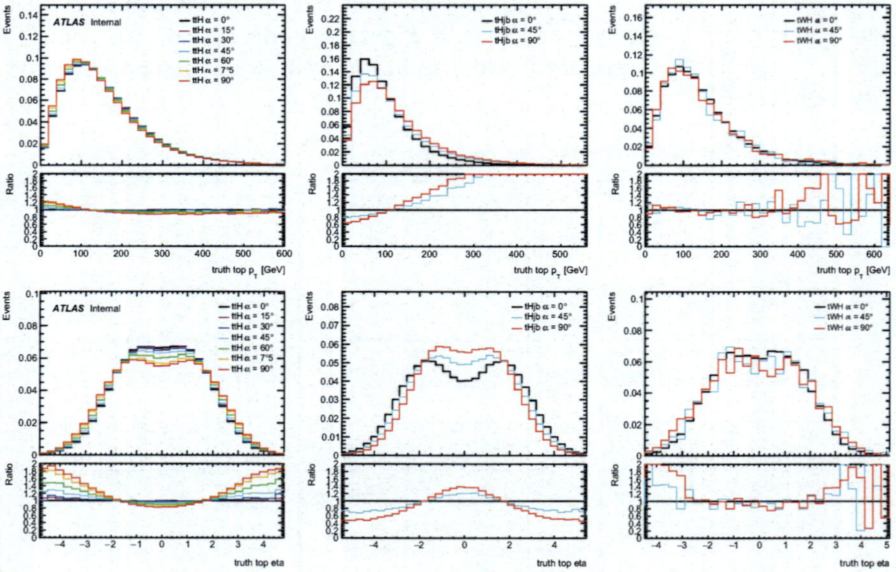

Fig. 11.3 Particle-level distributions of top quark p_T and η in $t\bar{t}H$ (left), $tHjb$ (center), and tWH (right) Monte Carlo for different values of the CP mixing angle α. The coupling strength κ_t is fixed to 1

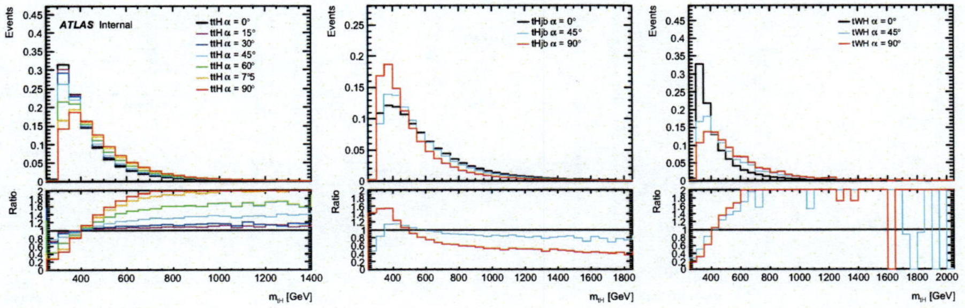

Fig. 11.4 Particle-level distributions of m_{tH} in $t\bar{t}H$ (left), $tHjb$ (center), and tWH (right) Monte Carlo for different values of the CP mixing angle α. The coupling strength κ_t is fixed to 1

11.3 Multivariate Categorization Strategy

As in the Standard Model $t\bar{t}H(\gamma\gamma)$ analysis (Chap. 9), events are preselected into two channels targeting different decays of the $t\bar{t}$ system: the hadronic channel targets the all-hadronic $t\bar{t}$ decay, and the leptonic channel targets final states with at least one semi-leptonic top decay. The preselection for both channels is unchanged with respect to the SM $t\bar{t}H(\gamma\gamma)$ analysis.

The $t\bar{t}H(\gamma\gamma)$ CP analysis categories are based on a partition of a two-dimensional BDT space in each channel (see the schematic in Fig. 11.5). The horizontal axis of the two-dimensional BDT plane is formed by the background rejection BDT discriminant, which separates $t\bar{t}H(\gamma\gamma)$ signal from background processes. The hadronic and leptonic background rejection BDTs are the same discriminants used in the SM $t\bar{t}H(\gamma\gamma)$ analysis and described in detail in Chap. 9.

The background rejection BDTs reject continuum background well, and acceptance at the signal-like end of the spectrum is high for both $t\bar{t}H$ and tH. Figures 11.6 and 11.7 show that there is weak dependence on CP mixing angle, and that this

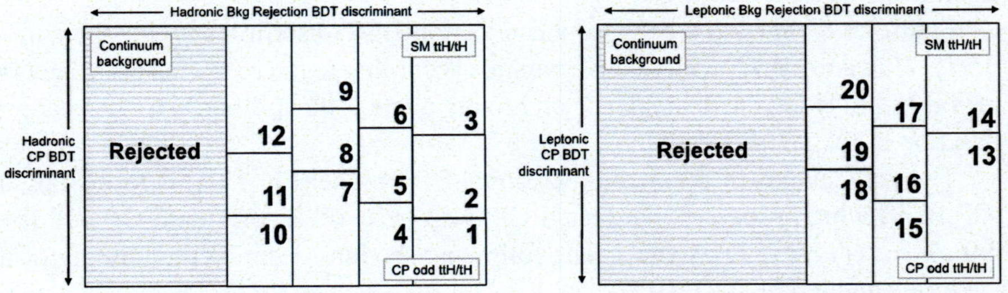

(a) Hadronic categories (b) Leptonic categories

Fig. 11.5 Diagram of the two-dimensional BDT plane and $t\bar{t}H(\gamma\gamma)$ CP categories in (left) the hadronic channel and (right) the leptonic channel. The horizontal axis represents the background rejection BDT score, and the vertical axis represents the CP BDT score. Vertical and horizontal lines denote the category boundaries, and the shaded regions contain rejected events

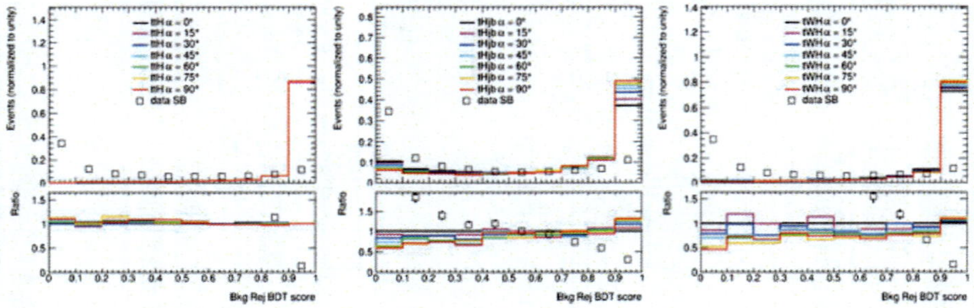

(a) $t\bar{t}H$ had (b) $tHjb$ had (c) tWH had

Fig. 11.6 Hadronic background rejection BDT score for $t\bar{t}H$, $tHjb$, and tWH for different values of the CP mixing angle. The coupling strength κ_t is fixed to 1. The open squares show TI data in the sideband region

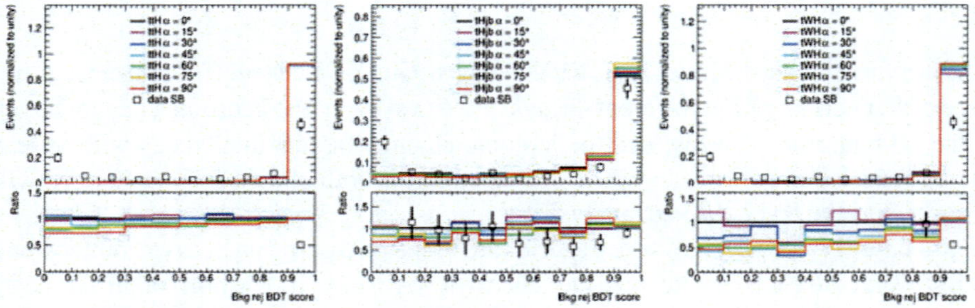

(a) $t\bar{t}H$ lep (b) $tHjb$ lep (c) tWH lep

Fig. 11.7 Leptonic background rejection BDT score for $t\bar{t}H$, $tHjb$, and tWH for different values of the CP mixing angle. The coupling strength κ_t is fixed to 1. The open squares show TI data in the sideband region

dependence is stronger in tH than in $t\bar{t}H$. The tight/isolated (TI) sideband data ($|m_{\gamma\gamma} - 125| > 5$) is overlaid, giving the approximate shape of the continuum background.

Figure 11.8 shows the background rejection BDT discriminants for the sum of $t\bar{t}H$, $tHjb$ and tWH, with relative weights according to the predicted cross sections in Table 5.6. Here, the dependence on mixing angle is due primarily to the changing fraction of the tH processes.

The vertical axis of the two-dimensional BDT plane (Fig. 11.5) is formed by the CP BDT, which separates simulated CP even ($\alpha = 0°$, $\kappa_t = 1$) and CP odd ($\alpha = 90°$, $\kappa_t = 1$) $t\bar{t}H + tH$ signals. The following sections describe the training of the hadronic and leptonic CP BDTs.

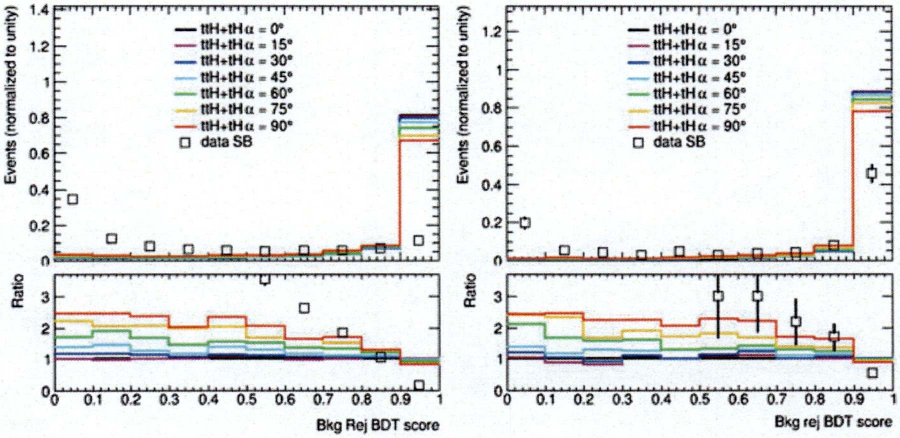

(a) Hadronic $t\bar{t}H + tH$ (b) Leptonic $t\bar{t}H + tH$

Fig. 11.8 Background rejection BDT score for the sum of $t\bar{t}H$, $tHjb$, and tWH for different values of the CP mixing angle in **a** the hadronic channel and **b** the leptonic channel. The coupling strength κ_t is fixed to 1. The open squares show TI data in the sideband region

11.3.1 Hadronic Channel

Following the preselection in Chap. 9, hadronic channel events must contain zero leptons (e or μ) and at least three jets, one or more of which is b-tagged at the 77% working point.

The hadronic CP BDT is trained to separate CP even and CP odd signal events passing hadronic preselection using XGBoost [4]. The training samples are constructed from MG5_aMC@NLO + Pythia8 Monte Carlo samples of $t\bar{t}H$, $tHjb$, and tWH. The composition of $t\bar{t}H$, $tHjb$, and tWH in each sample is determined by the predicted cross section of each process given in Table 5.6. The training is performed on 50% of each sample, while 25% is reserved for validation (including early stopping of the BDT and determination of category boundaries), and the final 25% is reserved for sensitivity evaluation.

The discriminating variables used in the hadronic CP BDT training are:

- p_T and η of the diphoton system (Higgs boson candidate)
- p_T, η, ϕ (with respect to the Higgs boson candidate), and Top Reco BDT score of the reconstructed top candidate and the second top candidate (see Chap. 8). In the case where no second top is reconstructed (either partially or fully), missing values are passed to XGBoost for the kinematics of the second top.
- Angles $\Delta\eta$ and $\Delta\phi$ between the top candidates. In the case where no second top is reconstructed (either partially or fully), missing values are passed to XGBoost.
- Two-object invariant masses $m_{t_1 H}$, and $m_{t_1 t_2}$. In the case where no second top is reconstructed (either partially or fully), the value of m_{t_1} is passed to XGBoost for the $m_{t_1 t_2}$ variable.

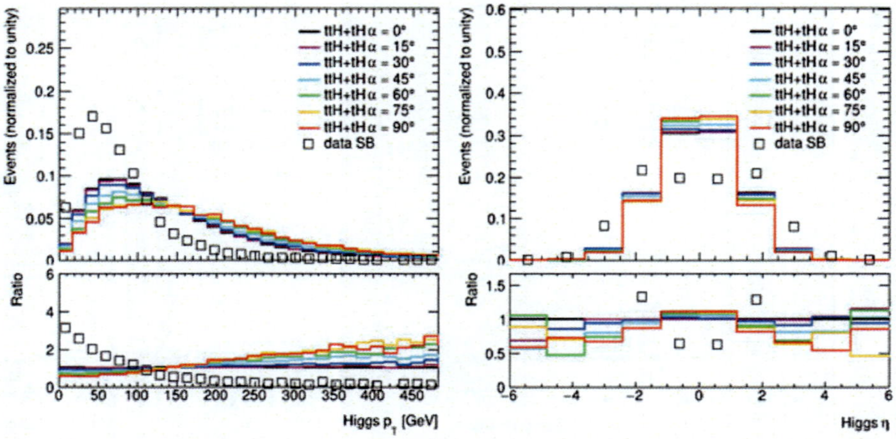

Fig. 11.9 Hadronic CP BDT training variables: $p_T^{\gamma\gamma}$ and $\eta^{\gamma\gamma}$. The open squares show TI data in the sideband region

- $H_T = \sum_{\text{jet } j} p_T^j$
- The minimum ΔR between a photon and a jet in the event
- The second-smallest ΔR between a photon and a jet in the event
- Jet multiplicity and b-jet multiplicity (77% working point)
- Missing E_T significance (Eq. 4.7)

Figs. 11.9–11.15 compare the Monte Carlo distribution of each training variable for different values of the mixing angle α, with the coupling strength κ_t fixed to 1.

The BDT hyper-parameters are chosen to maximize the area under the ROC curve evaluated on the validation set. The ROC AUC of the hadronic CP BDT is 78.39%. The distributions of the output BDT score for $t\bar{t}H$, $tHjb$ and tWH are shown separately for each process in Fig. 11.16.

Figure 11.17 shows the two-dimensional hadronic BDT plane in data passing hadronic preselection. For ease of visualization, the axes here represent transformations of the raw BDT scores. The inner (outer) contours capture 25% (50%) of the $t\bar{t}H + tH$ signal for the CP even and CP odd hypotheses. The projection onto the background rejection BDT and CP BDT axes are shown in the right-hand panels. Data and signal samples are normalized to unit area for a comparison of BDT shapes. It is clear that the background (the dominant component in the data) is concentrated at low background rejection BDT score, while the signal is concentrated at high background rejection BDT score. In addition, the CP even and CP odd signal hypotheses are separated by the CP BDT score. The SM expectation (CP even) is concentrated in the upper right corner of the 2D plane.

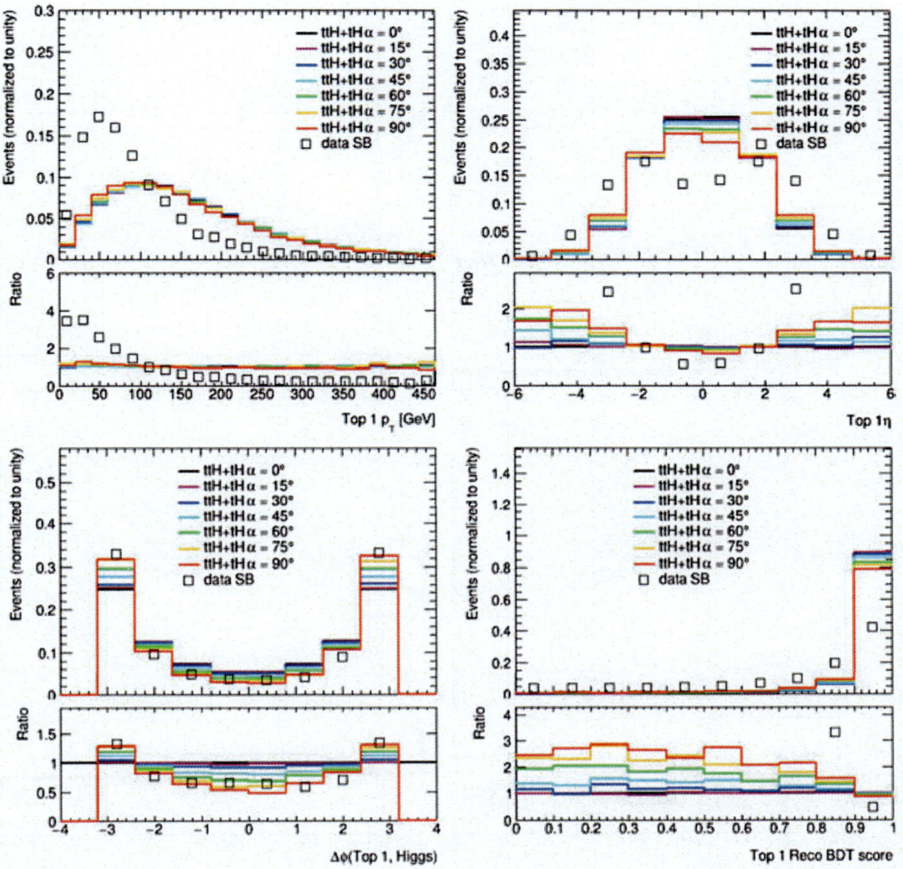

Fig. 11.10 Hadronic CP BDT training variables: p_T, η, ϕ with respect to the Higgs boson candidate, and Top Reco BDT score of the first reconstructed top quark. The open squares show data in the sideband region

11.3.2 Leptonic Channel

Again following the preselection in Chap. 9, leptonic channel events must contain at least one lepton (e or μ) and at least one b-jet.

The leptonic CP BDT is trained to separate CP even and CP odd events passing leptonic preselection using XGBoost [4]. The construction of training samples and the training variables used are analogous to those used in training of the hadronic CP BDT

The training samples are constructed from MG5_aMC@NLO + Pythia8 Monte Carlo samples of $t\bar{t}H$, $tHjb$, and tWH, with the composition of $t\bar{t}H$, $tHjb$, and tWH determined by the predicted cross sections in Table 5.6. The training is per-

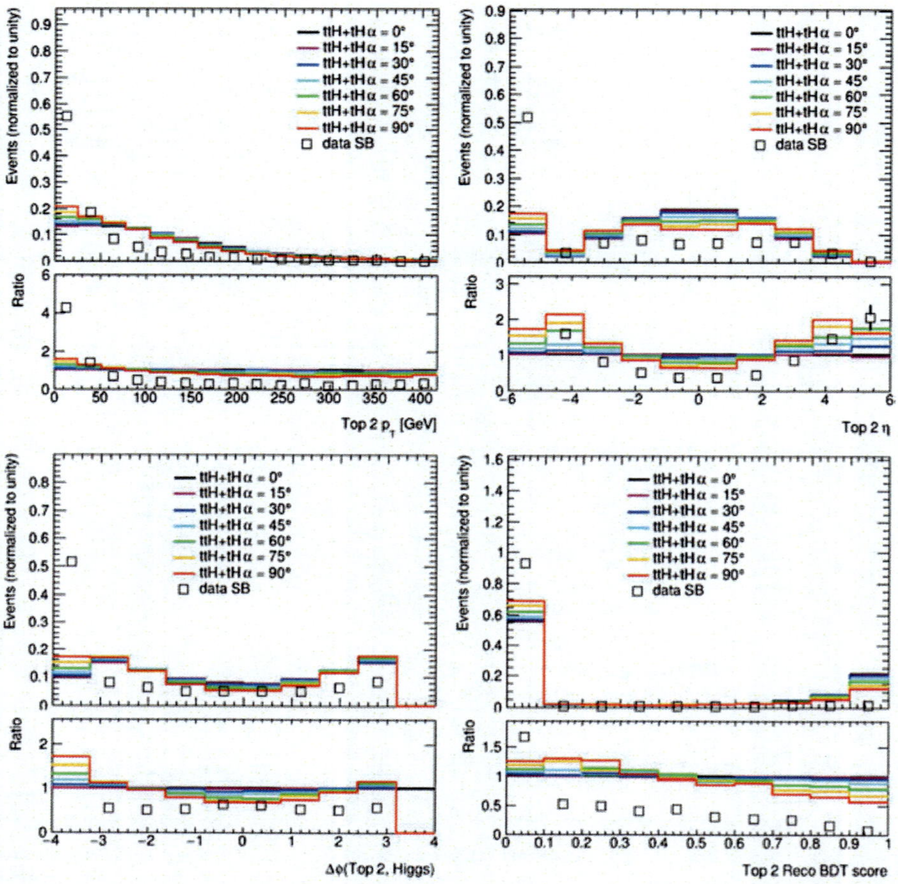

Fig. 11.11 Hadronic CP BDT training variables: p_T, η, ϕ with respect to the Higgs boson candidate, and Top Reco BDT score of the second reconstructed top quark. The underflow bins represent events where no second top is reconstructed (either partially or fully). The open squares show data in the sideband region

formed on 50% of each sample, while 25% is reserved for validation (including early stopping of the BDT and determination of category boundaries), and the final 25% is reserved for sensitivity evaluation.

The discriminating variables used in the leptonic CP BDT training are the same as those used in the hadronic training (though top quarks are reconstructed differently, as described in Chap. 8). In dilepton events (where no top quarks are reconstructed), missing values are passed to XGBoost for all top-related variables. Figures 11.18–11.24 compare the Monte Carlo distribution of each training variable for different values of the mixing angle α, with the coupling strength κ_t fixed to 1.

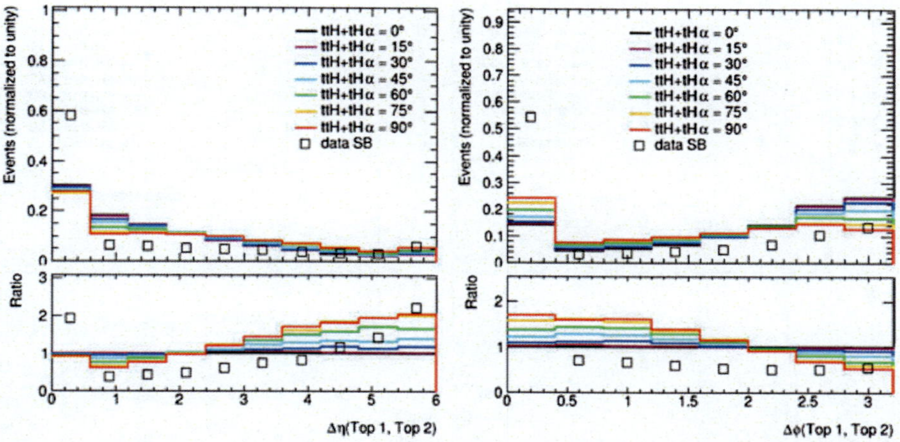

Fig. 11.12 Hadronic CP BDT training variables: $\Delta\eta(t_1, t_2)$ and $\Delta\phi(t_1, t_2)$. The underflow bins represent events where no second top is reconstructed (either partially or fully). The open squares show TI data in the sideband region

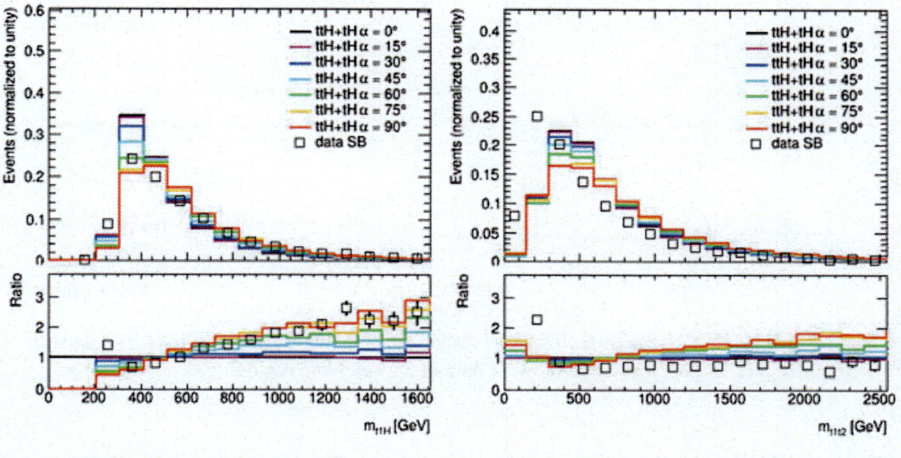

Fig. 11.13 Hadronic CP BDT training variables: $m_{t_1 H}$ and $m_{t_1 t_2}$. The open squares show TI data in the sideband region

The BDT hyper-parameters are chosen to maximize the area under the ROC curve. The ROC AUC for the leptonic CP BDT is 76.69%. The distributions of the output BDT score for $t\bar{t}H$, $tHjb$ and tWH are shown separately for each process in Fig. 11.25.

Figure 11.26 shows the two-dimensional leptonic BDT plane in data passing leptonic preselection. For ease of visualization, the axes here represent transformations of the raw BDT scores. The inner (outer) contours capture 25% (50%) of the $t\bar{t}H + tH$ signal for the CP even and CP odd hypotheses. The projection onto the background rejection BDT and CP BDT axes are shown in the right-hand panels.

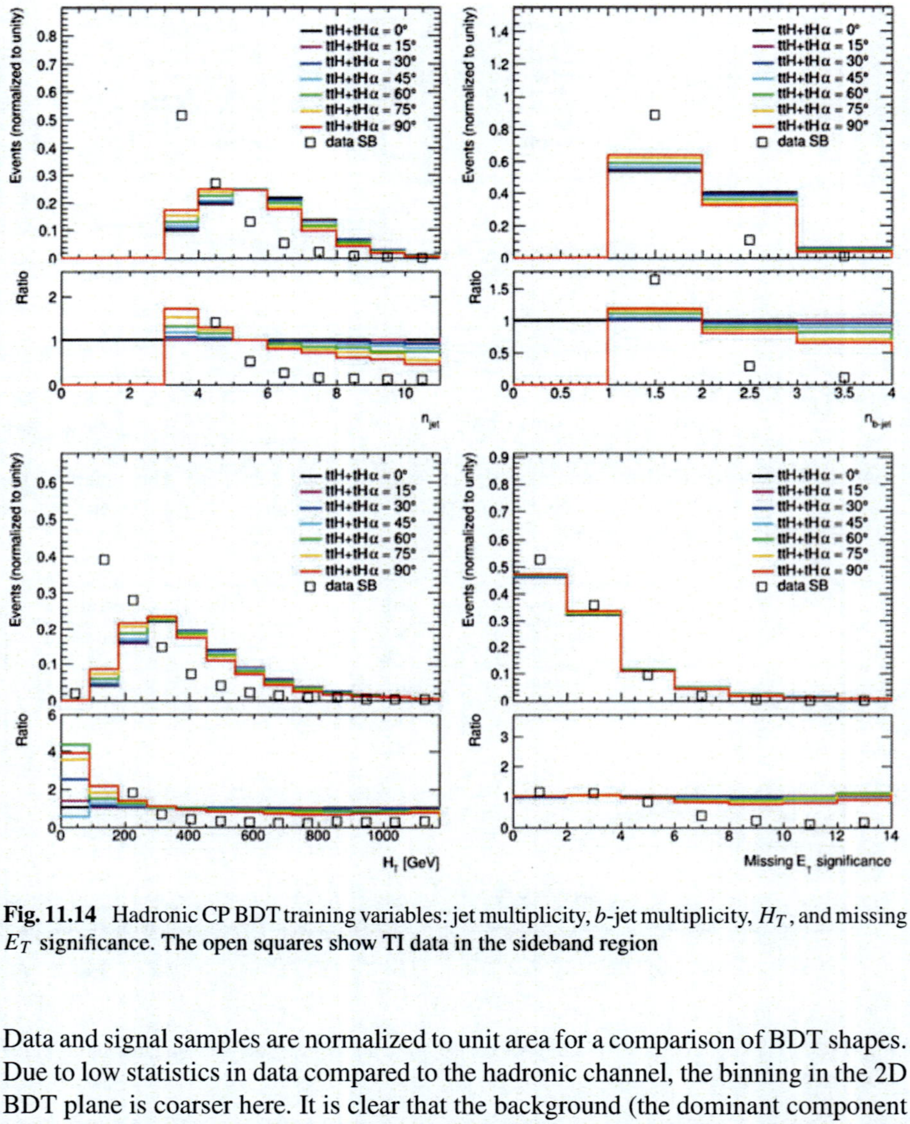

Fig. 11.14 Hadronic CP BDT training variables: jet multiplicity, b-jet multiplicity, H_T, and missing E_T significance. The open squares show TI data in the sideband region

Data and signal samples are normalized to unit area for a comparison of BDT shapes. Due to low statistics in data compared to the hadronic channel, the binning in the 2D BDT plane is coarser here. It is clear that the background (the dominant component in the data) is concentrated at low background rejection BDT score, while the signal is concentrated at high background rejection BDT score. In addition, the CP even and CP odd signal hypotheses are separated by the CP BDT score. The SM expectation (CP even) is concentrated in the upper right corner of the 2D plane.

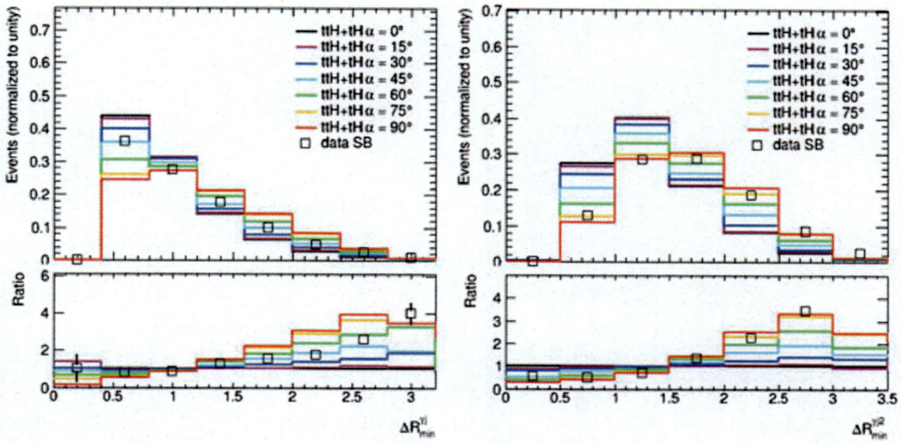

Fig. 11.15 Hadronic CP BDT training variables: $\Delta R^{\gamma j}_{\text{min}}$ and $\Delta R^{\gamma j2}_{\text{min}}$. The open squares show TI data in the sideband region

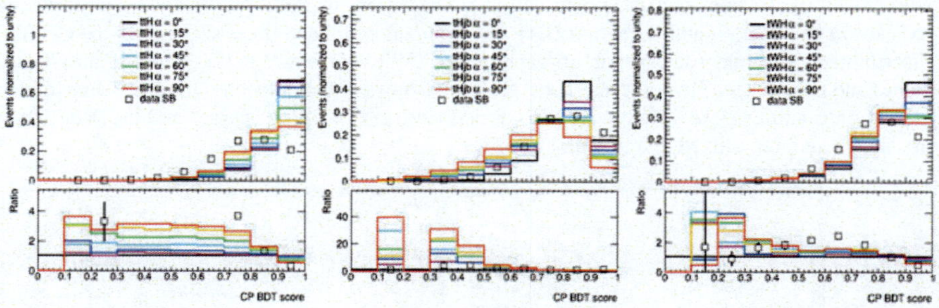

(a) $t\bar{t}H$ had (b) $tHjb$ had (c) tWH had

Fig. 11.16 Hadronic CP BDT score for $t\bar{t}H$, $tHjb$, and tWH for different values of the CP mixing angle. The coupling strength κ_t is fixed to 1. The open squares show TI data in the sideband region

11.4 Determination of Category Boundaries

The optimal choice of category boundaries maximizes both the expected $t\bar{t}H + tH$ significance ($Z_{t\bar{t}H+tH}$) and the expected number-counting exclusion of the CP odd hypothesis ($Z_{CP}(90)$ from Eq. 11.2). A brute force scan is performed over all possible boundary combinations, and the boundary values are chosen to maximize $Z_{t\bar{t}H+tH}$ and $Z_{CP}(90)$. For boundary optimization, both figures of merit are evaluated on the validation set, since the cuts forming the category boundaries are considered hyper-parameters of the analysis.

Up to four (three) categories are created by placing cuts on the hadronic (leptonic) background rejection BDT discriminant. Events falling below the loosest cut are rejected. Within each background rejection category, up to three CP categories are defined. Merging of categories is allowed by overlapping of the category boundaries.

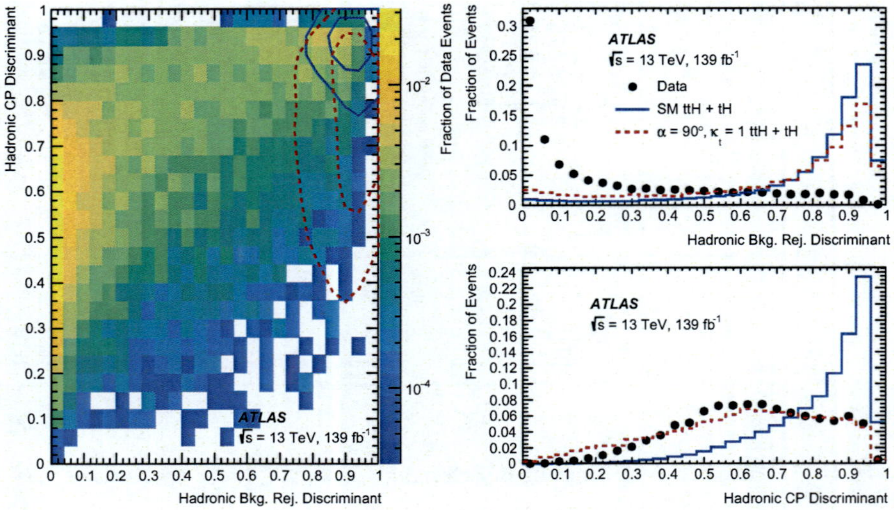

Fig. 11.17 Two-dimensional BDT distribution in data passing hadronic preselection. The horizontal axis shows the background Rejection BDT discriminant and the vertical axis shows the CP BDT discriminant. The inner (outer) contours capture 25% (50%) of the $t\bar{t}H + tH$ events for the CP even (blue) and CP odd (red) hypotheses. The right panel shows projections onto the BDT discriminant axes. Contributions from CP even (blue), CP odd (red) $t\bar{t}H + tH$ processes and the data (black) are shown and normalized to unit area

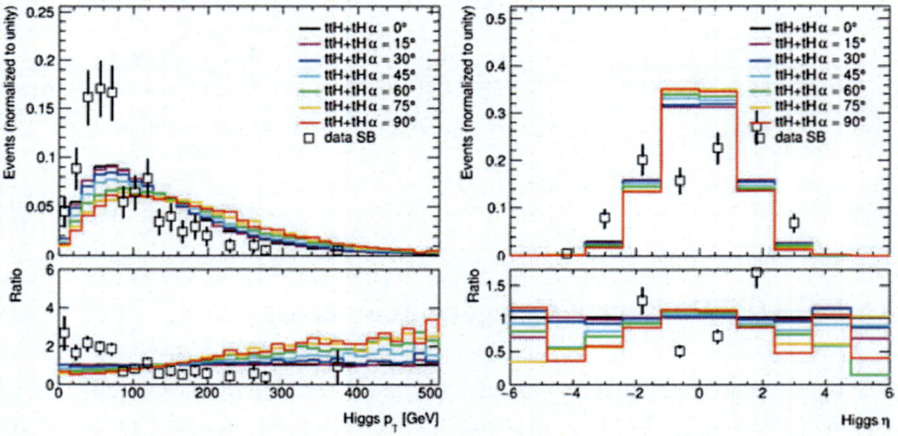

Fig. 11.18 Leptonic CP BDT training variables: $p_T^{\gamma\gamma}$ and $\eta^{\gamma\gamma}$. The open squares show TI data in the sideband region

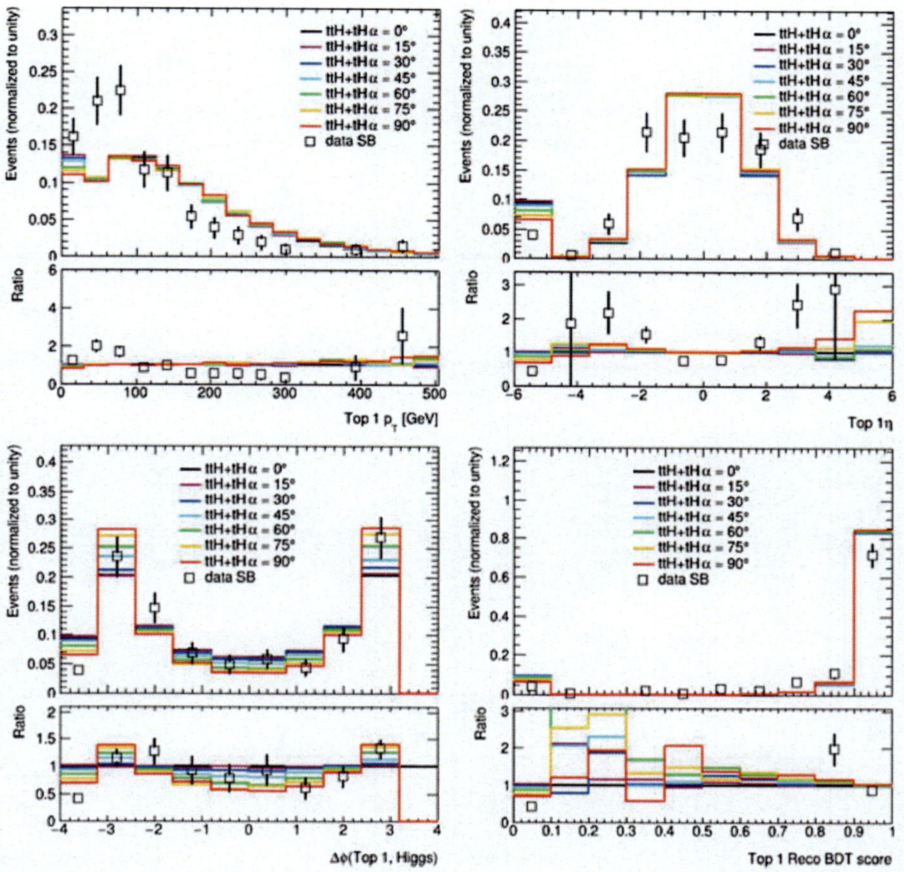

Fig. 11.19 Leptonic CP BDT training variables: p_T, η, ϕ with respect to the Higgs boson candidate, and Top Reco BDT score of the first reconstructed top quark. The underflow bin corresponds to events containing $\geq$ 2 leptons, where no top is reconstructed. The open squares show TI data in the sideband region

The maximum allowed number of categories is twelve (nine) for hadronic (leptonic). The selected categorization contains twelve hadronic and eight leptonic categories: two rather than three CP BDT categories are created in the tightest leptonic background rejection BDT region.

In order to have sufficient data in each category to perform the maximum likelihood fit described in Chap. 6, a minimum of 0.8 continuum background events (predicted from NTI data) is required in each category in the range $123 < m_{\gamma\gamma} < 127$ GeV. This corresponds to $\sim$ 10 events in the full fit range of 105–160 GeV.

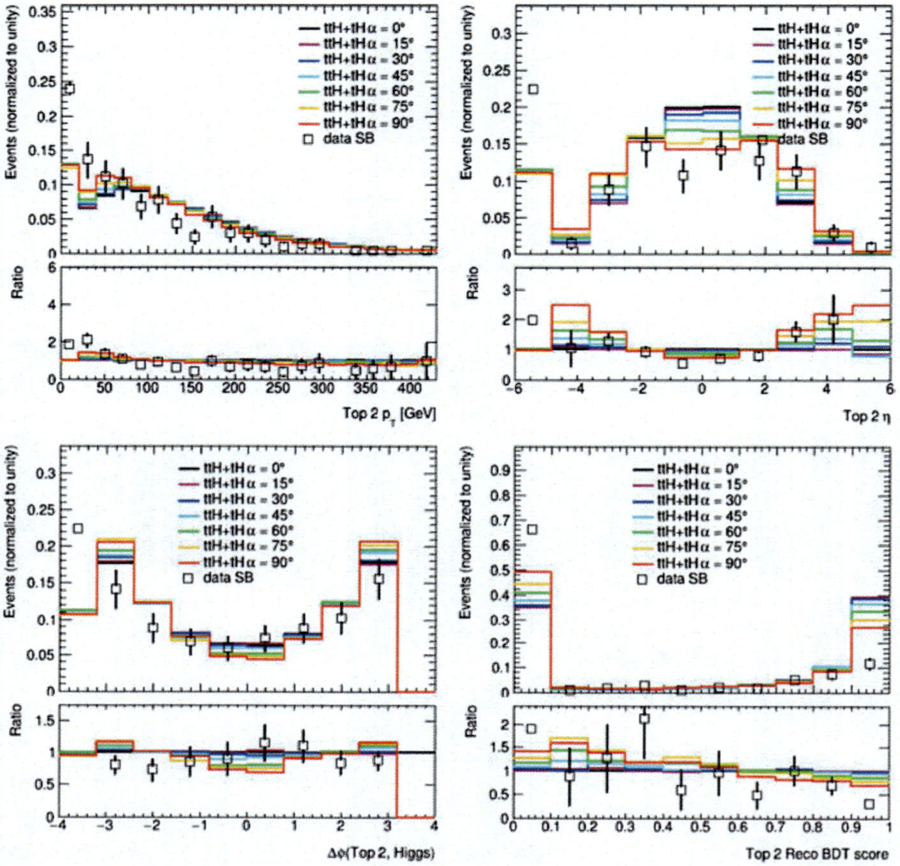

Fig. 11.20 Leptonic CP BDT training variables: p_T, η, ϕ with respect to the Higgs boson candidate, and Top Reco BDT score of the second reconstructed top quark. The underflow bins in p_T, η, and ϕ represent events containing ≥ 2 leptons or where no second top is reconstructed (either partially or fully). The underflow bin in Top 2 Reco BDT score represents events where the second top is not fully reconstructed. The open squares show TI data in the sideband region

The results of the brute force scan are shown in Fig. 11.27. It is clear that the same set of boundaries does not necessarily maximize both $Z_{t\bar{t}H+tH}$ and $Z_{CP}(90)$. The final category boundaries are selected by allowing $Z_{CP}(90)$ to be lower than the maximal value by up to 0.15σ: this corresponds to a small drop of 1° in the number-counting 95% confidence limit on the mixing angle α.

The chosen boundaries are listed in Table 11.1. The categories are labeled with numbers 1–20 according to Fig. 11.5. Categories 1–12 are in the hadronic channel, and categories 13–20 are in the leptonic channel. The most CP even-like categories are those with high CP BDT score (3, 6, 9, 12, 14, 17, 20), and the most CP odd-like categories are those with low CP BDT score (1, 4, 7, 10, 13, 16, 19).

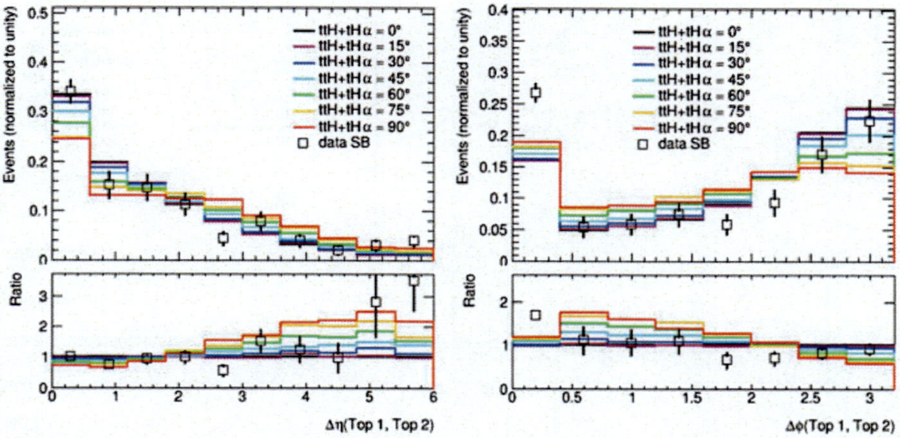

Fig. 11.21 Leptonic CP BDT training variables: $\Delta\eta(t_1, t_2)$ and $\Delta\phi(t_1, t_2)$. The underflow bins represent events containing ≥ 2 leptons or where no second top is reconstructed (either partially or fully). The open squares show TI data in the sideband region

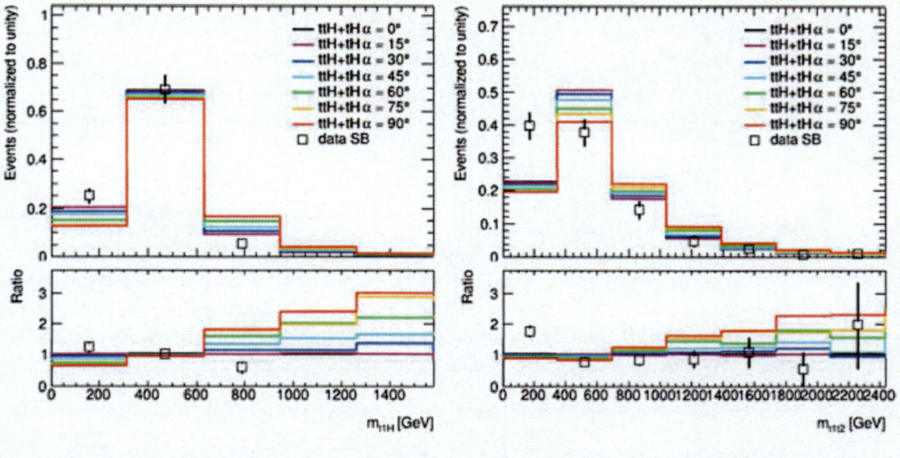

Fig. 11.22 Leptonic CP BDT training variables: $m_{t_1 H}$ and $m_{t_1 t_2}$. The underflow bins contain events with ≥ 2 leptons, where no top is reconstructed. The open squares show TI data in the sideband region

The expected significance of $t\bar{t}H$ and tH signal in the CP even and CP odd scenarios is shown in Table 11.2. This Table also includes the expected exclusion of the CP odd and maximal CP mixing ($\alpha = 45°$, $\kappa_t = 1$) hypotheses, assuming SM observation.

The background-only statistical uncertainty δ (Eq. 11.3) is shown for each category in Table 11.3 and compared to the major theory uncertainties that will impact the result: these include the 100% heavy flavor uncertainty on ggF, as well as on the Underlying Event/Parton Showering uncertainties on $t\bar{t}H$, $tHjb$, and tWH signal.

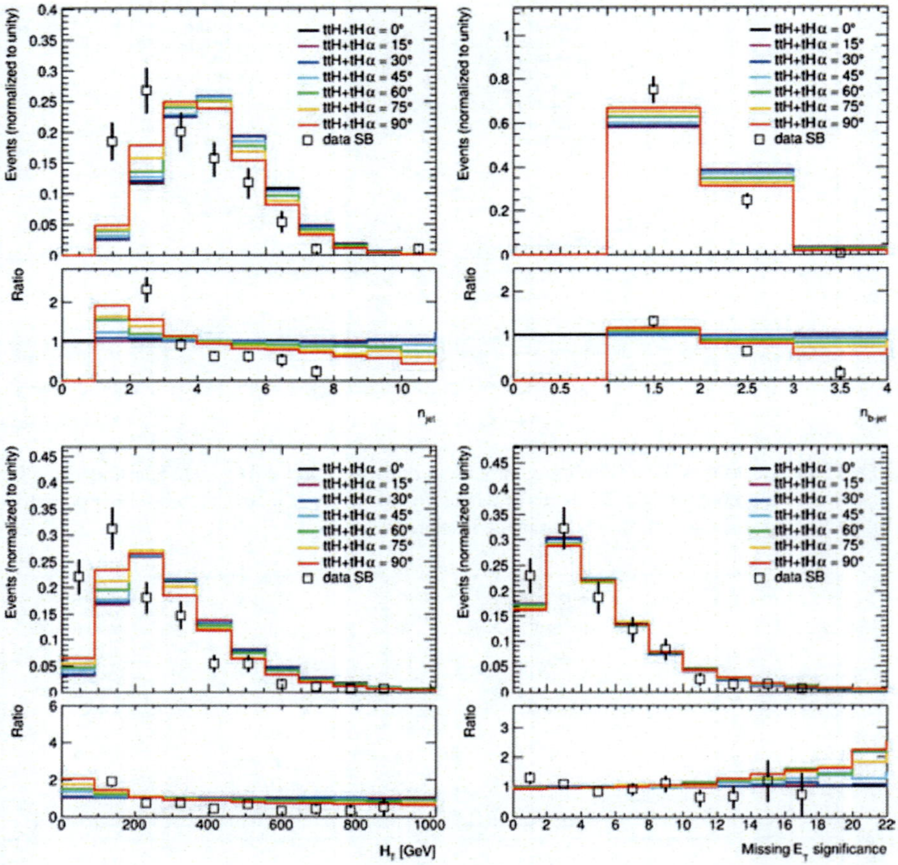

Fig. 11.23 Leptonic CP BDT training variables: jet multiplicity, *b*-jet multiplicity, H_T, and missing E_T significance. The open squares show TI data in the sideband region

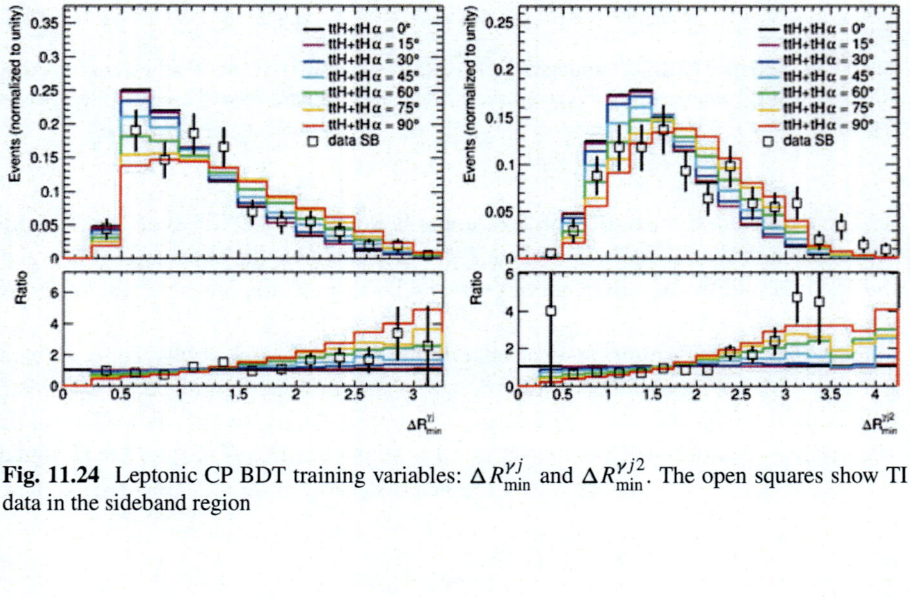

Fig. 11.24 Leptonic CP BDT training variables: $\Delta R^{\gamma j}_{\text{min}}$ and $\Delta R^{\gamma j2}_{\text{min}}$. The open squares show TI data in the sideband region

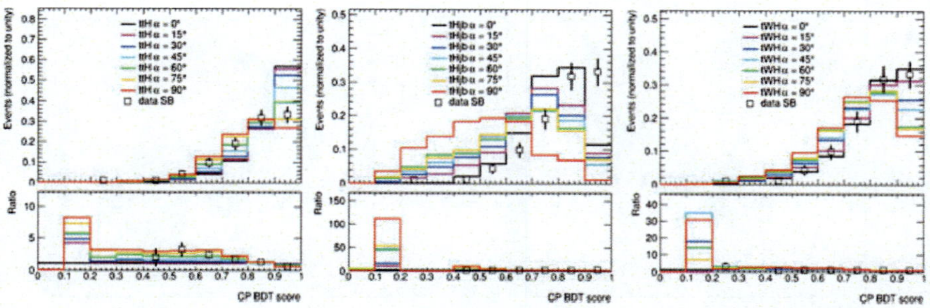

(a) $t\bar{t}H$ lep (b) $tHjb$ lep (c) tWH lep

Fig. 11.25 Leptonic CP BDT score for $t\bar{t}H$, $tHjb$, and tWH for different values of the CP mixing angle. The open squares show TI data in the sideband region

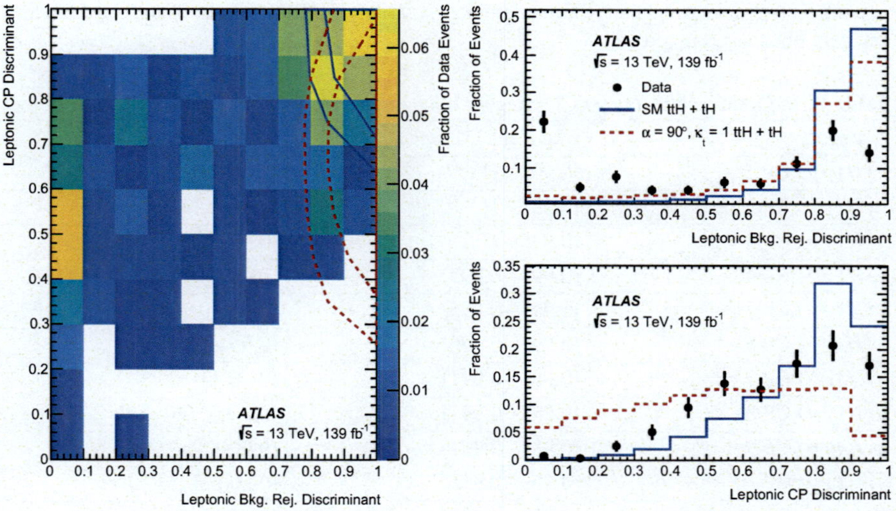

Fig. 11.26 Two-dimensional BDT distribution in data passing leptonic preselection. The horizontal axis shows the background rejection BDT discriminant and the vertical axis shows the CP BDT discriminant. The inner (outer) contours capture 25% (50%) of the $t\bar{t}H + tH$ signal events for the CP even (blue) and CP odd (red) hypotheses. The right panel shows projections onto the BDT discriminant axes. Contributions from CP even (blue), CP odd (red) $t\bar{t}H + tH$ processes and the data (black) are shown and normalized to unit area

All of these systematics are found to be much less than δ, and are therefore expected to have negligible impact on the final results.

The number-counting rejection $Z_{CP}(\alpha)$ as a function of mixing angle is shown in Fig. 11.28. The 95% confidence limit is shown as a horizontal dashed line. The limit is calculated to be $\alpha = 59.0°$ when background is estimated from the NTI sidebands, and $\alpha = 64.4°$ when background is estimated from the TI sidebands.

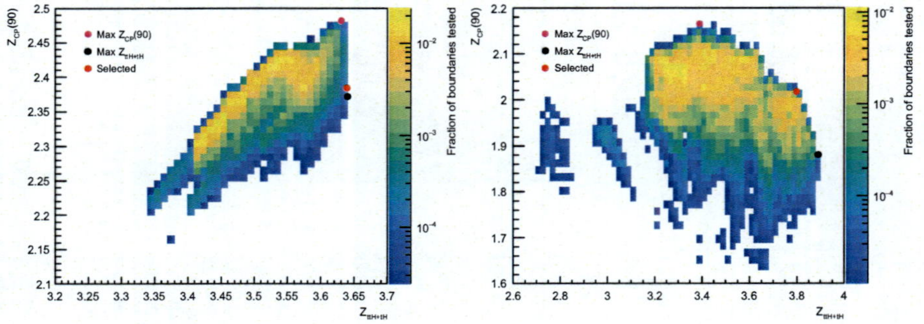

(a) Hadronic channel (b) Leptonic channel

Fig. 11.27 Summary of the CP odd exclusion $Z_{CP}(90)$ and SM $t\bar{t}H + tH$ significance $Z_{t\bar{t}H+tH}$ obtained by each set of boundaries in the brute force scan. The maximum exclusion of the CP odd hypothesis is shown in magenta, the maximum SM $t\bar{t}H + tH$ significance in black, and the selected category boundaries in red

Table 11.1 Optimized $t\bar{t}H(\gamma\gamma)$ CP category boundaries

Category	Bkg Rej BDT score	CP BDT score
$t\bar{t}H(\gamma\gamma)$ CP 1	[0.995–1]	[0–0.90]
$t\bar{t}H(\gamma\gamma)$ CP 2	[0.995–1]	[0.90–0.97]
$t\bar{t}H(\gamma\gamma)$ CP 3	[0.995–1]	[0.97–1]
$t\bar{t}H(\gamma\gamma)$ CP 4	[0.991–0.995]	[0–0.88]
$t\bar{t}H(\gamma\gamma)$ CP 5	[0.991–0.995]	[0.88–0.96]
$t\bar{t}H(\gamma\gamma)$ CP 6	[0.991–0.995]	[0.96–1]
$t\bar{t}H(\gamma\gamma)$ CP 7	[0.981–0.991]	[0–0.84]
$t\bar{t}H(\gamma\gamma)$ CP 8	[0.981–0.991]	[0.84–0.96]
$t\bar{t}H(\gamma\gamma)$ CP 9	[0.981–0.991]	[0.96–1]
$t\bar{t}H(\gamma\gamma)$ CP 10	[0.909–0.981]	[0–0.61]
$t\bar{t}H(\gamma\gamma)$ CP 11	[0.909–0.981]	[0.61–0.86]
$t\bar{t}H(\gamma\gamma)$ CP 12	[0.909–0.981]	[0.86–1]
$t\bar{t}H(\gamma\gamma)$ CP 13	[0.988–1]	[0–0.91]
$t\bar{t}H(\gamma\gamma)$ CP 14	[0.988–1]	[0.91–1]
$t\bar{t}H(\gamma\gamma)$ CP 15	[0.915–0.988]	[0–0.82]
$t\bar{t}H(\gamma\gamma)$ CP 16	[0.915–0.988]	[0.82–0.93]
$t\bar{t}H(\gamma\gamma)$ CP 17	[0.915–0.988]	[0.93–1]
$t\bar{t}H(\gamma\gamma)$ CP 18	[0.252–0.915]	[0–0.72]
$t\bar{t}H(\gamma\gamma)$ CP 19	[0.252–0.915]	[0.72–0.86]
$t\bar{t}H(\gamma\gamma)$ CP 20	[0.252–0.915]	[0.86–1]

Table 11.2 Predicted values of the figures of merit for the twenty $t\bar{t}H(\gamma\gamma)$ CP categories, calculated from event yields in the $m_{\gamma\gamma}$ region 125 ± 2 GeV

$t\bar{t}H$ significance (if even)	5.00σ
$t\bar{t}H$ significance (if odd)	2.75σ
tH significance (if even)	0.321σ
tH significance (if odd)	2.23σ
$Z_{CP}(90°)$	3.02σ
$Z_{CP}(45°)$	1.25σ

Continuum background is calculated by scaling NTI data sidebands according to Eq. 5.1

Table 11.3 Background-only statistical uncertainty (δ) in each $t\bar{t}H(\gamma\gamma)$ CP category compared with the 100% heavy flavor uncertainty (n_{ggF}) and the UEPS uncertainties on $t\bar{t}H$, $tHjb$, and tWH

Category	δ	n_{ggF}	UEPS = $n^{\text{Herwig}} - n^{\text{Pythia}}$		
			$t\bar{t}H$	$tHjb$	tWH
$t\bar{t}H(\gamma\gamma)$ CP 1	1.462	0.247	–0.157	0.008	0.010
$t\bar{t}H(\gamma\gamma)$ CP 2	1.581	0.044	–0.049	–0.007	–0.005
$t\bar{t}H(\gamma\gamma)$ CP 3	1.136	0.002	0.098	0.000	0.000
$t\bar{t}H(\gamma\gamma)$ CP 4	1.275	0.208	–0.039	–0.043	–0.038
$t\bar{t}H(\gamma\gamma)$ CP 5	1.534	0.054	–0.008	–0.020	–0.006
$t\bar{t}H(\gamma\gamma)$ CP 6	1.554	0.009	0.080	–0.005	–0.007
$t\bar{t}H(\gamma\gamma)$ CP 7	1.597	0.448	–0.053	–0.020	–0.014
$t\bar{t}H(\gamma\gamma)$ CP 8	2.713	0.186	0.015	–0.010	–0.002
$t\bar{t}H(\gamma\gamma)$ CP 9	2.623	0.021	0.094	0.007	–0.001
$t\bar{t}H(\gamma\gamma)$ CP 10	1.363	0.557	–0.003	0.019	–0.012
$t\bar{t}H(\gamma\gamma)$ CP 11	3.858	1.250	–0.053	–0.038	0.026
$t\bar{t}H(\gamma\gamma)$ CP 12	7.860	0.600	0.007	–0.025	0.006
$t\bar{t}H(\gamma\gamma)$ CP 13	1.398	0.001	–0.149	–0.024	–0.006
$t\bar{t}H(\gamma\gamma)$ CP 14	1.640	0.001	0.156	– 0.008	0.001
$t\bar{t}H(\gamma\gamma)$ CP 15	1.168	0.004	–0.089	–0.024	–0.011
$t\bar{t}H(\gamma\gamma)$ CP 16	1.643	0.000	–0.015	–0.012	–0.014
$t\bar{t}H(\gamma\gamma)$ CP 17	1.697	0.001	–0.036	0.001	0.000
$t\bar{t}H(\gamma\gamma)$ CP 18	1.082	0.008	–0.016	0.032	–0.027
$t\bar{t}H(\gamma\gamma)$ CP 19	1.627	0.008	–0.024	0.022	–0.010
$t\bar{t}H(\gamma\gamma)$ CP 20	1.973	0.003	–0.014	–0.013	0.013

All columns are in units of events at 139 fb^{-1} and are calculated in the range $m_{\gamma\gamma} = 125 \pm 2$ GeV

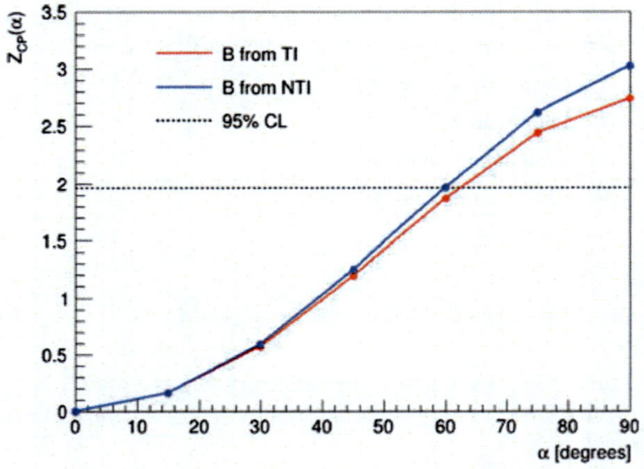

Fig. 11.28 Number-counting exclusion as a function of the CP mixing angle α. Continuum background yield is estimated from NTI sidebands (blue) and TI sidebands (red)

References

1. Ellis J et al (2014) Disentangling Higgs-top couplings in associated production. In: JHEP 04, p 004. https://doi.org/10.1007/JHEP04(2014)004. arXiv: 1312.5736 [hep-ph]
2. Boudjema F et al (2015) Lab-frame observables for probing the top-Higgs interaction. In: Phys. Rev. D92.1, p 015019. https://doi.org/10.1103/PhysRevD.92.015019. arXiv: 1501.03157 [hep-ph]
3. Amor Dos Santos S et al (2017) Probing the CP nature of the Higgs coupling in $t\bar{t}h$ eventsat the LHC. In: Phys. Rev. D96.1, p 013004. https://doi.org/10.1103/PhysRevD.96.013004. arXiv: 1704.03565 [hep-ph]
4. Chen T, Guestrin C (2016) XGBoost: a scalable tree boosting system. In: Proceedings of the 22nd ACM SIGKDD international conference on knowledge discovery and data mining. KDD '16. San Francisco, California, USA: ACM, 2016, pp 785–794. isbn: 978-1-4503-4232-2. https://doi.org/10.1145/2939672.2939785. http://doi.acm.org/10.1145/2939672.2939785

Chapter 12
Measurement of CP Properties in $t\bar{t}H$ and tH

A direct measurement of the CP properties of the Higgs-top interaction is performed with the full Run 2 ATLAS dataset of 139 fb^{-1} using the twenty $t\bar{t}H(\gamma\gamma)$ CP categories defined in the previous Chapter. This measurement focuses on the $t\bar{t}H$ and tH production modes in the $H \to \gamma\gamma$ decay channel, while remaining as insensitive as possible to the gluon fusion and $H \to \gamma\gamma$ loop processes. Section 12.1 describes the impact of CP violation in Higgs-top interactions on ggF and $H \to \gamma\gamma$ and two possibilities for their treatment in the statistical analysis. The second Section describes the $m_{\gamma\gamma}$ signal and background shapes in the $t\bar{t}H(\gamma\gamma)$ CP categories, as well as the derivation of the $t\bar{t}H$ and tH signal yield dependence on the mixing parameters α and κ_t. Section 12.3 discusses the major systematics, and the results are presented in Sect. 12.4.

12.1 Treatment of Loop Processes

The contribution of the top quark to the gluon fusion and $H \to \gamma\gamma$ loops gives these processes sensitivity to CP violation in the Higgs-top interaction. However, this sensitivity is only indirect, as many other models of BSM physics predict new loop contributions that change the rates and kinematics of these processes. In order to measure the effects of CP violation in Higgs-top interactions directly, the $t\bar{t}H(\gamma\gamma)$ CP analysis aims to be as insensitive to these loop processes as possible.

The $H \to \gamma\gamma$ decay kinematics are not expected to change with CP mixing angle, and it is demonstrated in Sect. 12.1.1 that the shape of the ggF yield in the $t\bar{t}H(\gamma\gamma)$ CP categories is independent of the mixing angle α. As a result, the introduction of CP mixing affects only the overall rate of ggF and $H \to \gamma\gamma$ in the analysis categories. The treatment of these rates with effective loop couplings is discussed in Sect. 12.1.2.

J. E. Dickinson, *ATLAS Measurements of the Higgs Boson Coupling to the Top Quark in the Higgs to Diphoton Decay Channel*, Springer Theses,
https://doi.org/10.1007/978-3-030-86368-5_12

12.1.1 Kinematics of Gluon Fusion

The ggF contribution to the hadronic $t\bar{t}H(\gamma\gamma)$ CP categories is studied in detail. The efficiency of the hadronic preselection on SM ggF is 0.5%, which is reduced to $< 0.1\%$ with the cut on the hadronic background rejection BDT discriminant at 0.909 (see Table 11.1). In the leptonic channel, ggF contamination is neglected altogether, since the efficiency of the leptonic preselection on SM ggF is $< 0.01\%$.

Figure 12.1 shows the hadronic background rejection BDT and CP BDT discriminants for multiple models of SM ggF (Powheg+Pythia8, Powheg+Herwig7, and MG5_aMC@NLO+ Pythia8), as well as two CP-violating scenarios (MG5_aMC@NLO+Pythia8). Though CP mixing in the Higgs-top coupling alters some kinematic variables in ggF (such as $\Delta\phi(j, j)$) [1], the BDTs in Fig. 12.1 (which are optimized on $t\bar{t}H$ and tH) are mostly insensitive to this. The background rejection BDT shape does not depend appreciably on the CP mixing angle α, and the CP BDT shape shows only a weak dependence.

In the statistical analysis of the $t\bar{t}H(\gamma\gamma)$ CP categories, several theoretical uncertainties are applied to allow migration of ggF events across categories. These uncertainties are derived from a comparison of SM Monte Carlo samples using different Monte Carlo generators and different parton showering algorithms (Table 12.4). With these conservative systematics applied, the shape of ggF yield in the CP BDT categories is assumed to be invariant under the mixing angle α.

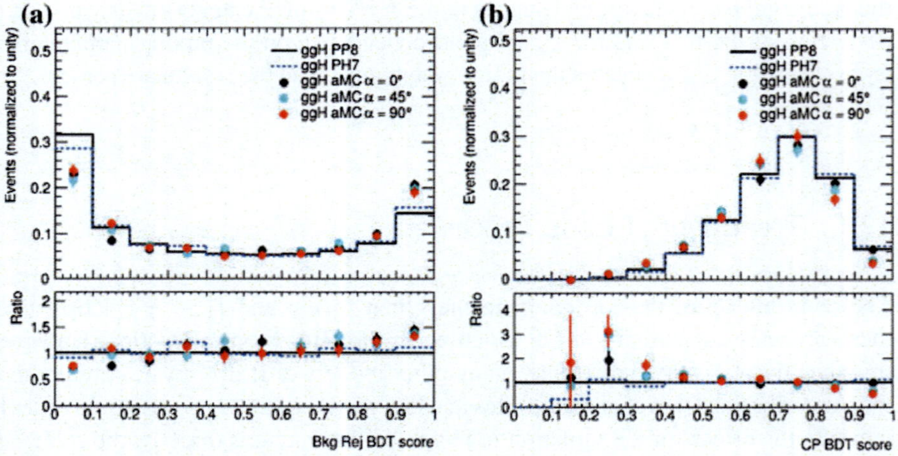

Fig. 12.1 Hadronic background rejection BDT score (**a**) and CP BDT score (**b**) for ggF Monte Carlo samples

12.1.2 Effective Loop Couplings

The overall rates of the ggF and $H \rightarrow \gamma\gamma$ processes are scaled by the square of the effective loop couplings κ_g and κ_γ, respectively:

$$\mu_{ggF} = \kappa_g^2 \tag{12.1}$$

$$\mu_{\gamma\gamma} = \kappa_\gamma^2 \tag{12.2}$$

Depending on the treatment of these loop couplings, the data are interpreted in the following two scenarios:

1. κ_g and κ_γ are constrained using data from the ATLAS combined Higgs coupling measurement [2].
2. κ_g and κ_γ are parameterized as function of κ_t and α and are varied coherently for $t\bar{t}H$, tH, ggF, and $H \rightarrow \gamma\gamma$.

Ideally, no model-dependent assumptions should be made about the ggF and $H \rightarrow \gamma\gamma$ processes: the first interpretation is therefore treated as the baseline. Details of each are included below.

Constraints from data

In order to constrain the ggF and $H \rightarrow \gamma\gamma$ processes, the dataset selected by the $t\bar{t}H(\gamma\gamma)$ CP analysis alone is insufficient. Since only one Higgs boson decay channel is selected, it is impossible to separate $H \rightarrow \gamma\gamma$ from the production mode cross section. And since the analysis rejects ggF as a background by design, the $t\bar{t}H(\gamma\gamma)$ CP categories are left with little sensitivity to this process. Therefore knowledge from other Higgs boson decay channels must be applied as well.

The ATLAS Higgs coupling combination [2] provides the measurement of κ_g and κ_γ shown in Fig. 12.2. The measured best fit couplings, errors, and correlation are re-derived with all $t\bar{t}H$ categories excluded from the combination. This ensures that the dataset used to constrain the ggF and $H \rightarrow \gamma\gamma$ rates is orthogonal to the $t\bar{t}H(\gamma\gamma)$ CP categories and independent of the measured $t\bar{t}H$ cross section. In addition, the systematic uncertainties between the re-derived combination and the $t\bar{t}H(\gamma\gamma)$ CP categories can be neglected.

The re-derived constraints on κ_g and κ_γ are:

$$\kappa_g = 1.034 \pm 0.067; \;\; \kappa_\gamma = 0.984 \pm 0.064; \;\; \rho(\kappa_g, \kappa_\gamma) = -0.47 \tag{12.3}$$

Since the ggF and $H \rightarrow \gamma\gamma$ measurements are dominated by high-statistics categories, the impact of removing the $t\bar{t}H$ categories is small.

Parameterize coherently with $t\bar{t}H$ and tH

Assuming that the only new physics impacting the ggF and $H \rightarrow \gamma\gamma$ loops is the model of CP violation described by Eq. 2.28, the effective loop couplings κ_g and κ_γ can be directly parameterized as a function of κ_t and α. The ggF and $H \rightarrow \gamma\gamma$ rates can then be varied coherently with $t\bar{t}H$ and tH production modes.

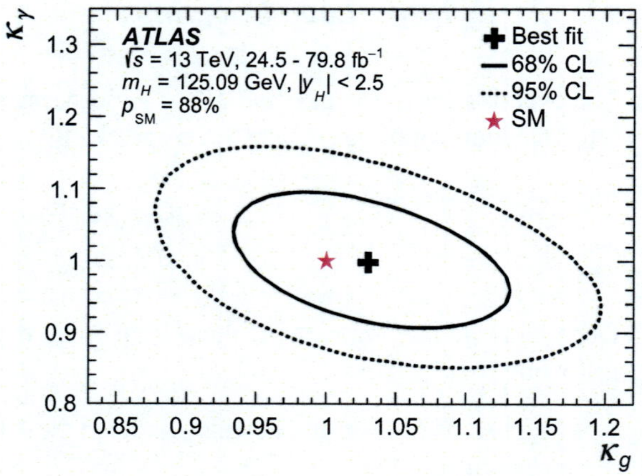

Fig. 12.2 Two-dimensional likelihood scan of κ_g and κ_γ from the ATLAS Higgs coupling combination [2]. The best fit value of $(\kappa_g, \kappa_\gamma)$ is shown with 1 and 2σ contours

The parameterization of κ_g and κ_γ derived in [3] can be re-written in terms of κ_t and α:

$$\kappa_g^2 = \kappa_t^2 c_\alpha^2 + 2.6\kappa_t^2 s_\alpha^2 + 0.11\kappa_t c_\alpha(\kappa_t c_\alpha - 1) \tag{12.4}$$

$$\kappa_\gamma^2 = (1.28 - 0.28\kappa_t c_\alpha)^2 + (0.43\kappa_t s_\alpha)^2 \tag{12.5}$$

where s_α and c_α represent the sine and cosine of the mixing angle α, respectively.

12.2 Signal and Background Model

12.2.1 Signal and Background Shapes

In each $t\bar{t}H(\gamma\gamma)$ CP category, the signal shape is determined by fitting a Double-Sided Crystal Ball (DSCB) function to a sample of inclusive Higgs boson Monte Carlo (see Sect. 6.2.1). The best fit DSCB parameters for all twenty categories are shown in Table 12.1, along with S_{90}, the smallest $m_{\gamma\gamma}$ range containing 90% of Higgs signal. The signal shape parameters are derived from the inclusive SM Higgs boson Monte Carlo sample. Using $t\bar{t}H + tH$ Monte Carlo with CP mixing angle $\alpha \neq 0°$ has negligible impact on the signal shape parameterization.

The functional form of the continuum background is extracted from templates constructed from $t\bar{t}\gamma\gamma$ Monte Carlo, since this process dominates the continuum background once the BDT selection has been applied. Because the $t\bar{t}\gamma\gamma$ background template in category 10 has exceptionally high statistical uncertainty ($\Delta_{MC} = 3.06$), the background shape and spurious signal are derived instead from a template built from NTI data ($\Delta_{MC} = 1.91$). In all other categories, the $t\bar{t}\gamma\gamma$ Monte Carlo template has a smaller statistical uncertainty than the NTI background template.

Table 12.1 Best fit parameter values for the signal $m_{\gamma\gamma}$ shape in each $t\bar{t}H(\gamma\gamma)$ CP category

Category	μ_{CB} [GeV]	σ_{CB} [GeV]	α_{low}	α_{high}	n_{low}	n_{high}	S_{90} [GeV]
$t\bar{t}H(\gamma\gamma)$ CP 1	125.10	1.16	1.6	1.5	6.2	15.5	4.3
$t\bar{t}H(\gamma\gamma)$ CP 2	125.12	1.35	1.8	1.5	4.1	15.5	5.2
$t\bar{t}H(\gamma\gamma)$ CP 3	125.27	1.64	1.7	1.7	4.5	11.9	6.0
$t\bar{t}H(\gamma\gamma)$ CP 4	125.13	1.21	1.6	1.5	4.8	5.7	4.7
$t\bar{t}H(\gamma\gamma)$ CP 5	125.17	1.50	1.7	1.5	6.9	15.0	5.6
$t\bar{t}H(\gamma\gamma)$ CP 6	125.17	1.64	1.6	1.4	8.0	35.9	6.3
$t\bar{t}H(\gamma\gamma)$ CP 7	125.09	1.21	1.6	1.5	5.6	11.6	4.7
$t\bar{t}H(\gamma\gamma)$ CP 8	125.18	1.58	1.6	1.7	5.8	7.6	5.9
$t\bar{t}H(\gamma\gamma)$ CP 9	125.21	1.72	1.7	1.7	4.4	12.2	6.3
$t\bar{t}H(\gamma\gamma)$ CP 10	125.10	1.29	1.8	1.6	13.6	7.9	4.5
$t\bar{t}H(\gamma\gamma)$ CP 11	125.15	1.55	1.7	1.6	4.7	7.6	5.4
$t\bar{t}H(\gamma\gamma)$ CP 12	125.21	1.70	1.5	1.5	8.5	13.3	6.5
$t\bar{t}H(\gamma\gamma)$ CP 13	125.11	1.29	1.6	1.5	6.8	11.1	5.0
$t\bar{t}H(\gamma\gamma)$ CP 14	125.17	1.61	1.7	1.5	5.2	15.0	6.1
$t\bar{t}H(\gamma\gamma)$ CP 15	125.15	1.45	1.6	1.6	5.7	6.0	5.4
$t\bar{t}H(\gamma\gamma)$ CP 16	125.19	1.70	1.7	1.7	6.7	14.4	6.2
$t\bar{t}H(\gamma\gamma)$ CP 17	125.20	1.71	1.5	1.5	7.9	14.4	6.7
$t\bar{t}H(\gamma\gamma)$ CP 18	125.03	1.68	1.8	1.6	18.7	12.7	5.7
$t\bar{t}H(\gamma\gamma)$ CP 19	125.17	1.76	2.2	1.5	1.4	10.2	6.6
$t\bar{t}H(\gamma\gamma)$ CP 20	125.14	1.78	1.6	1.6	10.8	14.7	6.6

Note that these are the fitted values for the MC samples at $m_H = 125$ GeV

The functional form of the continuum background and the spurious signal are listed in Table 12.2.

12.2.2 Signal Yield Dependence on Mixing Angle

Event yields and signal purities in the twenty CP BDT categories are shown in Fig. 12.3 for $\alpha = 0°$ (top row), $\alpha = 45°$ (middle row), and $\alpha = 90°$ (bottom row). The ggF event yields shown are from the MG5_aMC@NLO+Pythia8 Monte Carlo samples in which the ggF rate changes coherently with the CP mixing angle α. The changing composition of the $t\bar{t}H(\gamma\gamma)$ CP categories is clearly visible, particularly the increase in the tH rate with α.

To measure the mixing angle α and coupling strength κ_t, the signal yield must be parameterized as function of these two parameters. The signal yield in each category takes the same form as the $t\bar{t}H$ and tH cross sections in Chapter 2, with the parameters fitted independently in each category.

Table 12.2 Results of the spurious signal test for the $t\bar{t}H(\gamma\gamma)$ CP analysis categories

Category	Function	N_{sp}	χ^2/ndof	Prob(χ^2) [%]
$t\bar{t}H(\gamma\gamma)$ CP 1	Exponential	+0.426	0.621	90
$t\bar{t}H(\gamma\gamma)$ CP 2	Power Law	−0.297	0.71	82
$t\bar{t}H(\gamma\gamma)$ CP 3	Exponential	+0.260	1.69	3
$t\bar{t}H(\gamma\gamma)$ CP 4	Exponential	+0.311	0.672	74
$t\bar{t}H(\gamma\gamma)$ CP 5	Power Law	−0.566	1.15	29
$t\bar{t}H(\gamma\gamma)$ CP 6	Exponential	+0.335	0.573	94
$t\bar{t}H(\gamma\gamma)$ CP 7	Exponential	+0.644	1.91	1
$t\bar{t}H(\gamma\gamma)$ CP 8	Exponential	−1.23	0.677	85
$t\bar{t}H(\gamma\gamma)$ CP 9	Power Law	−0.264	0.742	79
$t\bar{t}H(\gamma\gamma)$ CP 10	Power Law	−1.31	0.795	62
$t\bar{t}H(\gamma\gamma)$ CP 11	Exponential	+4.50	1.18	26
$t\bar{t}H(\gamma\gamma)$ CP 12	Exponential	+3.05	0.981	48
$t\bar{t}H(\gamma\gamma)$ CP 13	Exponential	+0.199	0.981	48
$t\bar{t}H(\gamma\gamma)$ CP 14	Exponential	+0.129	0.996	46
$t\bar{t}H(\gamma\gamma)$ CP 15	Power Law	−0.205	1.06	39
$t\bar{t}H(\gamma\gamma)$ CP 16	Exponential	+0.398	0.841	66
$t\bar{t}H(\gamma\gamma)$ CP 17	Power Law	−0.393	1.38	12
$t\bar{t}H(\gamma\gamma)$ CP 18	Exponential	+0.761	0.91	57
$t\bar{t}H(\gamma\gamma)$ CP 19	Power Law	−0.498	0.718	81
$t\bar{t}H(\gamma\gamma)$ CP 20	Exponential	−0.135	0.618	90

The χ^2/ndof indicates the performance of a background only fit to the template in the $m_{\gamma\gamma}$ range 105–160 GeV, but is in general not a part of the spurious signal criteria

The $t\bar{t}H$ yield in category i is parameterized using the form of the cross section in Eq. 2.29:

$$y_i = A_i\kappa_t^2 c_\alpha^2 + B_i\kappa_t^2 s_\alpha^2 + E_i\kappa_t^2 s_\alpha c_\alpha \tag{12.6}$$

where s_α and c_α represent the sine and cosine of α, respectively. For a discussion of the diagrams corresponding to each term, refer to Sect. 2.4.1. The coefficients A, B, and E are determined in each category by fitting to the signal yield predicted by Monte Carlo sample at different mixing angles.

The inclusive parameterization of the $t\bar{t}H$ signal yield (sum of all twenty $t\bar{t}H(\gamma\gamma)$ CP categories) is shown in Fig. 12.4a, while Fig. 12.4b shows the fitted values of the coefficients A, B, and E in each category. As expected, interference between CP even and CP odd (coefficient E) is small.

For $tHjb$ and tWH, the interference between Higgs couplings to the top quark and the W boson must also be taken into account. The $tHjb$ and tWH yield in category i is parameterized using the form of the cross section in Eq. 2.30 (with different values of the coefficients for $tHjb$ and tWH):

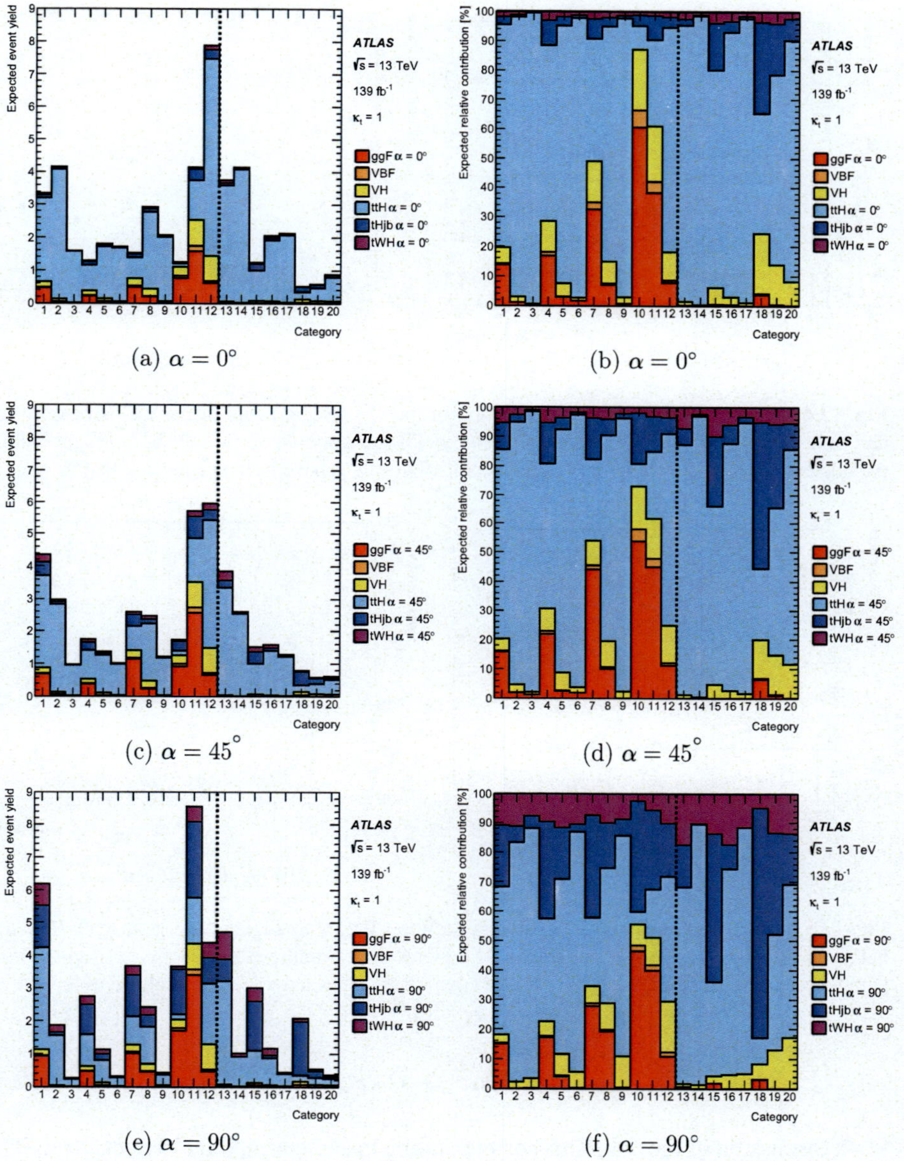

Fig. 12.3 (Left) Higgs event yields in the $t\bar{t}H(\gamma\gamma)$ CP categories at 139 fb^{-1} at different values of the CP mixing angle α. (Right) Purity of the Higgs yield in each category for different values of α. Yields are calculated in the smallest $m_{\gamma\gamma}$ window containing 90% of the fitted signal

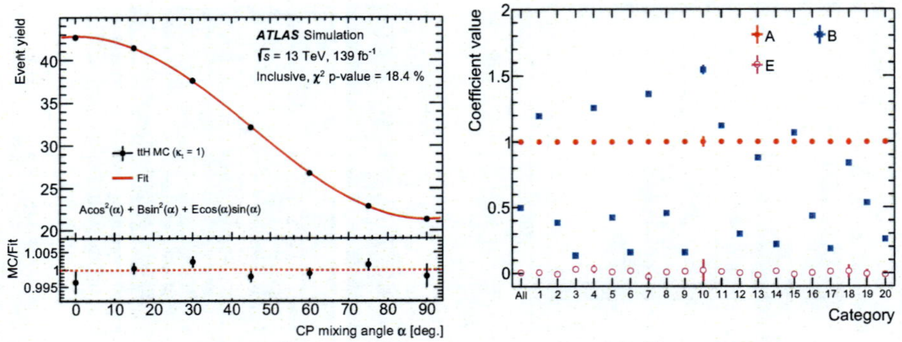

(a) Inclusive $t\bar{t}H$ fit (b) $t\bar{t}H$ **coefficients**

Fig. 12.4 (Left) Parameterization as function of the CP mixing angle of the inclusive $t\bar{t}H$ signal yield in the $t\bar{t}H(\gamma\gamma)$ CP categories. The yields are normalized to 139 fb^{-1}. (Right) Fitted parameterization coefficients for $t\bar{t}H$. The coefficient values have been normalized to SM

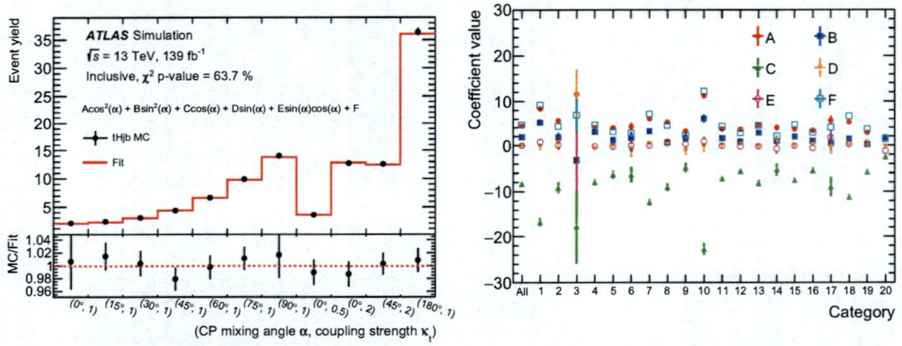

(a) Inclusive $tHjb$ fit (b) $tHjb$ **coefficients**

Fig. 12.5 (Left) Parameterization as function of the CP mixing angle of the inclusive $tHjb$ signal yield in the $t\bar{t}H(\gamma\gamma)$ CP categories. The yields are normalized to 139 fb^{-1}. (Right) Fitted parameterization coefficients for $tHjb$. The coefficient values have been normalized to SM

$$y_i = A_i\kappa_t^2 c_\alpha^2 + B_i\kappa_t^2 s_\alpha^2 + C_i\kappa_t c_\alpha + D_i\kappa_t s_\alpha + E_i\kappa_t^2 s_\alpha c_\alpha + F_i \quad (12.7)$$

For a discussion of the diagrams corresponding to each term, refer to Sect. 2.4.1. The coefficients A, B, C, D, E and F are determined in each category by fitting to the signal yield predicted by Monte Carlo. In the case of tH, the fit is performed over both the mixing angle α and the coupling strength κ_t in order to break the degeneracy among the parameters A, B, and F. Several $tHjb$ and tWH Monte Carlo samples with $\kappa_t \neq 1$ are generated for this purpose (see Sect. 5.2.2 for details).

The inclusive parameterizations of the $tHjb$ and tWH signal yield are shown in Figs. 12.5a and 12.6a, respectively. Figures 12.5b and 12.6b show the fitted values of the coefficients A, B, C, D, E, and F in each category. As expected, the interference between the CP even Higgs-top and Higgs-W couplings (coefficient C) is large and negative.

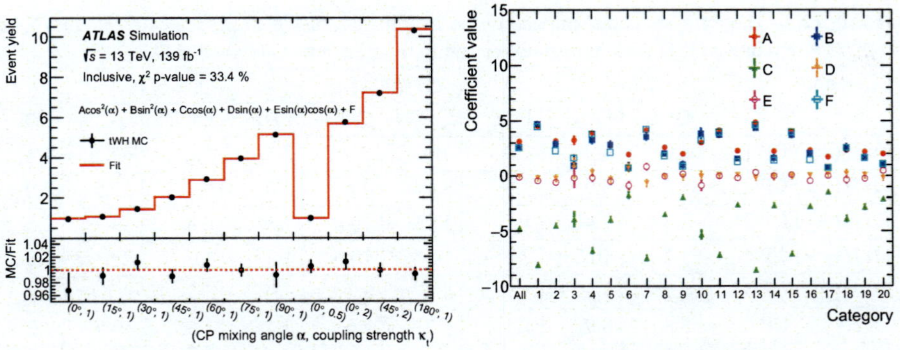

(a) Inclusive tWH fit (b) tWH cofficients

Fig. 12.6 (Left) Parameterization as function of the CP mixing angle of the inclusive tWH signal yield in the $t\bar{t}H(\gamma\gamma)$ CP categories. The yields are normalized to 139 fb^{-1}. (Right) Fitted parameterization coefficients for tWH. The coefficient values have been normalized to SM

Fig. 12.7 Inclusive parametrization of all processes dependent on the CP mixing angle α, with $\kappa_t = 1$. The yields are normalized to the Standard Model prediction

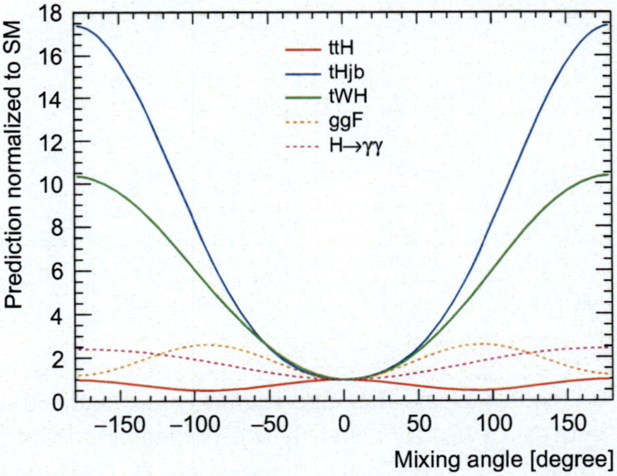

The signal parametrization is summarized in Fig. 12.7 for all processes whose rates depend on the CP mixing angle α. The parameterizations of ggF and $H \rightarrow \gamma\gamma$ from Eq. 12.4 are shown as dashed lines.

12.3 Systematic Uncertainties

The experimental systematics calculated for the $t\bar{t}H(\gamma\gamma)$ CP analysis are those summarized in Table 6.3. The theory systematics in Table 6.2 are calculated per process and per category for the $t\bar{t}H(\gamma\gamma)$ CP analysis, with magnitudes summarized in this Section.

Table 12.3 Theory systematics impacting the overall rate of signal processes ($t\bar{t}H$, $tHjb$, and tWH): the relative QCD scale uncertainties and PDF uncertainties (shown in %)

Category	$t\bar{t}H$ [%]			$tHjb$ [%]			tWH [%]		
	QCD+	QCD−	PDF	QCD+	QCD−	PDF	QCD+	QCD−	PDF
$t\bar{t}H(\gamma\gamma)$ CP 1	6.0	9.4	2.0	7.0	7.7	2.3	2.6	4.8	2.8
$t\bar{t}H(\gamma\gamma)$ CP 2	5.5	8.9	1.6	3.6	4.4	2.6	9.2	15.5	2.9
$t\bar{t}H(\gamma\gamma)$ CP 3	2.8	7.4	1.5	13.1	10.8	1.3	6.4	39.7	2.6
$t\bar{t}H(\gamma\gamma)$ CP 4	7.7	10.3	2.2	11.6	10.5	1.3	1.7	2.5	2.8
$t\bar{t}H(\gamma\gamma)$ CP 5	6.1	9.2	1.7	8.3	8.1	3.9	5.2	9.1	2.3
$t\bar{t}H(\gamma\gamma)$ CP 6	5.8	8.9	1.4	1.2	2.6	2.4	12.5	19.7	2.0
$t\bar{t}H(\gamma\gamma)$ CP 7	9.1	10.9	2.0	6.8	7.5	2.5	2.1	2.7	3.0
$t\bar{t}H(\gamma\gamma)$ CP 8	6.2	9.2	1.6	4.4	5.7	7.0	4.8	7.4	2.5
$t\bar{t}H(\gamma\gamma)$ CP 9	6.2	9.1	1.5	3.5	4.2	5.0	9.0	14.0	2.4
$t\bar{t}H(\gamma\gamma)$ CP 10	9.5	10.9	2.4	7.0	7.8	5.5	4.7	4.6	3.3
$t\bar{t}H(\gamma\gamma)$ CP 11	7.3	9.9	1.9	7.1	7.7	1.3	2.1	2.6	2.9
$t\bar{t}H(\gamma\gamma)$ CP 12	6.0	9.0	1.5	2.8	4.7	1.0	3.8	4.5	2.2
$t\bar{t}H(\gamma\gamma)$ CP 13	6.8	9.7	1.7	2.8	5.1	2.2	3.6	4.6	2.7
$t\bar{t}H(\gamma\gamma)$ CP 14	6.7	9.4	1.5	0.6	2.1	1.5	13.9	21.9	2.1
$t\bar{t}H(\gamma\gamma)$ CP 15	6.6	9.5	1.7	9.7	9.4	2.1	1.8	2.6	2.8
$t\bar{t}H(\gamma\gamma)$ CP 16	4.5	8.2	1.4	5.2	6.2	1.8	4.5	3.8	2.0
$t\bar{t}H(\gamma\gamma)$ CP 17	4.8	7.9	1.3	5.4	5.7	1.5	2.8	3.7	2.1
$t\bar{t}H(\gamma\gamma)$ CP 18	6.1	8.9	1.4	7.2	7.9	1.6	3.7	3.7	2.9
$t\bar{t}H(\gamma\gamma)$ CP 19	4.7	8.1	1.3	7.7	8.0	1.3	1.2	1.0	2.2
$t\bar{t}H(\gamma\gamma)$ CP 20	4.0	7.1	1.4	2.3	3.7	3.5	4.8	7.0	2.1

The QCD and PDF uncertainties are calculated in each category for each signal process ($t\bar{t}H$, $tHjb$, and tWH). The magnitudes of these uncertainties are shown in Table 12.3. The PDF uncertainties on $t\bar{t}H$, $tHjb$, and tWH are calculated according to the LHAPDF [4] prescription and treated as a single correlated nuisance parameter. In the case of ggF, VBF, and VH, a 100% heavy flavor uncertainty is applied to the normalization of each process to account for the uncertainty on the total cross section.

The Underlying Event/Parton Showering (UEPS) systematic is calculated by comparing the yield between SM samples interfaced with two different showering algorithms: Pythia8 and Herwig7. The UEPS uncertainty is calculated separately for each category and for each process ($t\bar{t}H$, $tHjb$, tWH, and ggF): the matrix element generator used for this comparison is MG5_aMC@NLO for $t\bar{t}H$ and tH, and Powheg for ggF. The magnitude of the UEPS uncertainty in each category is included in Table 12.4.

For $t\bar{t}H$ and ggF, an additional Monte Carlo generator uncertainty is applied, which accounts for differences in the predicted event yield per category from different matrix element calculations. For $t\bar{t}H$, this comparison is performed between the nominal MG5_aMC@NLO+Pythia8 and an alternative Powheg+Pythia8 sample. For ggF, the nominal sample is generated with Powheg+Pythia8 and the alternative

Table 12.4 Theory systematics impacting the migration of $t\bar{t}H$, $tHjb$, tWH, and ggF events across categories: the relative UEPS uncertainties and Monte Carlo generator uncertainties (shown in %)

Category	$t\bar{t}H$ [%]		$tHjb$ [%]	tWH [%]	ggF [%]	
	MC gen	UEPS	UEPS	UEPS	MC gen	UEPS
$t\bar{t}H(\gamma\gamma)$ CP 1	+1.4	−6.3	+9.1	+10.6	−87.0	−6.7
$t\bar{t}H(\gamma\gamma)$ CP 2	−4.3	−1.4	−19.5	−13.3	−51.5	+53.7
$t\bar{t}H(\gamma\gamma)$ CP 3	−9.7	+7.4	0.0	−7.4	0.0	−50.2
$t\bar{t}H(\gamma\gamma)$ CP 4	+2.5	−5.4	−40.7	−36.1	−16.0	+30.7
$t\bar{t}H(\gamma\gamma)$ CP 5	−5.6	−0.6	−45.2	−15.0	+28.1	+27.3
$t\bar{t}H(\gamma\gamma)$ CP 6	−7.2	+6.1	−24.1	−35.8	−77.2	−96.7
$t\bar{t}H(\gamma\gamma)$ CP 7	+5.3	−8.8	−21.5	−14.3	−12.4	−1.5
$t\bar{t}H(\gamma\gamma)$ CP 8	−5.7	+0.8	−12.3	−2.8	+0.4	+103.5
$t\bar{t}H(\gamma\gamma)$ CP 9	−7.8	+6.0	+47.9	−5.4	+7.0	+20.7
$t\bar{t}H(\gamma\gamma)$ CP 10	−1.8	−2.7	+44.1	−26.7	−38.2	+29.2
$t\bar{t}H(\gamma\gamma)$ CP 11	+2.2	−5.0	−12.0	+8.1	−18.8	+29.2
$t\bar{t}H(\gamma\gamma)$ CP 12	−3.1	+0.1	−11.1	+2.6	+19.0	+58.8
$t\bar{t}H(\gamma\gamma)$ CP 13	+3.5	−4.6	−35.0	−8.2	−96.9	+9.6
$t\bar{t}H(\gamma\gamma)$ CP 14	−2.6	+4.6	−37.3	+2.4	0.0	+58.6
$t\bar{t}H(\gamma\gamma)$ CP 15	+8.8	−10.9	−14.1	−6.5	−35.6	+7.1
$t\bar{t}H(\gamma\gamma)$ CP 16	+0.5	−1.0	−19.2	−23.9	0.0	0.0
$t\bar{t}H(\gamma\gamma)$ CP 17	+0.9	−2.2	+35.4	+15.7	0.0	−100.0
$t\bar{t}H(\gamma\gamma)$ CP 18	+7.5	−9.1	+21.5	−18.4	−203.2	+111.7
$t\bar{t}H(\gamma\gamma)$ CP 19	+5.6	−7.9	+27.7	−13.1	+94.3	−73.6
$t\bar{t}H(\gamma\gamma)$ CP 20	+3.4	−2.5	−24.0	+25.3	−86.6	−12.5

with MG5_aMC@NLO+Pythia8. The magnitude of the MC generator uncertainty in each category is included in Table 12.4.

12.4 Results

The diphoton mass spectra of the data in each of the twenty $t\bar{t}H(\gamma\gamma)$ CP categories are shown in Figs. 12.8, 12.9 and 12.10. Figure 12.8 shows the most CP even-like categories (those with high CP BDT score: 3, 6, 9, 12, 14, 17, 20), Fig. 12.9 shows the most CP odd-like categories (those with low CP BDT score: 1, 4, 7, 10, 13, 16, 19), and Fig. 12.10 shows the categories with intermediate CP BDT score (2, 5, 8, 11, 15, 18).

The reconstructed Higgs boson and top quark masses in data are shown in Fig. 12.11. The sum of all categories is shown with data in each category weighted

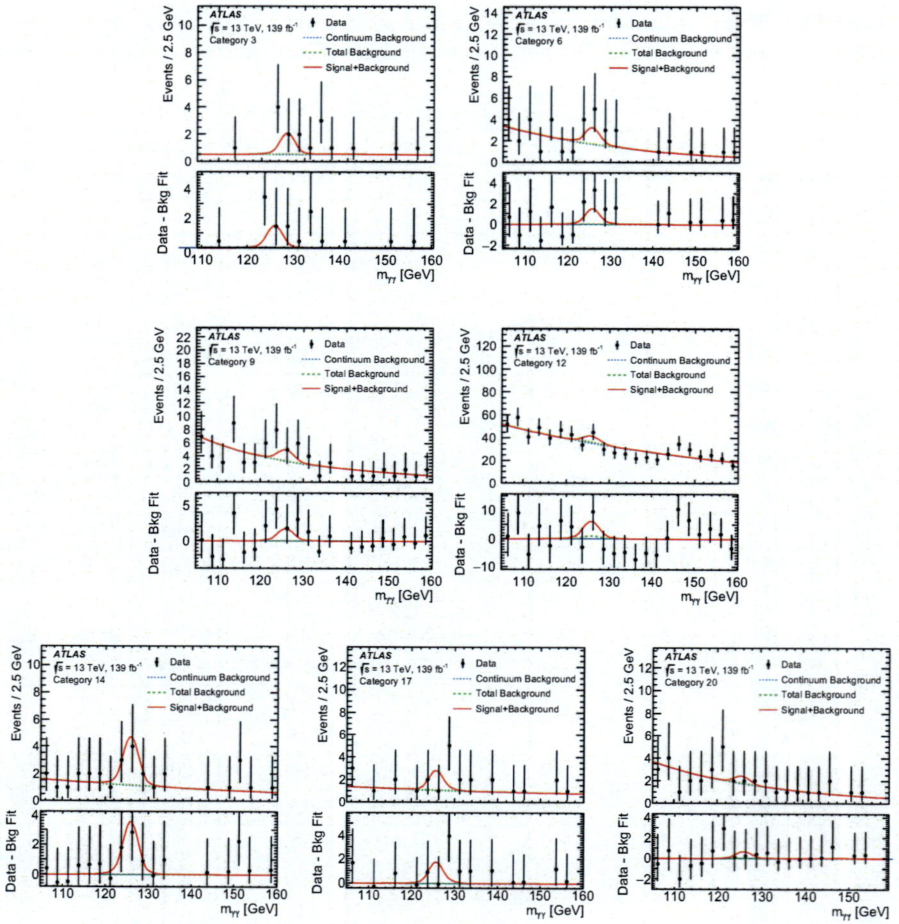

Fig. 12.8 Diphoton invariant mass spectrum in the $t\bar{t}H(\gamma\gamma)$ CP categories with high CP BDT score (CP even-like). The signal-plus-background fit is overlaid in red, the total background in dashed green, and the continuum background in dashed blue

by $\ln(1 + S/B)$, where S (B) is the fitted signal (background) in the smallest $m_{\gamma\gamma}$ window containing 90% of the signal. The fitted curves in the projection are similarly weighted. This weighting procedure is used to visually emphasize the contribution of the most powerful analysis categories (that is, the categories with highest signal-to-background ratio). A description of the calculation of errors on this plot is included in Sect. 6.5. A clear peak is visible at the Higgs boson mass and top quark mass, indicating the presence of both particles in the events in the most sensitive categories.

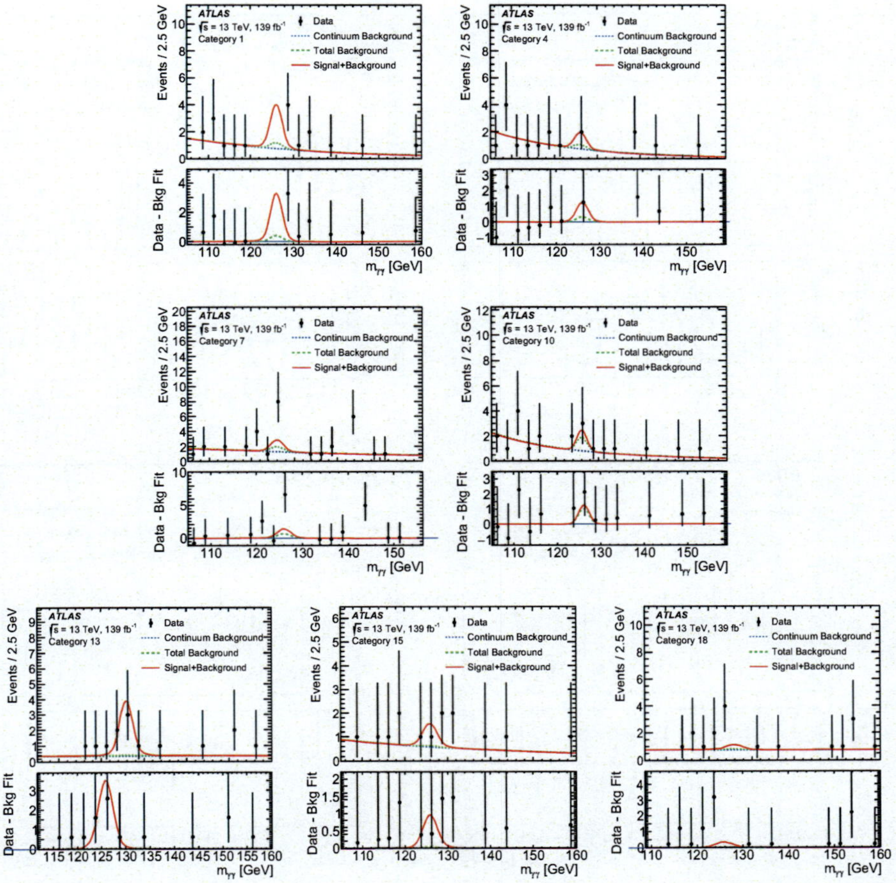

Fig. 12.9 Diphoton invariant mass spectrum in the $t\bar{t}H(\gamma\gamma)$ CP categories with low CP BDT score (CP odd-like). The signal-plus-background fit is overlaid in red, the total background in dashed green, and the continuum background in dashed blue

12.4.1 Sensitivity to SM $t\bar{t}H$ and tH

The sensitivity to $t\bar{t}H$ is evaluated by setting $\alpha = 0°$ and fixing the yield of all non-$t\bar{t}H$ production processes (including $tHjb$ and tWH) to the SM expectation. Theory and experimental uncertainties are retained on all event yields. The data is compared to the null hypothesis where $\mu_{t\bar{t}H} = 0$. The observed $t\bar{t}H$ significance is 5.2σ (compared to 4.4σ expected), and the measured signal strength is

$$\mu_{t\bar{t}H} = 1.43^{+0.39}_{-0.34} = 1.43^{+0.33}_{-0.31}(\text{stat.})^{+0.13}_{-0.11}(\text{exp.})^{+0.17}_{-0.09}(\text{th.}) \tag{12.8}$$

An upper limit on the cross section of tH production ($tHjb + tWH$) is derived by setting $\alpha = 0°$ and fixing the cross sections of all Higgs boson processes except

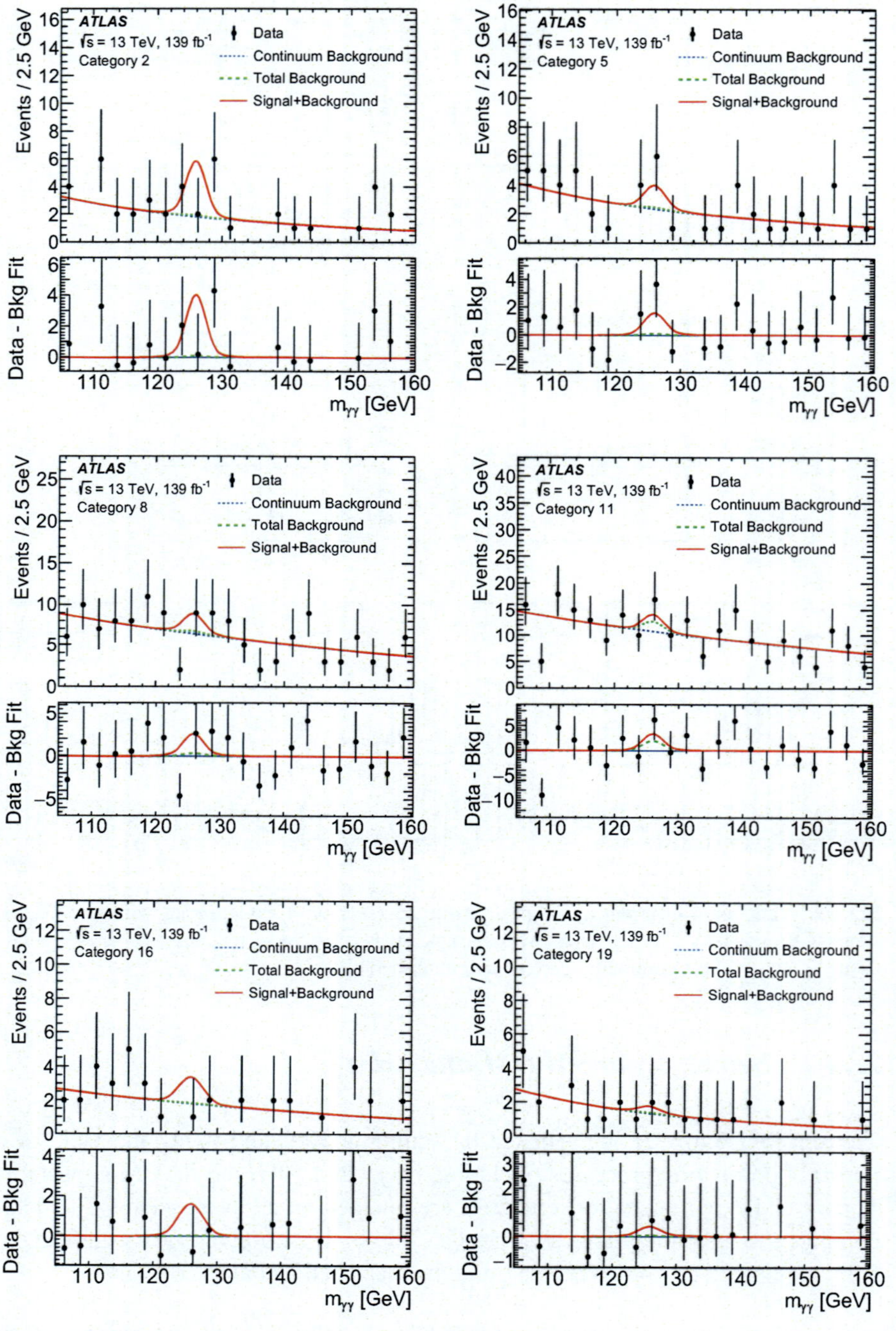

Fig. 12.10 Diphoton invariant mass spectrum in the $t\bar{t}H(\gamma\gamma)$ CP categories with intermediate CP BDT score. The signal-plus-background fit is overlaid in red, the total background in dashed green, and the continuum background in dashed blue

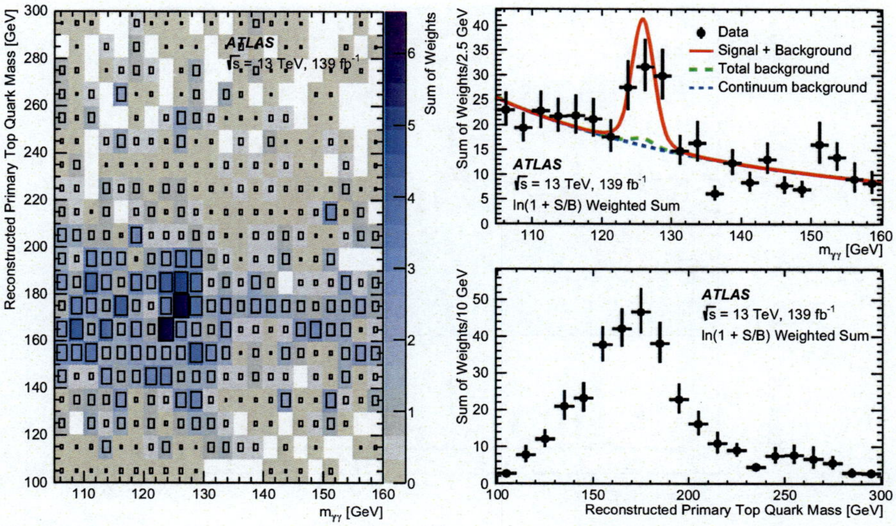

Fig. 12.11 Distribution of reconstructed top quark mass and reconstructed Higgs boson mass in the $t\bar{t}H(\gamma\gamma)$ CP analysis categories. The right panels show the projections onto the Higgs boson mass and top quark mass axes. In the upper panel, the fitted continuum background (blue), the total fitted background including non-$t\bar{t}H/tH$ Higgs production (green), and the signal-plus-background fit (red) are shown. All categories are shown, with data in each category weighted by $\ln(1 + S/B)$ of the category. The fitted curves in the projection are similarly weighted

$t\bar{t}H$, $tHjb$, and tWH to their SM expectations. A fit is performed where the tH and $t\bar{t}H$ signal strengths are left free.

The limit on the tH cross section is calculated using the CL_s procedure [5]. The observed upper limit on the tH cross section is 11.6 times the SM expectation (compared to 11.7 expected). At the time of writing, this limit is the strongest available on the tH process [6].

12.4.2 Constraints on CP Mixing

A fit is performed to the data in all twenty $t\bar{t}H(\gamma\gamma)$ CP categories where the CP mixing angle α and the coupling strength κ_t are left free, and the effective loop couplings κ_g and κ_γ are constrained by combined ATLAS Higgs data (Eq. 12.3). The fitted signal and background yields are shown in Fig. 12.12, calculated in the smallest $m_{\gamma\gamma}$ window containing 90% of the $t\bar{t}H + tH$ signal. The bottom panel compares the fitted signal and data minus the fitted background. Using the method described in [7], the compatibility between the observed data and the fit in Fig. 12.12 is found to have a p-value of 35%.

Table 12.5 shows the expected signal yield for three CP mixing hypotheses: Standard Model ($\alpha = 0°$, $\kappa_t = 1$), maximal CP mixing ($\alpha = 45°$, $\kappa_t = 1$), and fully CP

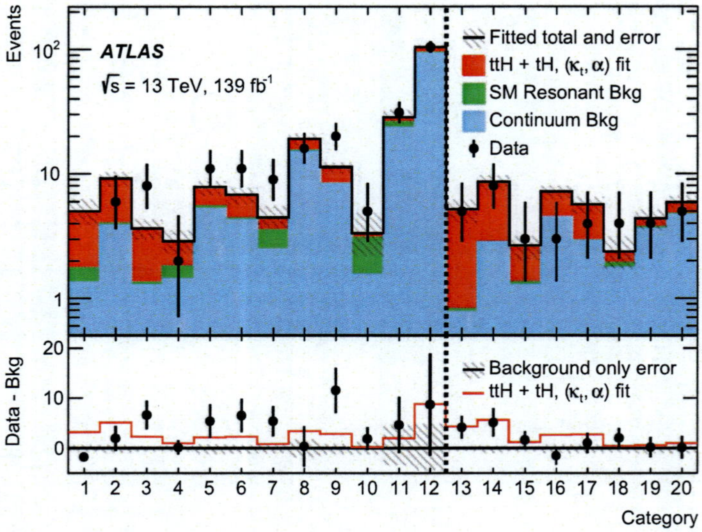

Fig. 12.12 Fitted signal and background yields in the $t\bar{t}H(\gamma\gamma)$ CP analysis categories, calculated in the smallest $m_{\gamma\gamma}$ window containing 90% of the $t\bar{t}H + tH$ signal. Note that the uncertainties in the hashed bands are correlated with the statistical error on the data. The bottom panel compares the fitted signal and data minus the fitted background

odd ($\alpha = 90°$, $\kappa_t = 1$). The expected background yields and observed data are also shown.

A comparison is performed between the data-minus-background and the extracted signal from different CP mixing angles. These fits are performed with α fixed and κ_t free. In order to obtain high enough statistics for a visual interpretation, categories are merged in to three bins: CP even-like (categories 3, 6, 9, 12, 14, 17, 20), CP odd-like (1, 4, 7, 10, 13, 16, 19), intermediate (2, 5, 8, 11, 15, 18). The results are shown in Fig. 12.13. The data clearly favor the $\alpha = 0°$ hypothesis over $\alpha = 90°$.

The two-dimensional likelihood scan of $\kappa_t \cos\alpha$ and $\kappa_t \sin\alpha$ is shown in Fig. 12.14a. The one-dimensional likelihood scan over CP mixing angle α (with the coupling strength κ_t profiled) is shown in Fig. 12.14b. Systematic uncertainties have negligible impact on the limit. Assuming that the SM (CP even) hypothesis is true, the expected 95% confidence limit is $|\alpha| > 63°$, and the observed data exclude $|\alpha| > 43°$ at 95% confidence level. The probability to observe a limit on α that is stronger than the expectation is 11% (evaluated on 500 toys).

The CP odd hypothesis ($\alpha = 90°$) is rejected at the 3.9σ level, and $\alpha = 180°$ is rejected at the 2.5σ level. The measured mixing angle (in degrees) is

$$\alpha = 0.0 \,^{+25.1}_{-21.2}\,(\text{stat.})\,^{+2.8}_{-1.7}\,(\text{syst.}) \tag{12.9}$$

Table 12.5 Observed and expected $t\bar{t}H$ and $tH = tHjb + tWH$ yields per $t\bar{t}H(\gamma\gamma)$ CP category, calculated in the smallest $m_{\gamma\gamma}$ window containing 90% of the fitted signal

Category	Expected $\kappa_t = 1$, $\alpha = 0°$		Expected $\kappa_t = 1$, $\alpha = 45°$		Expected $\kappa_t = 1$, $\alpha = 90°$		Resonant background	Continuum background	Data
	$t\bar{t}H$	tH	$t\bar{t}H$	tH	$t\bar{t}H$	tH			
$t\bar{t}H(\gamma\gamma)$ CP 1	2.1	0.14	2.3	0.55	2.5	1.67	0.40	1.4	0
$t\bar{t}H(\gamma\gamma)$ CP 2	3.5	0.067	2.4	0.12	1.3	0.377	0.12	3.9	6
$t\bar{t}H(\gamma\gamma)$ CP 3	1.6	0.007	0.93	0.015	0.21	0.040	0.060	1.3	8
$t\bar{t}H(\gamma\gamma)$ CP 4	0.63	0.12	0.72	0.28	0.79	0.85	0.37	1.5	2
$t\bar{t}H(\gamma\gamma)$ CP 5	1.5	0.063	1.1	0.091	0.63	0.28	0.23	5.4	11
$t\bar{t}H(\gamma\gamma)$ CP 6	1.6	0.026	0.95	0.023	0.26	0.071	0.10	4.4	11
$t\bar{t}H(\gamma\gamma)$ CP 7	0.51	0.12	0.60	0.39	0.70	1.2	1.1	2.5	9
$t\bar{t}H(\gamma\gamma)$ CP 8	2.3	0.12	1.7	0.22	1.04	0.59	0.42	15.2	16
$t\bar{t}H(\gamma\gamma)$ CP 9	1.9	0.046	1.14	0.045	0.31	0.11	0.11	8.3	20
$t\bar{t}H(\gamma\gamma)$ CP 10	0.094	0.064	0.12	0.30	0.15	1.0	1.5	1.6	5
$t\bar{t}H(\gamma\gamma)$ CP 11	1.1	0.38	1.2	0.80	1.22	2.3	2.6	23.7	31
$t\bar{t}H(\gamma\gamma)$ CP 12	5.8	0.39	3.7	0.52	1.72	1.4	1.8	93.5	104
$t\bar{t}H(\gamma\gamma)$ CP 13	2.9	0.15	2.7	0.42	2.55	1.2	0.042	0.79	5
$t\bar{t}H(\gamma\gamma)$ CP 14	3.9	0.052	2.4	0.073	0.86	0.17	0.013	2.9	8
$t\bar{t}H(\gamma\gamma)$ CP 15	0.73	0.19	0.75	0.46	0.78	1.3	0.075	1.3	3
$t\bar{t}H(\gamma\gamma)$ CP 16	1.7	0.13	1.3	0.17	0.76	0.50	–0.030	4.5	3
$t\bar{t}H(\gamma\gamma)$ CP 17	1.9	0.048	1.1	0.057	0.35	0.11	0.024	2.9	4
$t\bar{t}H(\gamma\gamma)$ CP 18	0.18	0.15	0.17	0.36	0.15	1.2	0.17	1.8	4
$t\bar{t}H(\gamma\gamma)$ CP 19	0.32	0.12	0.25	0.18	0.17	0.45	0.092	3.7	4
$t\bar{t}H(\gamma\gamma)$ CP 20	0.68	0.079	0.43	0.079	0.18	0.21	0.076	4.8	5

The expected yields assume $\kappa_t = 1$ for various α

12.4.2.1 Parametrize Loop Processes in Terms of κ_t and α

Assuming there is no new physics other than CP mixing in the Higgs-top interaction, the effective loop coupling strengths κ_g and κ_γ are directly parameterized as functions of κ_t and α using Eq. 12.4. This result is more model-dependent than the primary interpretation above. Here, the constraints from combined ATLAS Higgs data are not used, but only data in the twenty $t\bar{t}H(\gamma\gamma)$ CP analysis categories.

The two-dimensional likelihood scan of $\kappa_t \cos\alpha$ and $\kappa_t \sin\alpha$ is shown in Fig. 12.15a. The entire negative branch of $\kappa_t \cos\alpha$ is rejected at above 3σ. However, high values of κ_t, up to $\kappa_t \sim 6$, are allowed since no knowledge of the constraints from ggF and $H \to \gamma\gamma$ are applied in this interpretation. The region of allowed values contains a hole at $\kappa_t = 4.7$: near this value of κ_t, the $H \to \gamma\gamma$ branching fraction drops to zero (see Fig. 12.16), a scenario excluded by the observation of $\mu_{t\bar{t}H} > 0$ in the $t\bar{t}H(\gamma\gamma)$ CP analysis categories. The one-dimensional likelihood scan of α with κ_t profiled is shown in Figure 12.15b. Systematic uncertainties have negligible impact on the limit.

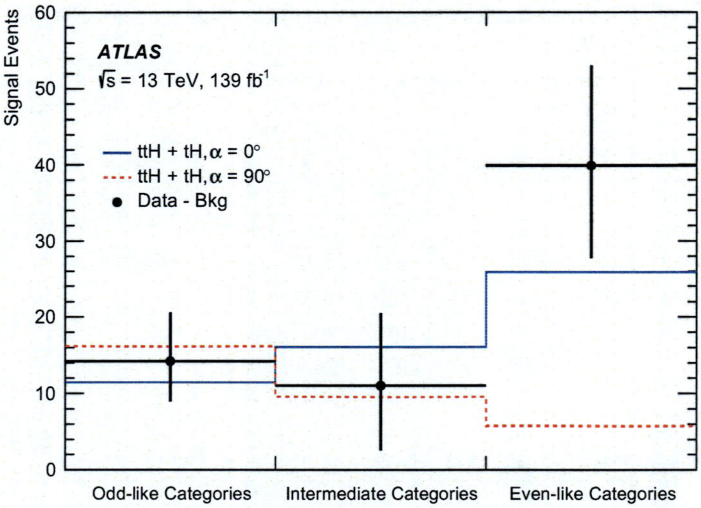

Fig. 12.13 Extracted $t\bar{t}H + tH$ signal yields in the $t\bar{t}H(\gamma\gamma)$ CP categories for $\alpha = 0°$ (red), $\alpha = 90°$ (blue), and in data. Yields are calculated in the smallest $m_{\gamma\gamma}$ interval containing 90% of the signal. The twenty categories are merged into three bins: CP odd-like, Intermediate, and CP even-like. $t\bar{t}H + tH$ signal yields are extracted fits performed with α fixed and κ_t free. The data is shown with fitted background subtracted

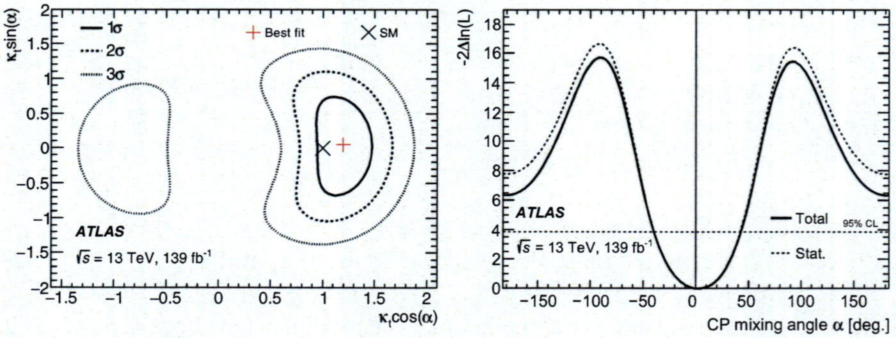

Fig. 12.14 (Left) The 1, 2, and 3σ exclusion contours in the ($\kappa_t \cos\alpha$, $\kappa_t \sin\alpha$) plane, calculated with κ_g and κ_γ constrained by ATLAS combined Higgs boson data. (Right) One-dimensional likelihood scan over the CP mixing angle α. The coupling strength κ_t is profiled

Assuming that the SM (CP even) hypothesis is true, the expected 95% confidence limit is $|\alpha| > 56°$. The observed data exclude $|\alpha| > 43°$ at 95% confidence level, and the CP odd hypothesis is rejected at the 4.0σ level. The measured mixing angle (in degrees) is

$$\alpha = 0.1^{+26.6}_{-23.2}(\text{stat.})^{+7.4}_{-5.2}(\text{syst.}) \tag{12.10}$$

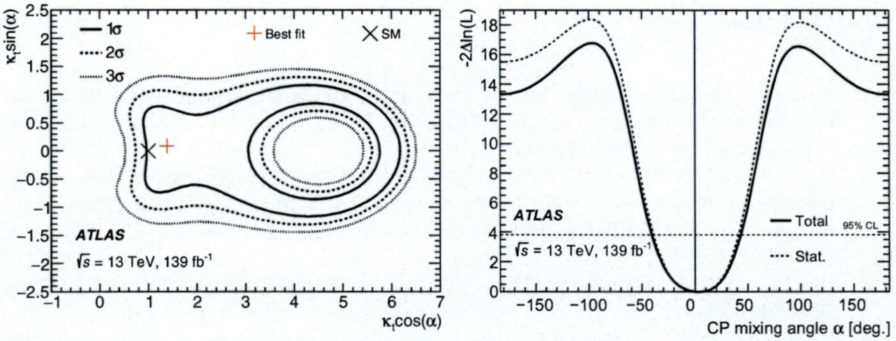

Fig. 12.15 (Left) The 1, 2, and 3σ exclusion contours in the ($\kappa_t \cos\alpha$, $\kappa_t \sin\alpha$) plane, with κ_g and κ_γ parameterized in terms of κ_t and α according to Eq. 12.4. (Right) One-dimensional likelihood scan over the CP mixing angle α. The coupling strength κ_t is profiled

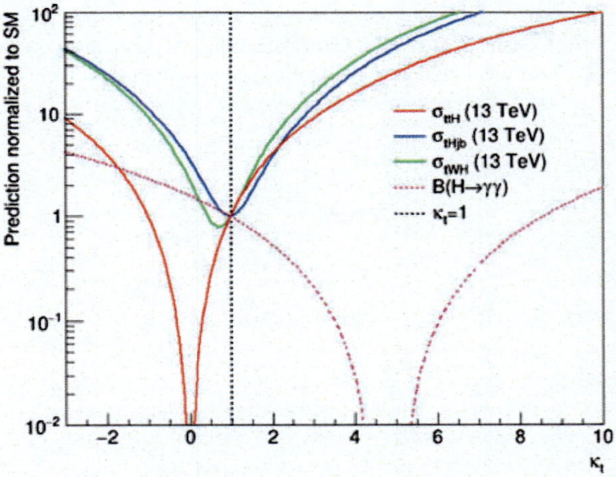

Fig. 12.16 The $H \to \gamma\gamma$ branching fraction and SM $t\bar{t}H$ and tH cross sections as a function of κ_t. The drop in the $H \to \gamma\gamma$ branching fraction corresponds to the inner excluded region in Fig. 12.15

The region $\kappa_t \cos\alpha > 1.5$ has been definitively excluded (> 95% CL) by ATLAS combined Higgs data [2], and an interpretation combining this full dataset with the parameterization in Eq. 12.4 would yield a stronger limit than either presented here. In particular, it would eliminate the region $\kappa_t \cos\alpha > 1.5$ from the second interpretation, which is not excluded by data in the $t\bar{t}H(\gamma\gamma)$ CP categories alone. However, this limit would be dominated by the statistical power of ggF (and $H \to \gamma\gamma$, to a lesser extent) rather than direct Higgs-top production. This analysis is therefore left to a future combination of ATLAS Higgs data.

References

1. Demartin F et al (2014) Higgs characterisation at NLO in QCD: CP properties of the top-quark Yukawa interaction. In: Eur. Phys. J. C74.9, p 3065. http://orcid.org/10.1140/epjc/s10052-014-3065-2. arXiv: 1407.5089 [hep-ph]
2. Aad G et al (2020) Combined measurements of Higgs boson production and decay using up to 80 fb-1 of proton-proton collision data at vs = 13 TeV collected with the ATLAS experiment. In: Phys. Rev. D 101.1, p 012002. https://doi.org/10.1103/PhysRevD.101.012002. arXiv: 1909.02845 [hep-ex]
3. Ellis J et al (2014) Disentangling Higgs-top couplings in associated production. In: JHEP 04, p 004. https://doi.org/10.1007/JHEP04(2014)004. arXiv: 1312.5736 [hep-ph]
4. Buckley A et al (2015) LHAPDF6: parton density access in the LHC precision era. In: Eur. Phys. J. C75, p 132. https://doi.org/10.1140/epjc/s10052-015-3318-8. arXiv: 1412.7420 [hep-ph]
5. Read AL (2002) Presentation of search results: theCLstechnique. J Phys G: Nuclear Particle Phys 28(10):2693–2704. https://doi.org/10.1088/0954-3899/28/10/313
6. Sirunyan AM et al (2019) Search for associated production of a Higgs boson and a single top quark in proton-proton collisions at vs = TeV. In: Phys. Rev. D 99.9, p 092005. https://doi.org/10.1103/PhysRevD.99.092005. arXiv: 1811.09696 [hep-ex]
7. Cousins RD (2013) Generalization of Chisquare goodness-of-fit test for binned data using saturated models, with application to histograms. Tech. rep. UCLA, 2013. http://www.physics.ucla.edu/~cousins/stats/cousins_saturated.pdf4

Chapter 13
Higgs Couplings in $H \to \gamma\gamma$

This work has focused thus far on the interaction between the Higgs boson and the top quark, κ_t, in the Higgs to diphoton decay channel. However, the $H \to \gamma\gamma$ decay provides powerful sensitivity to many Higgs boson couplings in addition to κ_t. This Chapter presents the full suite of Higgs coupling measurements obtained by targeting all major Higgs boson production modes, including $t\bar{t}H$, with 79.8 fb^{-1} of 13 TeV ATLAS data. Section 13.1 describes the selection of diphoton events into categories targeting gluon fusion, vector boson fusion, and vector boson associated production. Section 13.2 details the parameterization of the signal and background $m_{\gamma\gamma}$ shapes, and Sect. 13.3 reviews the dominant systematic uncertainties. Results are presented in Sect. 13.4.

The inclusive Higgs boson cross section is measured, as well as the individual cross sections for four Higgs boson production modes: ggF, VBF, VH, and $t\bar{t}H + tH$. In addition, the rate of $H \to \gamma\gamma$ events is measured in the **Simplified Template Cross Section (STXS)** scheme [1, 2], which gives a coarse differential binning in kinematic variables like jet multiplicity, Higgs candidate p_T, and jet p_T. The STXS regions targeted are shown on the left-hand side of Fig. 13.1. The boxes denote STXS region measured, including merging of categories due to low statistics.

The measurements reported in this Chapter were supplied as input to the ATLAS Higgs combination [3], which included results from analyses targeting all Higgs boson production and decay modes.

13.1 Event Categorization

All events included in the $H \to \gamma\gamma$ categories are required to have two photons passing the identification and isolation criteria outlined in Sect. 4.2. The categories are designed to target different Higgs boson production modes, and the categories

J. E. Dickinson, *ATLAS Measurements of the Higgs Boson Coupling to the Top Quark in the Higgs to Diphoton Decay Channel*, Springer Theses,
https://doi.org/10.1007/978-3-030-86368-5_13

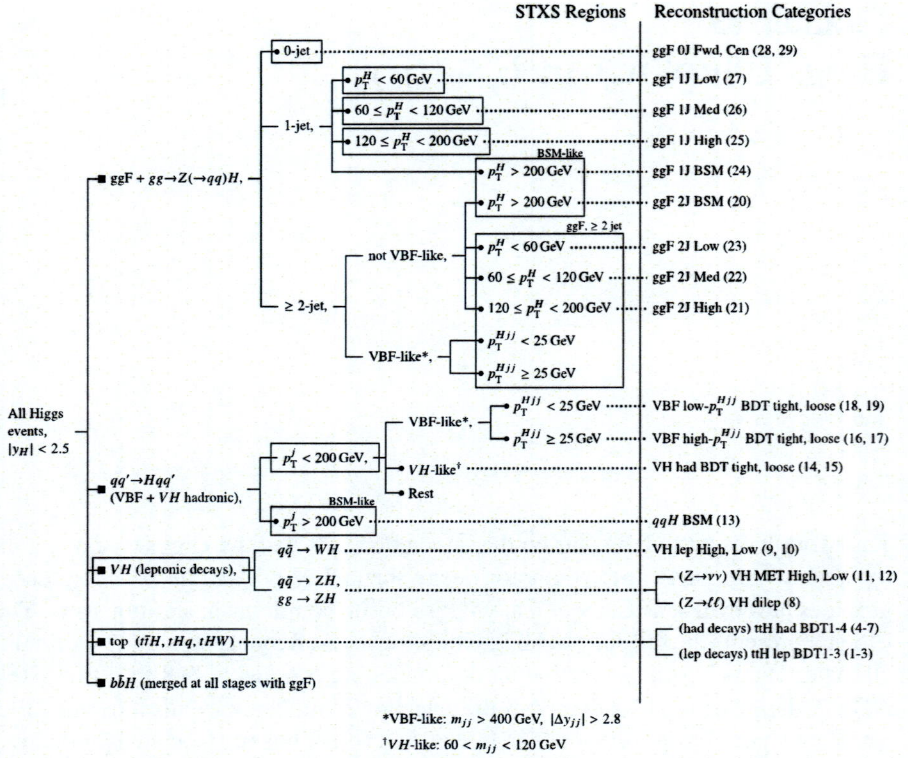

Fig. 13.1 Correspondence between Simplified Template Cross Section regions (left) and $H \to \gamma\gamma$ analysis categories (right)

targeting the rarest processes are filled first. Events failing selection for one production mode are considered for the next rarest mode. In total, the $H \to \gamma\gamma$ analysis includes:

1. Seven categories targeting $t\bar{t}H$
2. Five categories targeting leptonic VH
3. Seven categories targeting VBF and hadronic VH
4. Ten categories targeting gluon fusion

The seven $t\bar{t}H$ categories are those described in Chap. 9. The remaining categories are described in the following sub-sections.

Table 13.1 summarizes all 29 categories and their definitions. The STXS regions corresponding to each category are shown in Fig. 13.1, and the contribution from each STXS bin to the Higgs signal in each category is shown in Fig. 13.2. Figure 13.3 shows the contribution from each production mode to the total Higgs signal in each category.

Table 13.1 Summary of the 29 $H \to \gamma\gamma$ analysis categories

Category label	Selection
ttH lep BDT1	$N_{\text{lep}} \geq 1,\ N_{b\text{-jet}} \geq 1,\ \text{BDT}_{\text{ttHlep}} > 0.987$
ttH lep BDT2	$N_{\text{lep}} \geq 1,\ N_{b\text{-jet}} \geq 1,\ 0.942 < \text{BDT}_{\text{ttHlep}} < 0.987$
ttH lep BDT3	$N_{\text{lep}} \geq 1,\ N_{b\text{-jet}} \geq 1,\ 0.705 < \text{BDT}_{\text{ttHlep}} < 0.942$
ttH had BDT1	$N_{\text{lep}} = 0,\ N_{\text{jets}} \geq 3,\ N_{b\text{-jet}} \geq 1,\ \text{BDT}_{\text{ttHhad}} > 0.996$
ttH had BDT2	$N_{\text{lep}} = 0,\ N_{\text{jets}} \geq 3,\ N_{b\text{-jet}} \geq 1,\ 0.991 < \text{BDT}_{\text{ttHhad}} < 0.996$
ttH had BDT3	$N_{\text{lep}} = 0,\ N_{\text{jets}} \geq 3,\ N_{b\text{-jet}} \geq 1,\ 0.971 < \text{BDT}_{\text{ttHhad}} < 0.991$
ttH had BDT4	$N_{\text{lep}} = 0,\ N_{\text{jets}} \geq 3,\ N_{b\text{-jet}} \geq 1,\ 0.911 < \text{BDT}_{\text{ttHhad}} < 0.971$
VH dilep	$N_{\text{lep}} \geq 2,\ 70\,\text{GeV} \leq m_{\ell\ell} \leq 110\,\text{GeV}$
VH lep High	$N_{\text{lep}} = 1,\ \lvert m_{e\gamma} - 89\,\text{GeV}\rvert > 5\,\text{GeV},\ p_{\text{T}}^{\ell+E_{\text{T}}^{\text{miss}}} > 150\,\text{GeV}$
VH lep Low	$N_{\text{lep}} = 1,\ \lvert m_{e\gamma} - 89\,\text{GeV}\rvert > 5\,\text{GeV},\ p_{\text{T}}^{\ell+E_{\text{T}}^{\text{miss}}} < 150\,\text{GeV},\ E_{\text{T}}^{\text{miss}}$ significance > 1
VH MET High	$150\,\text{GeV} < E_{\text{T}}^{\text{miss}} < 250\,\text{GeV},\ E_{\text{T}}^{\text{miss}}$ significance > 9 or $E_{\text{T}}^{\text{miss}} > 250\,\text{GeV}$
VH MET Low	$80\,\text{GeV} < E_{\text{T}}^{\text{miss}} < 150\,\text{GeV},\ E_{\text{T}}^{\text{miss}}$ significance > 8
qqH BSM	$N_{\text{jets}} \geq 2,\ p_{\text{T,j1}} > 200\,\text{GeV}$
VH had BDT tight	$60\,\text{GeV} < m_{\text{jj}} < 120\,\text{GeV},\ \ \text{BDT}_{\text{VH}} > 0.78$
VH had BDT loose	$60\,\text{GeV} < m_{\text{jj}} < 120\,\text{GeV},\ 0.35 < \text{BDT}_{\text{VH}} < 0.78$
VBF high-p_{T}^{Hjj} BDT tight	$\lvert\Delta\eta_{jj}\rvert > 2,\ \lvert\eta_{\gamma\gamma} - 0.5(\eta_{\text{j1}} + \eta_{\text{j2}})\rvert < 5,\ p_{\text{T}}^{Hjj} > 25\,\text{GeV},\ \ \text{BDT}_{\text{VBF}}^{\text{high}} > 0.47$
VBF high-p_{T}^{Hjj} BDT loose	$\lvert\Delta\eta_{jj}\rvert > 2,\ \lvert\eta_{\gamma\gamma} - 0.5(\eta_{\text{j1}} + \eta_{\text{j2}})\rvert < 5,\ p_{\text{T}}^{Hjj} > 25\,\text{GeV},\ -0.32 < \text{BDT}_{\text{VBF}}^{\text{high}} < 0.47$
VBF low-p_{T}^{Hjj} BDT tight	$\lvert\Delta\eta_{jj}\rvert > 2,\ \lvert\eta_{\gamma\gamma} - 0.5(\eta_{\text{j1}} + \eta_{\text{j2}})\rvert < 5,\ p_{\text{T}}^{Hjj} < 25\,\text{GeV},\ \ \text{BDT}_{\text{VBF}}^{\text{low}} > 0.87$
VBF low-p_{T}^{Hjj} BDT loose	$\lvert\Delta\eta_{jj}\rvert > 2,\ \lvert\eta_{\gamma\gamma} - 0.5(\eta_{\text{j1}} + \eta_{\text{j2}})\rvert < 5,\ p_{\text{T}}^{Hjj} < 25\,\text{GeV},\ 0.26 < \text{BDT}_{\text{VBF}}^{\text{low}} < 0.87$
ggF 2J BSM	$N_{\text{jets}} \geq 2,\ p_{\text{T}}^{\gamma\gamma} \geq 200\,\text{GeV}$
ggF 2J High	$N_{\text{jets}} \geq 2,\ p_{\text{T}}^{\gamma\gamma} \in [120, 200]$ GeV
ggF 2J Med	$N_{\text{jets}} \geq 2,\ p_{\text{T}}^{\gamma\gamma} \in [60, 120]$ GeV
ggF 2J Low	$N_{\text{jets}} \geq 2,\ p_{\text{T}}^{\gamma\gamma} \in [0, 60]$ GeV
ggF 1J BSM	$N_{\text{jets}} = 1,\ p_{\text{T}}^{\gamma\gamma} \geq 200\,\text{GeV}$
ggF 1J High	$N_{\text{jets}} = 1,\ p_{\text{T}}^{\gamma\gamma} \in [120, 200]$ GeV
ggF 1J Med	$N_{\text{jets}} = 1,\ p_{\text{T}}^{\gamma\gamma} \in [60, 120]$ GeV
ggF 1J Low	$N_{\text{jets}} = 1,\ p_{\text{T}}^{\gamma\gamma} \in [0, 60]$ GeV
ggF 0J Fwd	$N_{\text{jets}} = 0$, one photon with $\lvert\eta\rvert > 0.95$
ggF 0J Cen	$N_{\text{jets}} = 0$, two photons with $\lvert\eta\rvert \leq 0.95$

13.1.1 Leptonic VH

The VH dilepton category targets $qq/qg \to ZH$ production where the Z boson decays to charged leptons: events in this category must contain two same-flavor opposite-sign leptons with dilepton invariant mass $m_{\ell\ell}$ between 70 and 110 GeV.

The VH single lepton categories target $qq/qg \to WH$ production where the W boson decays to a charged lepton (e or μ) and a neutrino. Events in these categories must contain exactly one electron or muon. Events where the electron and either of the two photons have $m_{e\gamma} \in [84, 94]$ GeV are vetoed. Two categories are defined by dividing events by the missing E_T significance and the transverse momentum of the lepton plus missing E_T system: low ($p_T^{\ell} + E_T^{\text{miss}} < 150$ GeV and E_T^{miss} significance > 1.0) and high ($p_T^{\ell} + E_T^{\text{miss}} > 150$ GeV).

Events with zero leptons (e or μ) fall into the VH MET categories if they contain high missing E_T. These categories target $qq/qg \to ZH$ production where the Z

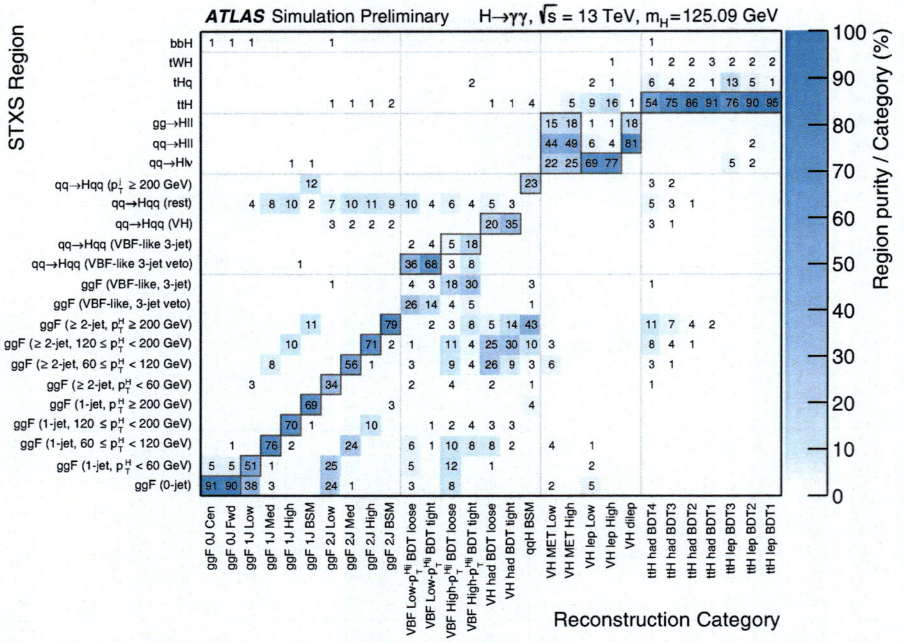

Fig. 13.2 Breakdown of expected Higgs signal by STXS process (y-axis) in each analysis category (x-axis). The regions bounded in light gray contain categories corresponding to the same Higgs production mode. The entries outlined in dark gray indicate the target process(es) corresponding to each category. The categories are generally very pure in the target process, though there is some contamination (e.g. between ggF and VBF categories)

boson decays to neutrinos. Two such categories are defined: low MET (80 GeV $< E_T^{\text{miss}} <$ 150 GeV and E_T^{miss} significance $>$ 8.0), or high MET ($E_T^{\text{miss}} >$ 150 GeV and E_T^{miss} significance $>$ 9.0, or $E_T^{\text{miss}} >$ 250 GeV).

13.1.2 Vector Boson Fusion and Hadronic VH

The qqH BSM category contains events with at least two jets and very high leading jet p_T ($>$ 200 GeV).

Hadronic VH events target WH and ZH production where the vector boson decays hadronically. Events in these categories must contain two jets with dijet invariant mass 60 GeV $< m_{jj} <$ 120 GeV. A BDT is trained to separate simulated VH events from a background sample constructed from simulated diphoton background and data-driven estimates of photon + jet and dijet backgrounds from NTI data. The BDT uses training variables $\Delta\phi$ and Δy between the diphoton and dijet systems, $\cos\theta^*_{\gamma\gamma,jj}$, dijet invariant mass, diphoton p_T orthogonal to the diphoton thrust axis in the transverse plane, p_T of the diphoton system, $H_T + E_T^{\gamma 1} + E_T^{\gamma 2}$, minimum

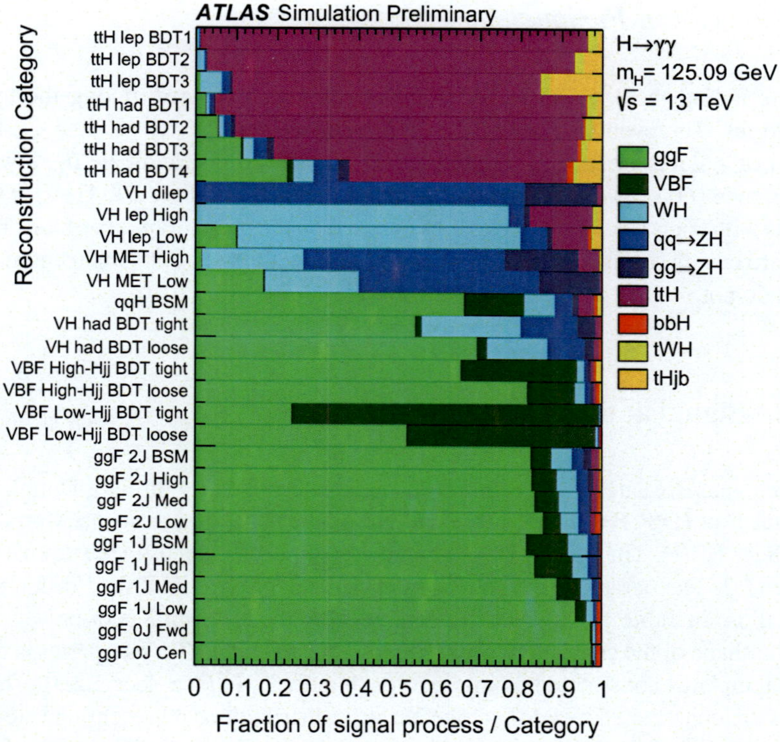

Fig. 13.3 Breakdown of expected Higgs signal by production mode in each analysis category. The categories are generally very pure in the target process, though there is some contamination (e.g. between ggF and VBF categories)

ΔR between a photon and a jet, the dijet transverse momentum, and Zeppenfeld η ($|\eta_{\gamma\gamma} - \frac{1}{2}(\eta_{j1} + \eta_{j2})|$). Using the BDT output as the discriminating variable, events are classified into two categories: tight (BDT score > 0.78) and loose (BDT score < 0.78).

In the four VBF categories, events are required to contain at least two jets. The two jets with highest p_T must be separated by a pseudorapidity difference $|\Delta\eta(j1, j2)| > 2.0$. In addition, events in the VBF categories must satisfy $|\eta_H - 0.5(\eta_{j1} + \eta_{j2})| < 5.0$, where η_H is the pseudorapidity of the Higgs candidate (diphoton system).

Another BDT is trained to separate simulated VBF events from a background sample constructed from simulated diphoton background and data-driven estimates of photon + jet and dijet backgrounds from NTI data. The BDT uses six training variables: dijet invariant mass, $\Delta\eta$ between the leading jets, $\Delta\phi$ between the diphoton and dijet systems, diphoton p_T orthogonal to the diphoton thrust axis in the transverse plane, minimum ΔR between a photon and a jet, and Zeppenfeld η. Four exclusive regions are defined based on the BDT discriminant and the transverse momentum of system composed of the Higgs candidate and two leading jets (p_T^{Hj1j2}). The cut values are listed in Table 13.1.

13.1.3 Gluon Fusion

Events failing the $t\bar{t}H$, VH, and VBF selection are divided among the ten ggF categories. The events are divided into bins of jet multiplicity (0, 1, ≥ 2). Events with 1 or ≥ 2 jets are further divided into four bins in Higgs candidate p_T: low (0–60 GeV), med (60–120 GeV), high (120–200 GeV), and BSM (> 200 GeV). The ggF events with 0 jets are split into two bins in based on the pseudorapidity of the photons: forward events (fwd) contain at least one photon with $|\eta| > 0.95$, and central events (cen) do not.

13.2 Signal and Background

In each analysis category, the signal shape is determined by fitting a Double-Sided Crystal Ball (DSCB) function to a sample of inclusive Higgs boson Monte Carlo (see Sect. 6.2.1). The best fit DSCB parameters in each category are summarized in Table 13.2. The parameter values for the $t\bar{t}H$ categories at 79.8 fb^{-1} differ slightly from those in Table 10.2, which are derived from the full Run 2 dataset.

The shape of the continuum background in the ggF and VBF categories is derived from templates constructed from diphoton Monte Carlo (see Sect. 5.2.3). The templates are constructed by adding together $m_{\gamma\gamma}$ distributions representing true diphoton ($\gamma\gamma$) events, single photon + jet (γj), and dijet (jj) events. The shape of the γj distribution is modeled using the data sidebands where exactly one photon fails the identification and/or isolation criteria. The shape of the jj distribution is modeled using the data sidebands where both photons fail the identification and/or isolation criteria. The relative fraction of $\gamma\gamma$, γj, and jj is calculated by a template fit of these three distributions to the TI data sidebands.

In the VH categories, background templates are constructed from $V + \gamma\gamma$ Monte Carlo generated with Sherpa following the setup used for the diphoton background sample described in Sect. 5.2.3. The Monte Carlo templates are normalized so that the number of sideband events is equal to that in the tight/isolated data. Statistics in the Monte Carlo sample are low in some VH categories, resulting in large spurious signal.

The background templates used for the $t\bar{t}H$ categories are described in Sect. 10.2.1. The functional form and spurious signal uncertainty in the $t\bar{t}H$ categories at 79.8 fb^{-1} differ slightly from those in Table 10.3, which are derived from the full Run 2 dataset.

The functional form of the continuum background and the spurious signal are listed in Table 13.3 for all the $H \rightarrow \gamma\gamma$ categories. The χ^2/ndof indicates the performance of a background only fit to the template in the $m_{\gamma\gamma}$ range 105–160 GeV, but is in general not a part of the spurious signal criteria.

Table 13.2 Best fit parameter values for the signal $m_{\gamma\gamma}$ shape in each $H \rightarrow \gamma\gamma$ category at 79.8 fb^{-1}

Category	μ_{CB} [GeV]	σ_{CB} [GeV]	α_{low}	α_{high}	n_{low}	n_{high}
ggH 0J cen	125.22	1.60	1.6	1.9	808.4	4.1
ggH 0J fwd	125.17	2.00	1.5	1.7	865.2	5.9
ggH 1J low	125.22	1.81	1.5	1.7	687.7	5.5
ggH 1J med	125.19	1.72	1.4	1.7	371.7	5.1
ggH 1J high	215.19	1.54	1.4	1.7	45.0	5.5
ggH 1J BSM	125.15	1.30	1.5	1.6	17.5	5.5
ggH 2J low	125.28	1.84	1.4	1.7	426.4	5.0
ggH 2J med	125.23	1.73	1.5	1.7	34.8	5.0
ggH 2J high	125.18	1.55	1.5	1.8	30.7	4.9
ggH 2J BSM	125.17	1.37	1.5	1.7	20.7	5.8
VBF low p_T^{Hjj}, loose	125.19	1.74	1.5	1.7	35.2	4.8
VBF low p_T^{Hjj}, tight	125.17	1.54	1.5	1.6	33.7	6.4
VBF high p_T^{Hjj}, loose	125.26	1.72	1.4	1.7	26.3	4.7
VBF high p_T^{Hjj}, tight	125.24	1.57	1.5	1.7	26.2	5.4
VH had loose	125.20	1.64	1.5	1.7	25.1	4.8
VH had tight	125.19	1.46	1.5	1.7	22.3	5.2
qqH BSM	125.15	1.33	1.5	1.6	17.6	5.9
VH MET low	125.19	1.70	1.5	1.5	6.1	5.4
VH MET high	125.19	1.48	1.6	1.7	11.0	6.1
VH lep low	125.19	1.75	1.5	1.7	23.1	4.7
VH lep high	125.14	1.38	1.4	1.6	13.8	5.0
VH dilep	125.20	1.59	1.8	1.7	6.0	4.2
ttH had 4	125.27	1.63	1.9	1.9	3.8	3.4
ttH had 3	125.27	1.59	2.0	1.8	3.8	4.3
ttH had 2	125.27	1.46	1.6	1.6	5.5	4.9
ttH had 1	125.24	1.32	1.7	1.7	6.5	4.5
ttH lep 3	125.26	1.73	1.9	1.7	3.6	4.9
ttH lep 2	125.24	1.68	1.9	1.8	4.2	4.4
ttH lep 1	125.22	1.45	1.7	1.6	5.3	4.9

13.3 Systematic Uncertainties

The experimental uncertainties included in the $H \rightarrow \gamma\gamma$ couplings measurement are those summarized in Sect. 6.4.2.

For each Higgs signal process, QCD scale variations and PDF uncertainties (31 eigenvector scheme) are evaluated separately in each category. The heavy flavor uncertainty in the $t\bar{t}H$ categories is retained, and UEPS uncertainties are evaluated on each Higgs production mode.

Table 13.3 Results of the spurious signal test for the $H \to \gamma\gamma$ categories at 79.8 fb^{-1}

Category	Function	N_{sp}	χ^2/ndof	Prob(χ^2) [%]
ggH 0J cen	ExpPoly2	−48.3	1.12	25.1
ggH 0J fwd	ExpPoly2	−228	1.1	29.6
ggH 1J low	ExpPoly2	+45.9	1.18	17.5
ggH 1J med	ExpPoly2	+40.7	1.59	0.445
ggH 1J high	PowerLaw	−9.79	1.03	41.6
ggH 1J BSM	Exponential	−2.63	1.23	12.1
ggH 2J low	ExpPoly2	−38.1	0.957	56.3
ggH 2J med	ExpPoly2	+21.7	1	47.1
ggH 2J high	PowerLaw	+7.43	1.28	8.27
ggH 2J BSM	Exponential	+0.426	1.11	26.8
VBF low p_T^{Hjj}, loose	PowerLaw	+10.3	1.05	38.2
VBF low p_T^{Hjj}, tight	Exponential	+5.16	0.922	63.7
VBF high p_T^{Hjj}, loose	PowerLaw	−12.2	0.895	69
VBF high p_T^{Hjj}, tight	PowerLaw	−17.4	1.37	3.75
VH had loose	PowerLaw	+6.45	1	47.3
VH had tight	Exponential	−1.81	1.07	33.1
qqH BSM	Exponential	−11.7	1.02	43.8
VH MET low	Exponential	+2.45	2.61	0.001
VH MET high	PowerLaw	+1.94	0.989	47.9
VH lep low	Exponential	−4.67	1	46.4
VH lep high	Exponential	−0.598	1.56	3.52
VH dilep	PowerLaw	+0.608	1.51	4.53
ttH had 4	Power Law	+0.96	0.98	
ttH had 3	Power Law	+0.63	0.88	
ttH had 2	Exponential	+0.36	0.72	
ttH had 1	Power Law	+0.11	0.86	
ttH lep 3	Exponential	+0.24	1.26	
ttH lep 2	Power Law	+0.33	0.79	
ttH lep 1	Power Law	+0.12	0.97	

The χ^2/ndof indicates the performance of a background only fit to the template in the $m_{\gamma\gamma}$ range 105–160 GeV, but is in general not a part of the spurious signal criteria

For the measurement of the ggF cross section, the UEPS uncertainty on ggF and the spurious signal in the 0-jet ggF categories dominate the systematic uncertainty, along with the photon identification efficiency. For the VBF cross section, the UEPS uncertainty on VBF and the experimental jet uncertainties are the dominant systematics. The dominant uncertainties for the measurement of the VH cross section are the spurious signal in the leptonic VH categories.

13.4 Results (79.8 fb^{-1})

The inclusive cross section is extracted from the fit with a single POI μ. For the inclusive fit, the number of Higgs boson signal events can be expressed as

$$N_S^c = \mu \times \sum_p \left[N_p^{SM,c}(1 + \sigma_L \theta_L) \prod_i (1 + \sigma_{\epsilon,p}^{i,c} \theta_\epsilon^i) \prod_j (1 + \sigma_{T,p}^{j,c} \theta_T^j) \right] \quad (13.1)$$

where the systematic uncertainty magnitudes σ and nuisance parameters θ have the same definitions given in Chap. 6.

The measured inclusive signal strength is

$$\mu = 1.06\ ^{+0.13}_{-0.12} = 1.06\ \pm 0.08\ \text{(stat)}\ ^{+0.10}_{-0.09}\ \text{(syst)} \quad (13.2)$$

This is in good agreement with the SM prediction of $1.00^{+0.13}_{-0.12}$. The likelihood scan for the inclusive cross section is shown in Fig. 13.4.

Figure 13.5 shows the weighted diphoton invariant mass spectrum summed over all 29 $H \to \gamma\gamma$ analysis categories. In order to visually enhance the contribution of the categories targeting each production mode, the data and fit in each category are weighted by $\ln(1 + S/B)$ where S (B) is the expected signal (background) in the smallest $m_{\gamma\gamma}$ window containing 90% of the Higgs signal. The signal here is taken to be all Higgs boson events, and the calculation of error bars in Fig. 13.5 follows the prescription outlined in Sect. 6.5.

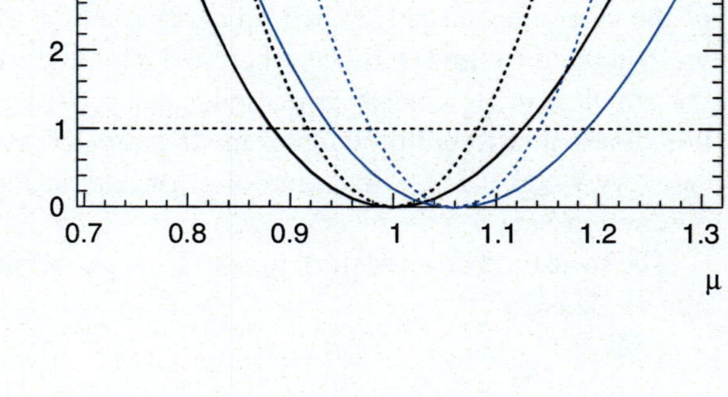

Fig. 13.4 Likelihood scan of the inclusive Higgs signal strength measured in $H \to \gamma\gamma$ with 79.8 fb^{-1}

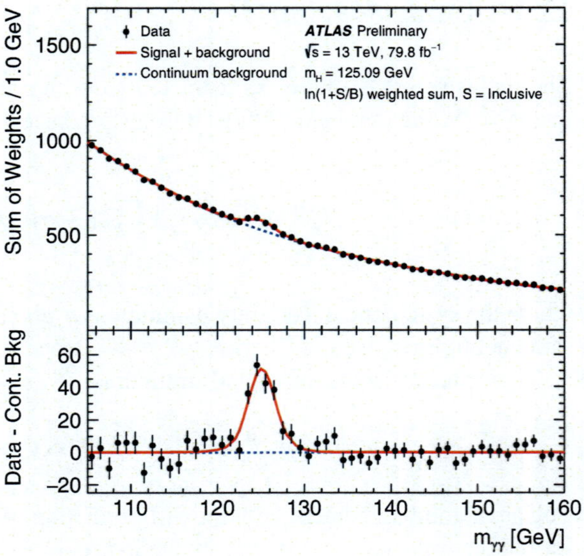

Fig. 13.5 Weighted diphoton invariant mass spectrum of data from all 29 $H \rightarrow \gamma\gamma$ analysis categories with 79.8 fb^{-1}. Each data event is weighted by $\ln(1 + S_c/B_c)$, where c is the category containing the event, and S_c and B_c are the fitted Higgs signal and continuum background in category c, respectively

13.4.1 Higgs Cross Section by Production Mode

The cross sections of individual Higgs boson production modes (p = ggF + $b\bar{b}H$, VBF, VH, $t\bar{t}H + tH$) are fitted simultaneously with POIs μ_p corresponding to each process:

$$N_S^c = \sum_p \left[\mu_p \times N_p^{SM,c}(1 + \sigma_L \theta_L) \prod_i (1 + \sigma_{\epsilon,p}^{i,c} \theta_\epsilon^i) \prod_j (1 + \sigma_{T,p}^{j,c} \theta_T^j) \right] \tag{13.3}$$

where the systematic uncertainty magnitudes σ and nuisance parameters θ have the same definitions given in Chap. 6.

Figure 13.6 shows the weighted diphoton invariant mass spectrum summed over all 29 $H \rightarrow \gamma\gamma$ analysis categories. In order to visually enhance the contribution of the categories targeting each production mode, the data and fit in each category are weighted by $\ln(1 + S/B)$ where S (B) is the expected signal (background) in the smallest $m_{\gamma\gamma}$ window containing 90% of the Higgs signal. In each panel of this figure, a different production mode is treated as signal in the calculation of the category weight. The calculation of error bars in Fig. 13.6 follows the prescription outlined in Sect. 6.5.

The measured cross section times $H \rightarrow \gamma\gamma$ branching fraction for each production mode is:

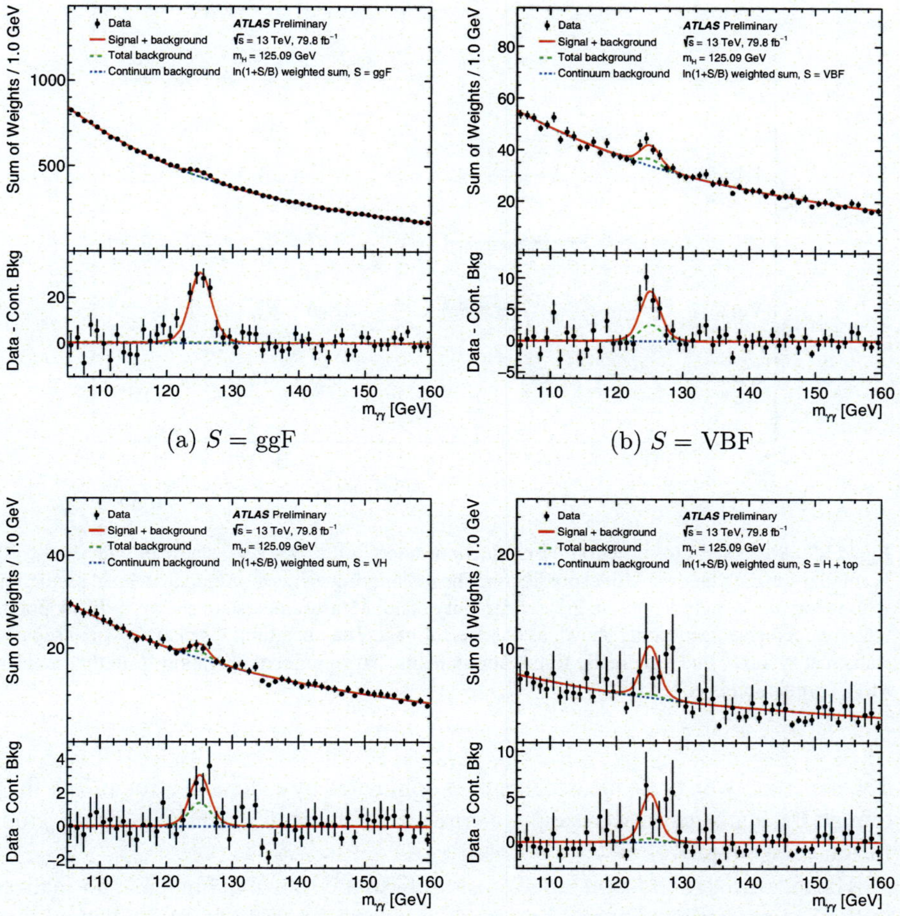

(a) $S = \text{ggF}$ (b) $S = \text{VBF}$

(c) $S = VH$ (d) $S = t\bar{t}H + tH$

Fig. 13.6 Weighted diphoton invariant mass spectrum of data from all 29 $H \to \gamma\gamma$ analysis categories with 79.8 fb^{-1}, with different Higgs boson processes treated as signal. Each data event is weighted by $\ln(1 + S_c/B_c)$, where c is the category containing the event. In subfigures (**a**), (**b**), (**c**), and (**d**), S_c corresponds to the fitted ggF, VBF, VH, and $t\bar{t}H + tH$ signal, respectively. The background B_c in each case is corresponds to the fitted continuum background plus the fitted Higgs background from non-signal processes

$$\sigma_{ggF} \times B_{\gamma\gamma} = 98.28\ ^{+14.71}_{-14.08}\ \text{fb} = 98.28\ \pm 11.18\ (\text{stat})\ ^{+9.15}_{-8.13}\ (\text{syst})\ \text{fb} \tag{13.4}$$

$$\sigma_{VBF} \times B_{\gamma\gamma} = 11.15\ ^{+3.38}_{-2.92}\ \text{fb} = 11.15\ ^{+2.56}_{-2.40}\ (\text{stat})\ ^{+2.24}_{-1.60}\ (\text{syst})\ \text{fb} \tag{13.5}$$

$$\sigma_{VH} \times B_{\gamma\gamma} = 4.84\ ^{+2.66}_{-2.47}\ \text{fb} = 4.84\ ^{+2.40}_{-2.26}\ (\text{stat})\ ^{+1.13}_{-1.00}\ (\text{syst})\ \text{fb} \tag{13.6}$$

$$\sigma_{t\bar{t}H+tH} \times B_{\gamma\gamma} = 1.50\ ^{+0.56}_{-0.48}\ \text{fb} = 1.50\ ^{+0.49}_{-0.44}\ (\text{stat})\ ^{+0.27}_{-0.19}\ (\text{syst})\ \text{fb} \tag{13.7}$$

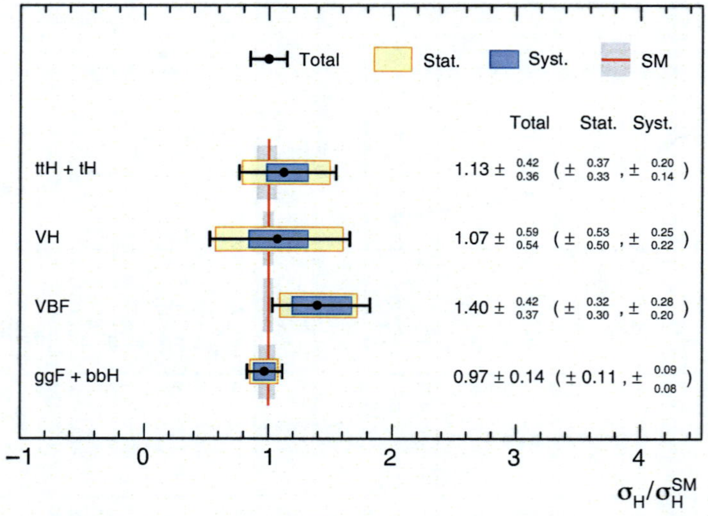

Fig. 13.7 Measured cross section by production mode in the $H \to \gamma\gamma$ channel with 79.8 fb^{-1}, normalized to the Standard Model prediction. The red line represents the Standard Model prediction, with theory error included in the gray uncertainty band. Measurements are shown as black points, with the inclusive uncertainty shown as black error bars. The uncertainty is also shown in separate statistical (yellow) and systematic (blue) components. No significant deviation from the Standard Model prediction is observed

The uncertainty on these measurements is dominated by statistical error at 79.8 fb^{-1}, though the statistical and systematic errors are similar in magnitude for the gluon fusion cross section.

The measured production mode cross sections are shown in Fig. 13.7, normalized to the SM prediction. The red line represents the Standard Model prediction, with theory error included in the gray uncertainty band. Measurements are shown as black points, with the inclusive uncertainty shown as black error bars. The uncertainty is also shown in separate statistical (yellow) and systematic (blue) components. The likelihood scan for each cross section measurement is shown in Fig. 13.8. No significant deviation from the Standard Model is observed.

The measured correlations between the production mode cross sections are displayed in Fig. 13.9. The correlations are overall small, with the correlation between the ggF and VBF modes being the largest.

The expected and observed significance of each production mode is reported in Table 13.4.

In addition to direct measurements of the production mode cross sections, ratios of the production mode cross sections to the ggF cross section are calculated. This allows for the cancellation of some common systematic uncertainties. The measured cross section ratios are shown in Fig. 13.10, normalized to the SM expectation. The red line represents the Standard Model prediction, with theory error included in the gray uncertainty band. Measurements are shown as black points, with the inclusive

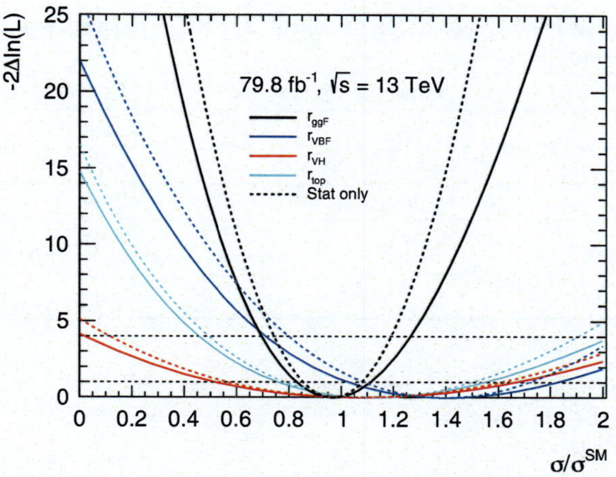

Fig. 13.8 Likelihood scan of Higgs production cross sections in the $H \rightarrow \gamma\gamma$ channel with 79.8 fb^{-1}. The ggF, VBF, VH, and $t\bar{t}H + tH$ production modes are shown in black, blue, cyan, and red respectively. The statistical only errors are indicated by dashed lines. No significant deviation from the Standard Model is observed

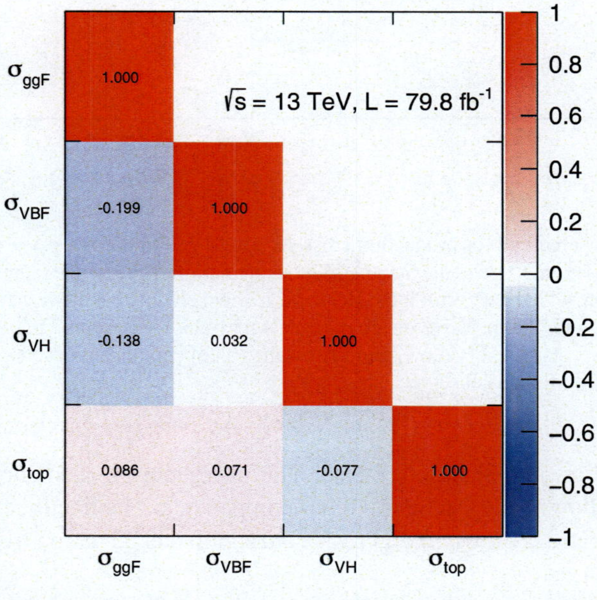

Fig. 13.9 Measured correlations between production mode cross sections in the $H \rightarrow \gamma\gamma$ channel with 79.8 fb^{-1}

Table 13.4 Expected and observed significance of each Higgs production mode in the $H \to \gamma\gamma$ channel with 79.8 fb^{-1}

Production mode	Expected significance	Observed significance
ggF	7.9σ	7.6σ
VBF	3.6σ	4.7σ
VH	1.9σ	2.0σ
$t\bar{t}H + tH$	3.6σ	3.8σ

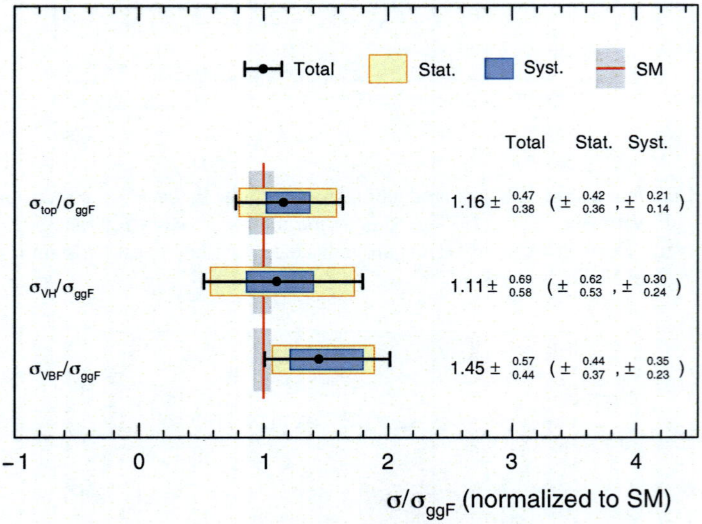

Fig. 13.10 Measured ratio of production cross section to ggF in the $H \to \gamma\gamma$ channel with 79.8 fb^{-1}. The ratio is normalized to the Standard Model prediction. The red line represents the Standard Model prediction, with theory error included in the gray uncertainty band. Measurements are shown as black points, with the inclusive uncertainty shown as black error bars. The uncertainty is also shown in separate statistical (yellow) and systematic (blue) components. No significant deviation from the Standard Model is observed

uncertainty shown as black error bars. The uncertainty is also shown in separate statistical (yellow) and systematic (blue) components. The likelihood scan for each cross section ratio is shown in Fig. 13.11. No significant deviation from the Standard Model is observed.

13.4.2 Measurement of STXS

For the measurement of STXS rates, a POI μ_b is associated with each STXS bin b. The exception is the ggF and VBF BSM bins, where the sum is fitted with POI μ_{BSM} and the difference is profiled. All categories are fit simultaneously. The measured

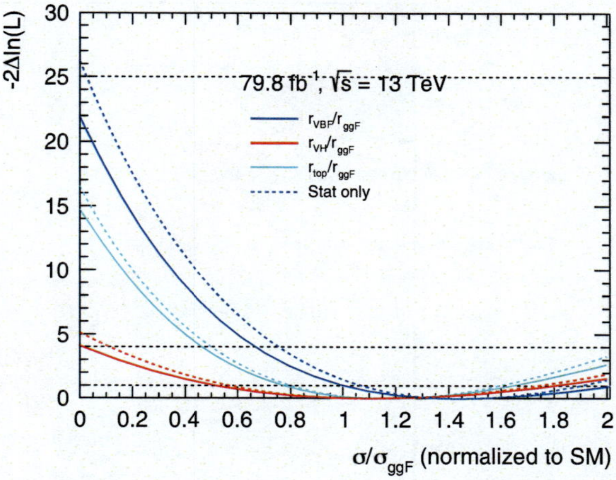

Fig. 13.11 Likelihood scan of Higgs production cross section ratio to ggF in the $H \rightarrow \gamma\gamma$ channel with 79.8 fb^{-1}. The ratios of VBF, VH, and $t\bar{t}H + tH$ to ggF are shown in blue, cyan, and red respectively. The statistical only errors are indicated by dashed lines. No significant deviation from the Standard Model is observed

STXS rates are shown in Fig. 13.12, normalized to the SM expectation. The red line represents the Standard Model prediction, with theory error included in the gray uncertainty band. Measurements are shown as black points, with the inclusive uncertainty shown as black error bars. The uncertainty is also shown in separate

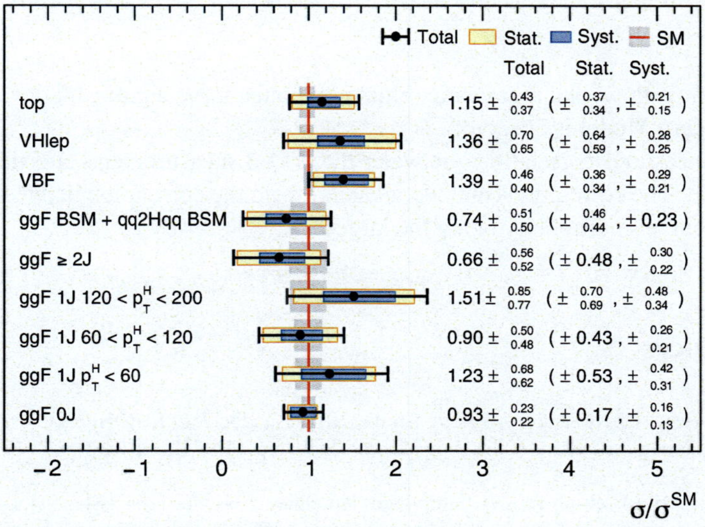

Fig. 13.12 Measured Simplified Template Cross Sections in the $H \rightarrow \gamma\gamma$ channel with 79.8 fb^{-1}, normalized to the Standard Model prediction. The red line represents the Standard Model prediction, with theory error included in the gray uncertainty band. Measurements are shown as black points, with the inclusive uncertainty shown as black error bars. The uncertainty is also shown in separate statistical (yellow) and systematic (blue) components. No significant deviation from the Standard Model prediction is observed

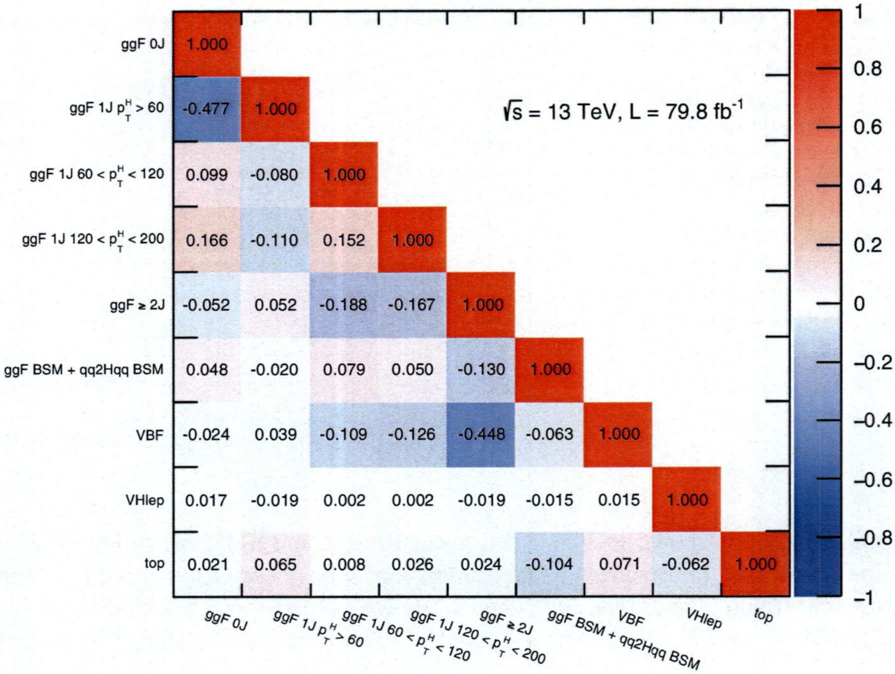

Fig. 13.13 Measured correlations between STXS bins in the $H \to \gamma\gamma$ channel with 79.8 fb^{-1}

statistical (yellow) and systematic (blue) components. No significant deviation from the Standard Model is observed.

The measured correlations between the STXS measurements are displayed in Fig. 13.13. The correlations are overall small, with the correlation between ggF bins of different jet multiplicity being the largest.

References

1. LHC Higgs Cross Section Working Group (2017) Handbook of LHC Higgs Cross Sections: 4. Deciphering the Nature of the Higgs Sector. Tech. rep. FERMILAB/CERN, 2017. https://cds.cern.ch/record/2227475
2. Delmastro M et al (2019) Simplified template cross sections—Stage 1.1. Tech. rep. arXiv:1906.02754. 14 pages, 3 figures. Geneva: CERN, 2019. https://cds.cern.ch/record/2669925
3. Aad G et al (2020) Combined measurements of Higgs boson production and decay using up to 80 fb-1 of proton-proton collision data at vs = 13 TeV collected with the ATLAS experiment. In: Phys. Rev. D 101.1 (2020), p 012002. https://doi.org/10.1103/PhysRevD.101.012002. arXiv:1909.02845 [hep-ex]

Chapter 14
Conclusions

During Run 2 of the Large Hadron Collider, protons were collided at $\sqrt{s} = 13$ TeV, the highest center of mass energy achieved by a collider to date. During this time, the ATLAS experiment collected a dataset of unprecedented size, corresponding to a luminosity of 139 fb^{-1}. Analysis of this dataset has provided fundamental insight into the nature of the Higgs boson and its interaction with the top quark, the heaviest particle in the Standard Model.

By performing a statistical combination of ATLAS $t\bar{t}H$ searches in multiple Higgs decay modes, Higgs boson production in association with a top-antitop quark pair was observed for the first time in the partial Run 2 dataset at the level of 6.3σ (5.1σ expected). This observation directly established the interaction of the Higgs boson with quarks and serves as powerful validation of the theory of Higgs-quark couplings in the SM.

Due to the excellent diphoton mass resolution of the ATLAS detector and the clean signature of the $H \rightarrow \gamma\gamma$ decay, this channel has proven to be one of the most powerful for $t\bar{t}H$ measurements. Using Machine Learning techniques, a sophisticated multivariate analysis was developed based on the momenta of jets, photons, and leptons. The resulting categories contain a high ratio of $t\bar{t}H$ signal to background. Top quarks were reconstructed with high probability in selected $t\bar{t}H(\gamma\gamma)$ events, verifying the presence of both a top quark and a Higgs boson in the final state. The measurement of the $t\bar{t}H$ cross section in the $H \rightarrow \gamma\gamma$ channel is among the most precise single channel measurements to date: a cross section times branching fraction of $\sigma_{t\bar{t}H} \times B_{\gamma\gamma} = 1.59\ ^{+0.43}_{-0.39}$ fb is measured using the full Run 2 ATLAS dataset.

With the $t\bar{t}H$ process directly established, a direct measurement of the CP properties of the Higgs-top interaction was then carried out in the $H \rightarrow \gamma\gamma$ decay channel. A CP-sensitive multivariate categorization was developed using reconstructed top

J. E. Dickinson, *ATLAS Measurements of the Higgs Boson Coupling to the Top Quark in the Higgs to Diphoton Decay Channel*, Springer Theses,
https://doi.org/10.1007/978-3-030-86368-5_14

quark variables, yielding a single-channel observed $t\bar{t}H$ significance of 5.2σ (4.4σ expected) and an observed upper limit on the tH cross section of 11.6 times the SM expectation (11.7 expected). The observed data excludes a fully CP odd Higgs-top coupling at the level of 3.9σ. The CP mixing angle is constrained to $|\alpha| > 43°$ at 95% confidence level ($|\alpha| > 63°$ expected).

A study of all Higgs production modes in the $H \rightarrow \gamma\gamma$ channel provides measurements of the cross sections of gluon fusion, vector boson fusion, and vector boson associated production, in addition to $t\bar{t}H + tH$. With 79.8 fb^{-1} of Run 2 LHC data, the cross section times $H \rightarrow \gamma\gamma$ branching ratio is measured with unprecedented precision, and measurements in the differential bins of the Simplified Template Cross Section scheme [1, 2] are possible. These measurements in the $H \rightarrow \gamma\gamma$ channel provide critical input to the combined ATLAS measurements of Higgs boson couplings [3].

With Run 2 brought to a successful conclusion, the LHC experiments (ATLAS included) are undergoing large-scale upgrades in preparation for the High-Luminosity LHC (HL-LHC), which will expand the existing proton–proton dataset by a factor of ten or more over the next few decades. The exploration of the Higgs boson and its couplings will continue at the HL-LHC through high precision and differential Higgs measurements, in addition to searches for new production and decay modes.

References

1. LHC Higgs Cross Section Working Group (2017) Handbook of LHC Higgs cross sections: 4. deciphering the nature of the Higgs sector. Tech. rep. FERMILAB/CERN, 2017. https://cds.cern.ch/record/2227475
2. Delmastro M et al (2019) Simplified template cross sections—Stage 1.1. Tech. rep. arXiv:1906.02754. 14 pages, 3 figures. Geneva: CERN, 2019. https://cds.cern.ch/record/2669925
3. Aad G et al (2020) Combined measurements of Higgs boson production and decay using up to 80 fb-1 of proton-proton collision data at vs = 13 TeV collected with the ATLAS experiment. In: Phys. Rev. D 101.1 (2020), p 012002. http://orcid.org/10.1103/PhysRevD.101.012002. arXiv:1909.02845 [hep-ex]

Appendix A
Search for TeV-Scale Gravity

The increase in proton–proton center of mass energy from $\sqrt{s} = 8$ TeV to 13 TeV in LHC Run 2 opened up a large phase space for new physics that had not been explored by previous experiments. This Appendix describes an early Run 2 search for signatures of TeV-scale gravity, such as microscopic black holes or string balls [1, 2], in final states with many high p_T jets [3].

The production of microscopic black holes and subsequent decay to jets are generated using the CHARYBDIS2 event generator [4]. Figure A.1 shows the distribution of jet multiplicity (n_{jet}) and scalar sum of jet p_T (H_T) for an example signal point: a model of rotating black holes in 6 extra dimensions, where the Planck scale in $4+6$ dimensions (M_D) is 2 TeV and the black hole threshold mass (M_{th}) is 7 TeV. This signal point is just beyond the limit on this process set by ATLAS using 8 TeV data [5]. The signal is concentrated around $H_T \sim 5.5$ TeV and $n_{\text{jet}} \sim 8$.

This search targets signals with high jet multiplicity and large H_T, such as that shown in Fig. A.1. Jets are required to have $p_T > 50$ GeV and $|\eta| < 2.8$. In order to ensure that the trigger is fully efficient in the selected region, each event must have $H_T > 1$ TeV and leading jet $p_T > 200$ GeV. The analysis is performed in six inclusive jet multiplicity bins: $n_{\text{jet}} \geq 3 - 8$.

The dominant background passing this selection is QCD multijet production, which is estimated using a data-driven method. In each n_{jet} bin, the H_T range is divided into three regions as shown in Fig. A.2: a Control Region (CR) at low H_T, a Validation Region (VR) at intermediate H_T, and a Signal Region (SR) at high H_T. The region boundaries C, V, and S are optimized using a Pythia8 multijet Monte Carlo sample that is normalized to data at low H_T:

1. C is chosen so that the background extrapolation uncertainty is minimized.
2. S is chosen so that the background extrapolation uncertainty in the SR is 0.5 events.
3. V is chosen so that 20 background events are predicted in the range $V < H_T < S$.

J. E. Dickinson, *ATLAS Measurements of the Higgs Boson Coupling to the Top Quark in the Higgs to Diphoton Decay Channel*, Springer Theses,
https://doi.org/10.1007/978-3-030-86368-5

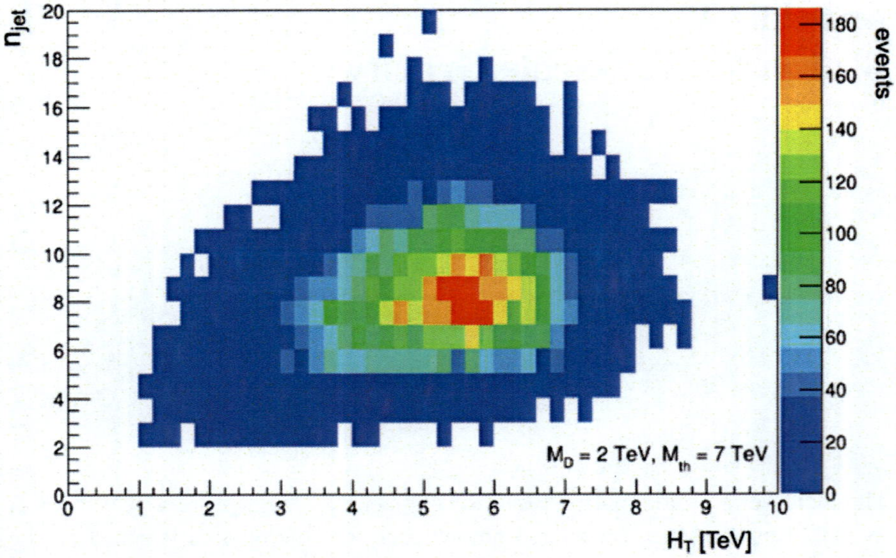

Fig. A.1 Distribution of H_T and n_{jet} for a rotating black hole signal in 6 extra dimensions, where the Planck scale in $4 + 6$ dimensions (M_D) is 2 TeV and the black hole threshold mass (M_{th}) is 7 TeV

Fig. A.2 Diagram of the control region, validation region, and signal region definitions

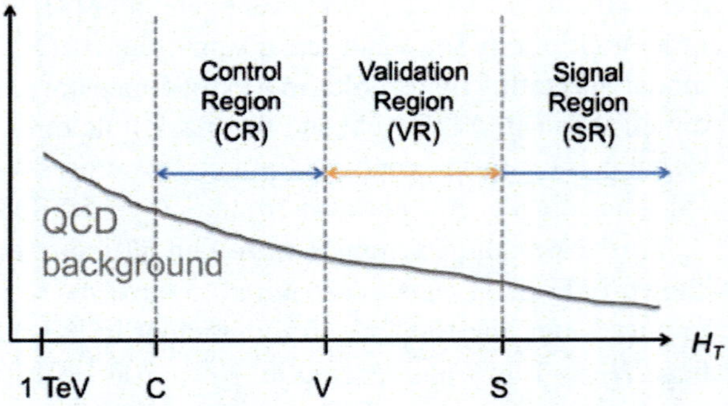

The multijet background is estimated separately in each n_{jet} bin by fitting to data in the Control Region. The functional forms considered for the shape of the multijet background are listed in Table A.1. To evaluate the performance of each fit function, pseudo-experiments (PEs) are generated from the Pythia8 multijet Monte Carlo. A candidate function qualifies if the fit converges and the function decreases monotonically in at least 95% of PEs. Qualifying functions are ranked based on their extrapolation uncertainty (determined using PEs). The function with smallest extrapolation uncertainty is selected as the baseline: extrapolation of this function to the VR and SR gives the central values of the background prediction. Figure A.3 shows the performance of the fit functions in one bin of the analysis.

Table A.1 Functional forms considered for the H_T distribution of the multijet background

	Functional form	p_1	p_2
1	$f_1(x) = \frac{p_0(1-x)^{p_1}}{x^{p_2}}$	(0,+∞)	(0,+∞)
2	$f_2(x) = p_0(1-x)^{p_1}e^{p_2x^2}$	(0,+∞)	(−∞, +∞)
3	$f_3(x) = p_0(1-x)^{p_1}x^{p_2x}$	(0,+∞)	(−∞, +∞)
4	$f_4(x) = p_0(1-x)^{p_1}x^{p_2 lnx}$	(0,+∞)	(−∞, +∞)
5	$f_5(x) = p_0(1-x)^{p_1}(1+x)^{p_2x}$	(0,+∞)	(0,+∞)
6	$f_6(x) = p_0(1-x)p_1(1+x)^{p_2 lnx}$	(0,+∞)	(0,+∞)
7	$f_7(x) = \frac{p_0}{x}(1-x)^{[p_1-p_2 lnx]}$	(0,+∞)	(0,+∞)
8	$f_8(x) = \frac{p_0}{x^2}(1-x)^{[p_1-p_2 lnx]}$	(0,+∞)	(0,+∞)
9	$f_9(x) = \frac{p_0(1-x^{1/3})^{p_1}}{x^{p_2}}$	(0,+∞)	(0,+∞)
10	$f_{10}(x) = p_0(1-x^{1/3})^{p_1}x^{p_2 lnx}$	(0,+∞)	(−∞, +∞)

The uncertainty on the background prediction has two components:

1. PE: extrapolation uncertainty calculated from the pseudo-experiments
2. DD: data-driven systematic uncertainty

The systematic uncertainty from pseudo-experiments is the difference in the number of events in the SR between the Monte Carlo prediction and the fit extrapolation. The data-driven systematic uncertainty is derived by extrapolating all qualifying fit functions to the VR and SR. This gives a projected background yield that is different from the baseline. Any qualifying function that does not agree with data in the VR within 95% confidence level is rejected. The data-driven systematic uncertainty is taken to be the maximum difference in the SR background projection between the baseline and the remaining fit functions.

Bootstrap Analysis Method

The analysis strategy outlined above relies on the assumption that there is no significant signal contamination in the Control Region. However, if the analysis is performed directly with a few inverse femtobarns of 13 TeV data, then there are allowed signal points that could impact the H_T spectrum in the CR and bias the background fit.

In order to ensure that the analysis is not biased by possible signal contamination, a **boostrap analysis method** is applied: the analysis is repeated and at several steps in integrated luminosity. If no new physics is observed at Step i, then Step $i+1$ (at about 10× higher luminosity) is safe from signal contamination in the CR. The Run 1 ATLAS exclusion limit is taken as a starting point, and the 13 TeV analysis is performed at four luminosity steps: 6.5 pb^{-1}, 74 pb^{-1}, 0.44 fb^{-1}, and 3.0 fb^{-1}. The total integrated luminosity used in the analysis is 3.6 fb^{-1}.

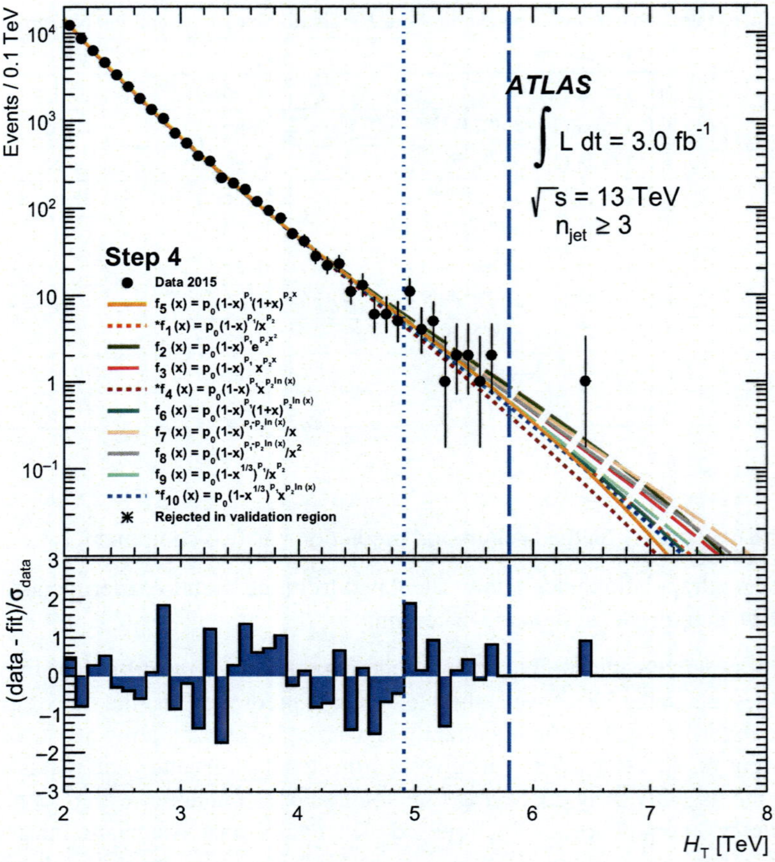

Fig. A.3 H_T distribution in data with 3.0 fb^{-1} in the $n_{\text{jet}} \geq 3$ bin, with qualifying fit functions overlaid. Functions rejected by data in the validation region are marked with an asterisk. The region boundary V (S) is shown as a blue dotted (dashed) vertical line

Figure A.4 shows the observed H_T distribution in data at each luminosity step in the $n_{\text{jet}} \geq 3$ bin. The region boundaries C, V, and S (overlaid in blue) are calculated independently for each n_{jet} bin and for each luminosity step. A signal point near the exclusion limit set by each step is shown in red.

The observed and predicted number of events in each region are shown in Table A.2, A.3, A.4, A.5 for each luminosity step and each jet multiplicity bin. No significant excess is observed. At Step 1, only the $n_{\text{jet}} \geq 3$ bin has enough statistics to derive a result. Statistics are too low to derive a result in the $n_{\text{jet}} \geq 8$ bin until Step 3.

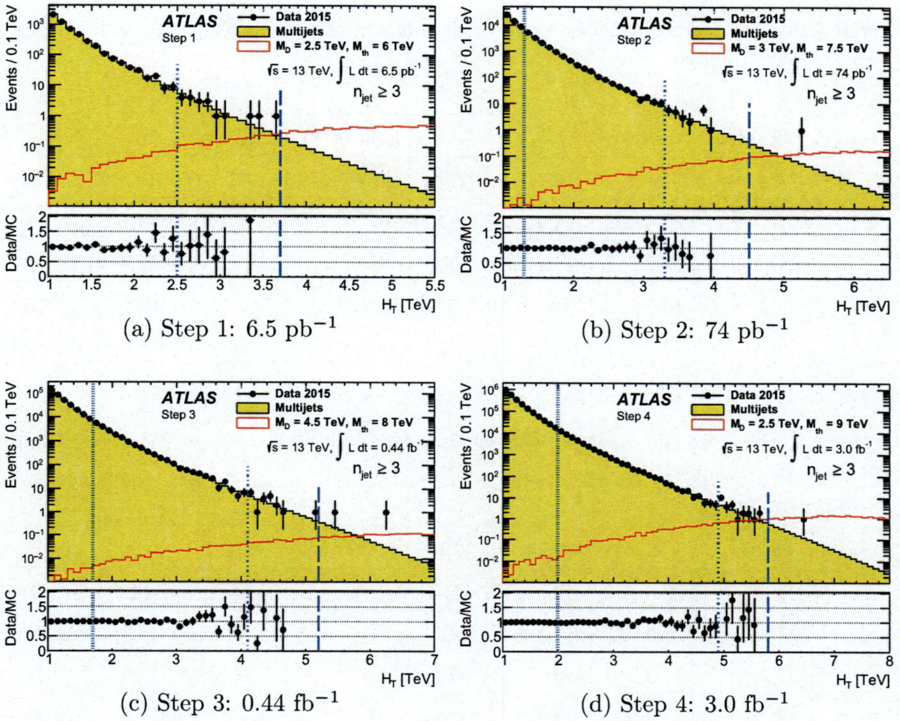

(a) Step 1: 6.5 pb^{-1} (b) Step 2: 74 pb^{-1}

(c) Step 3: 0.44 fb^{-1} (d) Step 4: 3.0 fb^{-1}

Fig. A.4 Comparison of data and multijet Monte Carlo at each step of the bootstrap analysis in the $n_{\text{jet}} \geq 3$ bin. The Monte Carlo prediction from Pythia8 (yellow) is normalized to data at low H_T. The region boundaries C, V, and S are overlaid in blue, and a signal point near the exclusion limit is shown in red for each luminosity step

Table A.2 Observed and expected event yields at each luminosity step in the bootstrap analysis, shown in bins of inclusive jet multiplicity

$n_{\text{jet}} \geq$	VR (obs)	VR (exp)	SR (obs)	SR (exp)
3	19	20.4 ± 4.4 (PE) ± 2.6 (DD)	0	0.65 ±0.46 (PE) ± 0.64 (DD)

Only the $n_{\text{jet}} \geq 3$ bin has enough statistics to derive a result

Limits on New Physics

Limits are derived on multiple signal models simulated using the CHARYBDIS2 event generator [4]. Figure A.5 shows the exclusion of rotating black holes in 6 extra dimensions in terms of the model parameters M_D and M_{th}. Figure A.6 shows the exclusion of string balls in terms of the string coupling (g_S), the string scale (M_S), and the string ball threshold mass (M_{th}).

A model-independent limit on the cross section of new physics at high H_T can be derived from the limit on the number of signal events N_S:

Table A.3 Observed and expected event yields at luminosity step 1 (6.5 pb^{-1}), shown in bins of inclusive jet multiplicity

(a) *Step 1: 6.5 pb*$^{-1}$

$n_{\text{jet}} \geq$	VR (obs)	VR (exp)	SR (obs)	SR (exp)
3	23	27.1 ± 3.7 (PE) ± 9.6 (DD)	1	1.42 ± 0.41 (PE)$^{+4:3}_{-1:42}$ (DD)
4	27	25.4 ± 3.2 (PE) ± 15.5 (DD)	0	1.62 ± 0.46(PE)$^{+9:2}_{-1:62}$(DD)
5	21	18.9 ± 2.9 (PE) ± 9.9 (DD)	0	1.32 ± 0.48(PE)$^{+5:1}_{-1:32}$(DD)
6	18	20.7 ± 3.3 (PE) ± 10.4 (DD)	0	1.19 ± 0.48(PE)$^{+13:3}_{-1:19}$(DD)
7	29	22.2 ± 3.7 (PE) ± 7.0 (DD)	0	0.81 ± 0.36 (PE) ± 0.60 (DD)

Table A.4 Observed and expected event yields at luminosity step 2 (74 pb^{-1}), shown in bins of inclusive jet multiplicity

(b) *Step 2: 74 pb*$^{-1}$

$n_{\text{jet}} \geq$	VR (obs)	VR (exp)	SR (obs)	SR (exp)
3	21	20.4 ± 2.7 (PE) ±10.5 (DD)	2	1.46 ± 0.42(PE)$^{+4:37}_{-1:46}$(DD)
4	23	29.9 ± 3.9 (PE) ± 8.1 (DD)	2	1.95 ± 0.46(PE)$^{+4:06}_{-1:95}$(DD)
5	17	21.4 ± 3.4 (PE) ± 7.1 (DD)	1	1.56 ± 0.51(PE)$^{+3:47}_{-1:56}$(DD)
6	19	28.3 ± 4.3 (PE) ± 6.3 (DD)	0	1.44 ± 0.40(PE)$^{+2:13}_{-1:44}$(DD)
7	28	24.7 ± 3.8 (PE) ± 4.5 (DD)	0	0.96 ± 0.39(PE)$^{+1:74}_{-0:96}$(DD)
8	25	31.8 ± 4.7 (PE) ± 1.4 (DD)	2	2.86 ± 0.40 (PE) ± 0.70 (DD)

Table A.5 Observed and expected event yields at luminosity step 3 (0.44 fb^{-1}), shown in bins of inclusive jet multiplicity

(c) *Step 3: 0.44 fb*$^{-1}$

$n_{\text{jet}} \geq$	VR (obs)	VR (exp)	SR (obs)	SR (exp)
3	28	19.5 ± 3.6 (PE) ±4.1 (DD)	1	2.10 ± 0.51 (PE) ± 1.78 (DD)
4	27	20.8 ± 2.3 (PE) ± 6.4 (DD)	2	2.36 ± 0.52 (PE) ± 2.12 (DD)
5	26	22.3 ± 2.6 (PE) ± 6.8 (DD)	2	1.95 ± 0.45(PE)$^{+2:10}_{-1:95}$ (DD)
6	20	20.3 ± 2.9 (PE) ± 5.4 (DD)	3	1.82 ± 0.49(PE)$^{+1:91}_{-1:82}$ (DD)
7	14	20.7 ± 4.1 (PE) ± 1.7 (DD)	0	0.53 ± 0.36 (PE) ± 0.22 (DD)
8	19	18.2 ± 4.9 (PE) ± 3.5 (DD)	0	0.43 ± 0.36 (PE) ± 0.26 (DD)

(d) *Step 4: 3.0 fb*$^{-1}$

At Step 1, only the $n_{\text{jet}} \geq 3$ bin has enough statistics to derive a result. Statistics are too low to derive a result in the $n_{\text{jet}} \geq 8$ bin until Step 3

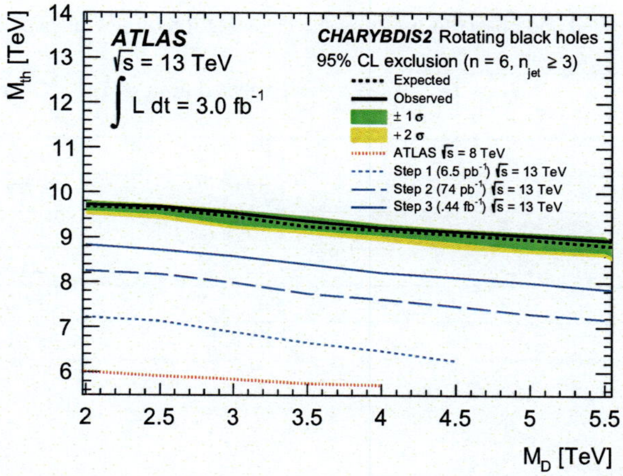

Fig. A.5 Limit on the production of rotating black holes in 6 extra dimensions generated using CHARYBDIS2. The Run 1 ATLAS limit is shown in red, and the limit obtained at each step in the bootstrap in the analysis is shown in blue

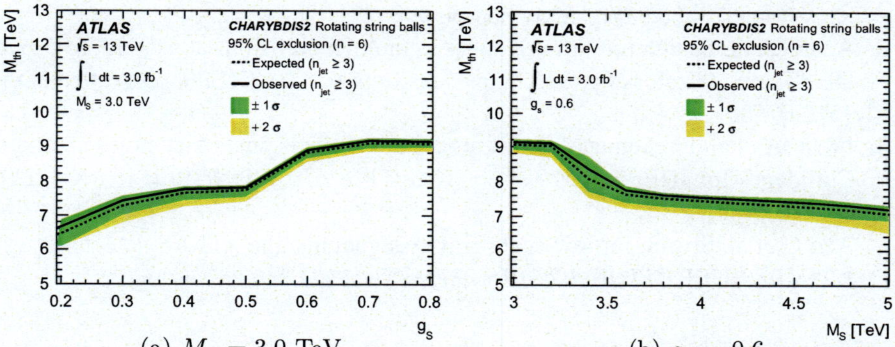

(a) $M_S = 3.0$ TeV (b) $g_S = 0.6$

Fig. A.6 Limit on the production of string balls generated using CHARYBDIS2

$$\sigma = \frac{N_S}{L} \frac{\epsilon^{\text{reco}}}{\epsilon^{\text{true}}} \tag{A.1}$$

Here, L is the integrated luminosity and ϵ^{true} (ϵ^{reco}) represents the analysis selection efficiency on true (reconstructed) signal events. The ratio of efficiencies $\epsilon^{\text{reco}}/\epsilon^{\text{true}}$ is model dependent, so a conservative estimate is used to obtain the cross section limit. The expected and observed cross section limits are reported in Table A.6 for each inclusive jet multiplicity bin. The H_T range of the Signal Region is also shown.

Table A.6 Expected and observed limits on the cross section of new physics in bins of inclusive jet multiplicity

$n_{\text{jet}} \geq$	$H_T > H_T^{min}$ (TeV)	Expected limit (fb)	Observed limit (fb)
3	5.8	$1.63^{+0.70}_{-0.57}$	1.33
4	5.6	$1.77^{+0.70}_{-0.57}$	1.77
5	5.5	$1.56^{+0.73}_{-0.50}$	1.75
6	5.3	$1.52^{+0.69}_{-0.50}$	2.15
7	5.4	$1.02^{+0.36}_{-0.0}$	1.02
8	5.1	$1.01^{+0.29}_{-0.0}$	1.01

References

1. Arkani-Hamed N, Dimopoulos S, Dvali GR. new dimensions at a millimeter. Phys Lett. https://doi.org/10.1016/S0370-2693(98)00466-3. arXiv:hep-ph/9803315.
2. Randall L, Sundrum R (1999) A Large mass hierarchy mension. Phys Rev Lett 83:3370–3373. 83.3370. arXiv:hep-ph/9905221.
3. Aad G et al. Search for strong gravity in multijet collisions at vs = 13 TeV using the ATLAS detector at p. 026. https://doi.org/10.1007/JHEP03(2016)026.arXiv: 1512.02586
4. Frost JA et al. Phenomenology of production dimensional black holes at Hadron Colliders'. In: JHEP https://doi.org/1088/1126-6708/2009/10/014. arXiv:0904. 0979[hep-ph].
5. Aad G et al. Search for low-scale gravity signatures the ATLAS detector at vs = 8 TeV. In: JHEP 07 JHEP07(2015)032. arXiv:1503.08988[hep-ex].

Appendix B
Multijet Modeling with MG5_aMC@NLO

Multijet final states constitute a major background in many ATLAS analyses, especially searches for Beyond the Standard Model physics (see for example Appendix A). Because pure QCD processes are challenging to model in Monte Carlo, background predictions of multijet processes are often derived using data-driven methods. In order for multijet Monte Carlo samples to be useful for analysis, a thorough understanding of their inputs and features is essential.

This Appendix presents the development of the first leading order $2 \rightarrow 4$ multijet sample generated for use in ATLAS at $\sqrt{s} = 13$ TeV. The MG5_aMC@NLO generator is selected for this sample because of its ability to perform leading order multileg and NLO calculations, and because of its interface to multiple showering packages. For references associated with each Monte Carlo generator software, see Table 5.3.

Throughout this Appendix, the MG5_aMC@NLO sample is compared to a leading order $2 \rightarrow 2$ Pythia sample generated with version 8.1.86. This sample uses PDF set NNPDF23 [1] and the A14 parameter tune, and EvtGen for bottom and charm hadron decays. This Pythia sample is found to agree well with data in important jet distributions, including leading and inclusive jet p_T [2].

The MG5_aMC@NLO multijet sample is generated using MG5_aMC@NLO version 2.3.3. The contribution from diagrams with up to four partons in the final state is calculated at leading order in α_S. Additional jets are generated by the parton shower, which is performed by Pythia 8.2.12. The matching/merging is done according to the CKKW-L algorithm [3] with a merging scale of 30 GeV. The EvtGen program is used for bottom and charm hadron decays.

The NNPDF30 PDF set [4] is selected for the baseline sample with the A14 parameter tune [5]. The CT10NLO PDF [6] set is chosen as an alternative PDF set, and a small test sample is generated in order to examine the impact of the choice of PDF on jet variables. The LHAPDF PDF reweighting procedure [7] is applied to this test sample, and the reweighting is found to close within 15%. The total effect of PDF choice on the MG5_aMC@NLO sample is taken to be the envelope of these

J. E. Dickinson, *ATLAS Measurements of the Higgs Boson Coupling to the Top Quark in the Higgs to Diphoton Decay Channel*, Springer Theses,
https://doi.org/10.1007/978-3-030-86368-5

Table B.1 H_T slice boundaries used for the MG5_aMC@NLO multijet Monte Carlo sample

Slice name	HT1	HT2	HT3	HT4	HT5	HT6
Min H_T [GeV]	0	100	200	300	500	700
Max H_T [GeV]	100	200	300	500	700	1000
Slice name	HT7	HT8	HT9	HT10	HT11	HT12
Min H_T [GeV]	1000	1500	2000	4000	6000	8000
Max H_T [GeV]	1500	2000	4000	6000	8000	13,000

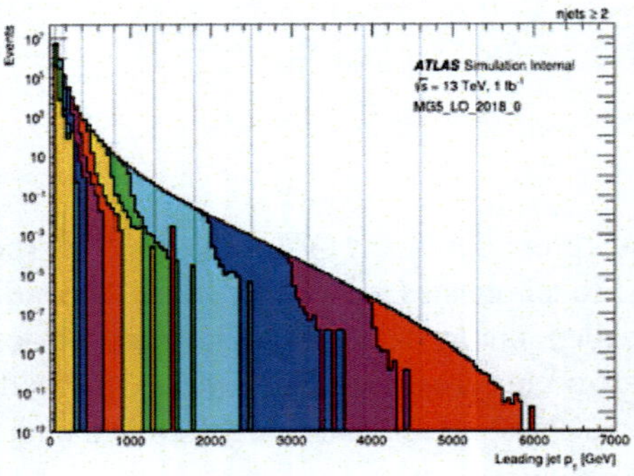

(a) Leading jet p_T

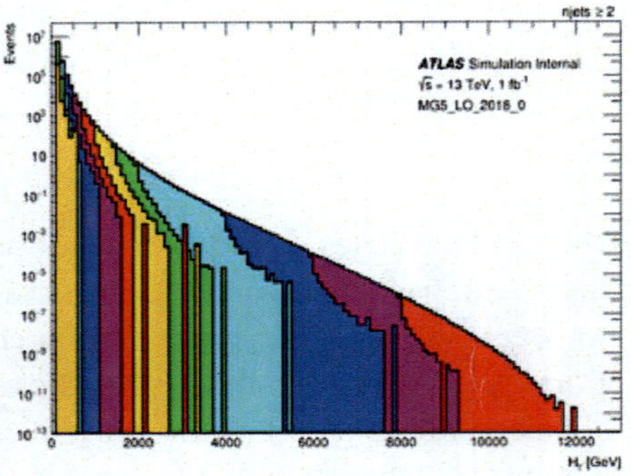

(b) H_T

Fig. B.1 Leading jet p_T and H_T for the MG5_aMC@NLO multijet sample sliced in parton-level H_T. Each slice is represented in a different color. The dashed black lines mark the boundaries of the H_T slices

reweighted distributions. This gives an uncertainty on the total jet cross section of about 10–20% for jet $p_T < 4.5$ TeV. The impact of PDF choice grows large in the high p_T tail.

Slicing

The MG5_aMC@NLO sample is composed of twelve regions (or "slices") in parton-level H_T (scalar sum p_T). Slicing at the parton level rather than particle level removes the need for filters with low efficiency and therefore long generation times. The contribution from each H_T slice to the full p_T and H_T distributions can be seen in Fig. B.1. The boundaries of the H_T slices are listed in Table B.1.

The H_T sliced sample agrees well with an H_T inclusive test sample, and it has been verified that the distribution is smooth at the slice boundaries (Fig. B.2).

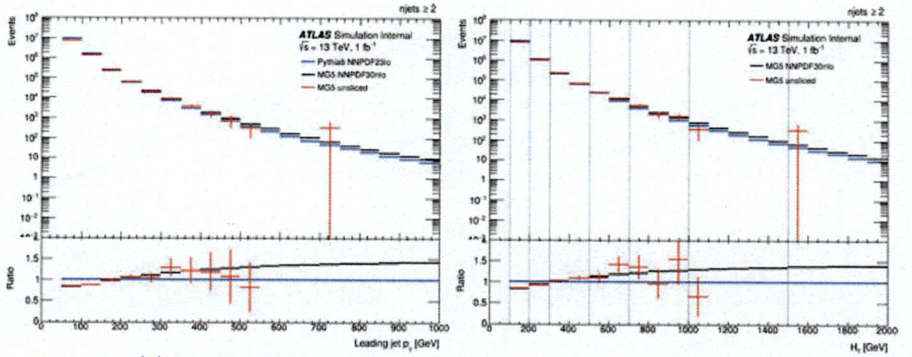

(a) Leading jet p_T (b) H_T

Fig. B.2 Leading jet p_T and H_T for the MG5_aMC@NLO multijet sample sliced in parton-level H_T (black) and unsliced (red). The ratio in the bottom panel is to the Pythia8 sample. Agreement is good in the region where the unsliced sample has sufficient statistics, including at H_T slice boundaries (vertical dashed lines)

Scale Variations

Two forms are tested for the renormalization and factorization scales μ_R and μ_F:

- S_0: the transverse mass of the $2 \to 2$ system resulting from k_T clustering (the default for LO generation in MG5_aMC@NLO)
- S_1: the total transverse energy in the event:

$$S_1 = \sum_{i=1}^{N} \frac{E_i p_{T,i}}{\sqrt{p_{x,i}^2 + p_{y,i}^2 + p_{z,i}^2}} \tag{B.1}$$

The renormalization and factorization scales (μ_R and μ_F) are varied independently up and down by a factor of two to provide an uncertainty band on each distribution. The total scale uncertainty on the MadGraph sample is taken to be the envelope of these eight variations. The impact of scale variations on the jet cross section is dominated by μ_R rather than μ_F. These scale variations give an uncertainty on the total jet cross section of about 40%.

The leading jet p_T and H_T distributions are shown in Fig. B.3 using scales S_0 and S_1. The sample using scale S_1 agrees better in shape with the Pythia8 sample in both variables. In addition, the uncertainty due to scale variations is smaller on the sample with scale S_1. This form of the scale is therefore chosen for the final MG5_aMC@NLO sample.

Test samples with $2 \to 2$, $2 \to 3$, and $2 \to 4$ partons in the matrix element were generated using scale S_1. Figure B.4 shows that the scale variations on the $2 \to 4$ multileg sample (light gray band) cover the differences between these calculations.

Variables sensitive to the third jet (Fig. B.5) can be used to probe the differences between the $2 \to 2$ and $2 \to 3, 4$ samples. The $2 \to 3+$ samples have a harder p_T

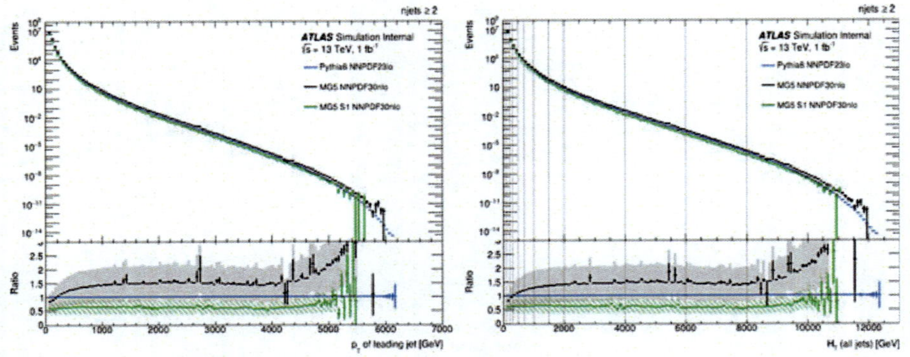

(a) Leading jet p_T (b) H_T

Fig. B.3 Leading jet p_T and H_T using two different forms of the scale in MG5_aMC@NLO multijet Monte Carlo. The default scale S_0 is shown in black and S_1 in green. The uncertainty bands show the envelope of distributions with μ_R and μ_F varied up and down by a factor of two. The ratio is shown to the Pythia8 sample. All jets have $p_T > 50$ GeV and $|\eta| < 5.0$

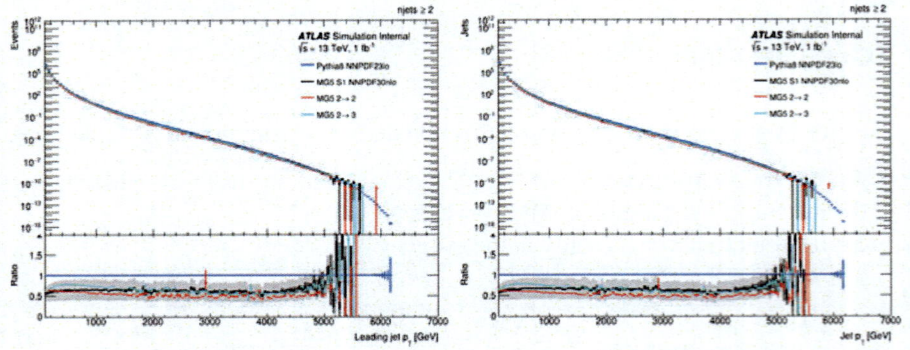

(a) Leading jet p_T (b) Inclusive jet p_T

Fig. B.4 Leading jet p_T and inclusive jet p_T for MG5_aMC@NLO multijet samples using scale S_1. The light gray band shows scale variations on the $2 \rightarrow 4$ sample, and the dark gray band shows the PDF uncertainty on this sample. The ratio is shown to the Pythia8 sample

spectrum for the third jet than $2 \rightarrow 2$, as well as a higher average jet multiplicity. All MG5_aMC@NLO samples show good agreement in the angular separation between the second and third jets ($\Delta\phi(j_2, j_3)$), but there is a large discrepancy with Pythia. The angle between the plane formed by the leading jet momentum/beam and the plane formed by the second/third leading jet momentum (ψ^*) [8] exhibits shape differences between the $2 \rightarrow 2$ and $2 \rightarrow 3+$ samples, and the MG5_aMC@NLO $2 \rightarrow 2$ sample is most similar in shape to Pythia.

In an optimal multijet Monte Carlo setup, the data provide a guide. Additional measurements at $\sqrt{s} = 13$ TeV, particularly measurements of variables sensitive to the third jet, will shed further light on the relative performance of the available multijet samples. Such comparisons are underway, and this MG5_aMC@NLO sample

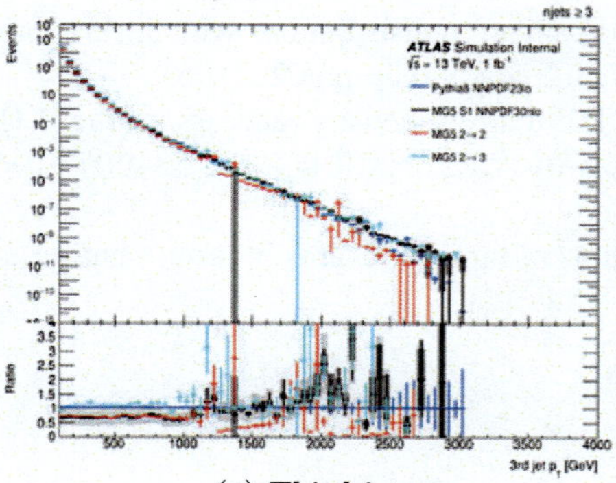

(a) Third jet p_T

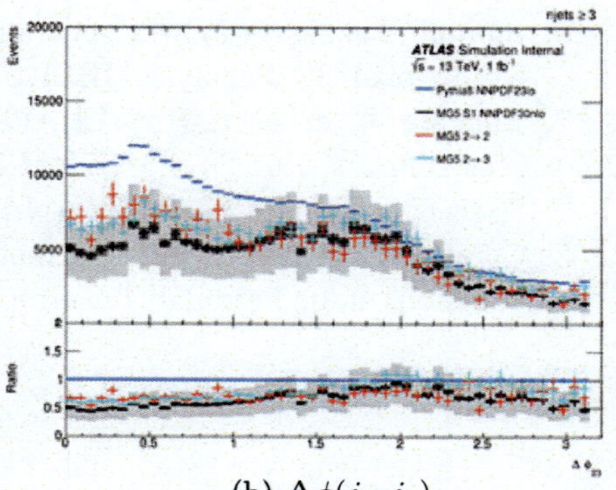

(b) $\Delta\phi(j_2, j_3)$

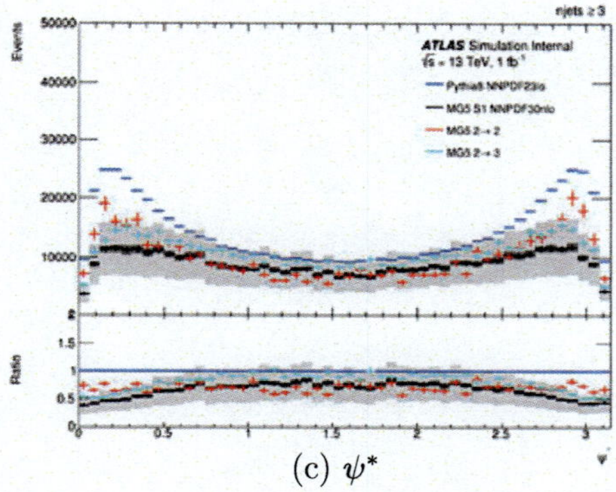

(c) ψ^*

Fig. B.5 Variables sensitive to the third jet for MadGraph multijet Monte Carlo samples. The light gray band shows scale variations on the 2 → 4 MadGraph sample, and the dark gray band shows the impact of PDF choice on this sample. The ratio is shown to the Pythia8 sample

is included in a comprehensive comparison of ATLAS multijet Monte Carlo samples [2].

References

1. Ball RD et al (2013) Parton distributions with LHC data. Nucl Phys B867:244–289. https://doi.org/10.1016/j.nuclphysb.2012.10.003. arXiv: 1207.1303 [hep-ph].
2. Multijet simulation for 13 TeV ATLAS Analyses. Tech. rep. ATL-PHYS-PUB-2019- 017. Geneva: CERN, 2019. https://cds.cern.ch/record/2672252.
3. Kuhn R, Catania S, Krauss F, Webber BR (2001) QCD matrix elements + Parton showers. JHEP 11:63. https://doi.org/10.1088/1126-6708/2001/11/063.
4. Ball RD et al (2015) Parton distributions for the LHC Run II. JHEP 04:040. https://doi.org/10.1007/JHEP04(2015)040. arXiv: 1410.8849 [hep-ph].
5. Summary of ATLAS Pythia 8 tunes. Tech. rep. ATL-PHYS-PUB-2012-003. Geneva: CERN, 2012. https://cds.cern.ch/record/1474107.

6. Pumplin J et al (2002) New generation of parton distributions with uncertainties from global QCD analysis. JHEP 07:012. arXiv: hep-ph/0201195.
7. Buckley A et al (2015) LHAPDF6: parton density access in the LHC precision era. Eur Phys J C75:132. https://doi.org/10.1140/epjc/s10052-015-3318-8. arXiv: 1412.7420 [hep-ph].
8. Carey RW (1989) Angular distributions of three jet events in proton-antiproton collisions at the Fermilab Tevatron.

Appendix C
Radiation Damage in the IBL

This Appendix presents two offline studies of the effects of radiation damage on the ATLAS Insertable B-Layer (IBL). Before describing these studies in detail, a description of the IBL layout and electronics are provided that is beyond the scope of Chap. 3. Section C.1 reports measurements of the rate of Single Event Upset (SEU) in the IBL obtained by reading back pixel configuration latches. Section C.2 describes a measurement of leakage current in the IBL and includes projections for Run 3.

The IBL is composed of silicon pixel sensors with two different designs:

1. **Planar sensors** are used in the central region ($|z| < 24$ cm). These sensors are of the same design used in other layers of the Pixel detector. Each planar sensor consists of 26,880 pixels, each with surface area of $250 \times 50\ \mu\text{m}^2$ and thickness of 200 μm.
2. **3D sensors** are used in the forward region. This new sensor design is used for the first time in ATLAS [1]. Each 3D sensor has a surface area of $250 \times 50\ \mu\text{m}^2$ and thickness of 230 μm.

The readout chip used in the IBL is the FE-I4B, which is built in 130 nm CMOS technology [2, 3]. Each planar IBL module is composed of two FE-I4B chips and one planar sensor, and each 3D module is composed of a single FE-I4B chip and one 3D sensor. Each pixel in the sensor is electrically bonded to one channel of an FE-I4B chip to form a module.

Fourteen carbon fiber staves support the IBL modules. Each stave is 64.3 cm long and 3 cm wide, and is oriented along the beam direction. Figure C.1 shows the orientation of the IBL staves and the numbering scheme used to label them. Each stave holds twelve planar and eight 3D modules. Four 3D modules are located on the A side ($|z| > 0$ cm) and four on the C side ($|z| < 0$ cm). In total, there are 32 FE-I4B readout chips on each stave.

J. E. Dickinson, *ATLAS Measurements of the Higgs Boson Coupling to the Top Quark in the Higgs to Diphoton Decay Channel*, Springer Theses,
https://doi.org/10.1007/978-3-030-86368-5

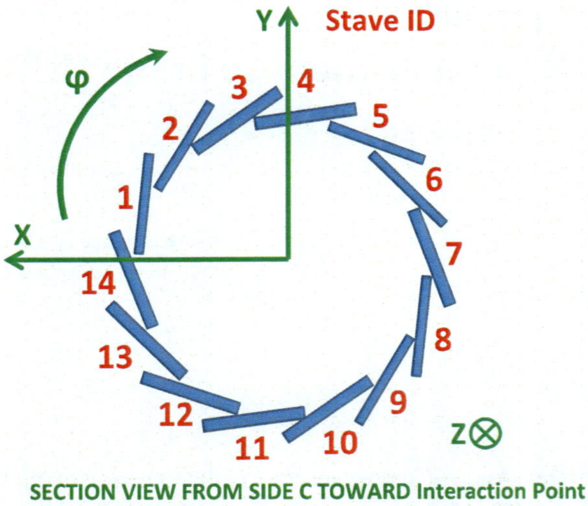

Fig. C.1 Position of IBL staves in the $\hat{x} \times \hat{y}$ plane [4]

C.1 Single Event Upset

Single Event Upset (SEU) is an effect of ionizing radiation where charge injected into a circuit's memory cell(s) leads to a flip in one or more logic states. At the LHC, SEUs are caused by recoil nuclei and hadronic showers resulting from the interaction of particles in the detector material near the memory cells affected [5]. Such interactions are difficult to model, but an understanding of SEU rates is essential for detector operations. This Section presents an in-situ measurement of the probability of bit flips in the radiation hard memory cells known as Dual Interlocked CElls, or DICE latches [6, 7], which store the configuration of each pixel in the IBL.

Each DICE latch contains thirteen bits, which correspond to the following configuration settings:

- 0: Output enable
- 1–5: Threshold tuning DAC (TDAC) value. 1 = most significant bit (MSB)
- 6–7: Selection of charge injection capacitor
- 8: HitBus (input to logical OR of all pixel discriminator outputs in the matrix)
- 9–12: Time-over-threshold (ToT) tuning DAC (FDAC) value. ToT represents the length of time that single pixel discriminator is over threshold in counts of an externally supplied clock (nominally 40 MHz). 12 = most significant bit (MSB)

During data taking, more than 99% of pixels are enabled: that is, they have a value of one stored in bit 0. Both capacitor selection bits and the HitBus bit also store values of one. The 5-bit TDAC stores an average value of fifteen, and the 4-bit FDAC stores an average value of seven.

The SEU rate is measured by copying the values of the pixel latches (which are radiation hard) into a register which is not radiation hard. After allowing a certain fluence of particles to traverse the device, the latch values are re-copied, read back,

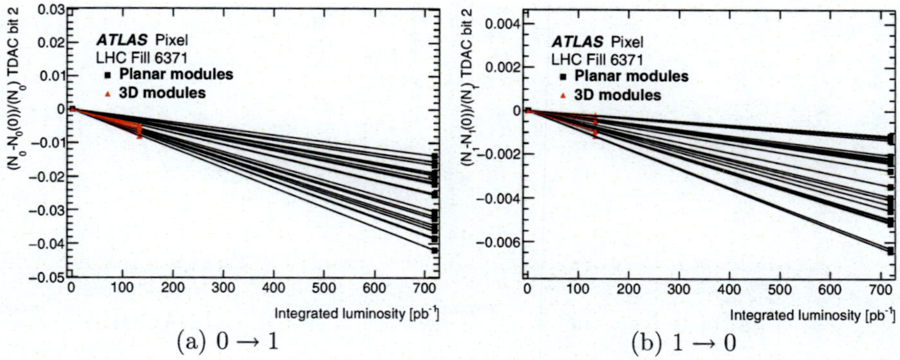

Fig. C.2 Fraction of pixels in which TDAC-bit2 flips, measured at the beginning of LHC fill 6371 and after about 150 pb^{-1} of integrated luminosity for 3D modules (red) and after the end of LHC fill for planar modules (black)

and compared to their original values. The fraction of pixels in which the bit state flips after taking data of integrated luminosity L depends on the probability of $0 \rightarrow 1$ ($\sigma_{0\rightarrow 1}$) and $1 \rightarrow 0$ ($\sigma_{1\rightarrow 0}$) transitions due to SEU:

$$\frac{N_1(0) - N_1}{N_1} = \sigma_{1\rightarrow 0} \times L \tag{C.1}$$

$$\frac{N_0(0) - N_0}{N_0} = \sigma_{0\rightarrow 1} \times L \tag{C.2}$$

Figure C.2 shows the change in the fraction of ones (**a**) and zeros (**b**) for a sample bit (the second most significant TDAC bit) during LHC fill 6371 in 2017. The first point is zero by construction. Because the register cannot be read back while the FE-I4B chip is receiving trigger signals, there are only two points for the planar modules (black): one before the start of collisions, and the other after beam dump. For the 3D modules (red), the register is read back after about 150 pb^{-1} of data taking.

From read back measurements like those shown in Fig. C.2, the SEU probabilities $\sigma_{0\rightarrow 1}$ and $\sigma_{1\rightarrow 0}$ can be extracted for each of the thirteen pixel latches. The resulting probabilities are shown in Fig. C.3 for the three bits that are most important for operations: the output enable bit, the TDAC most significant bit, and the FDAC most significant bit. The probability $\sigma_{0\rightarrow 1}$ is not calculated for the output enable bit because most pixels are enabled initially.

For the TDAC and FDAC most significant bits, the $0 \rightarrow 1$ transition is about five times as frequent as the $1 \rightarrow 0$ transition. For additional studies of this asymmetry, refer to [4].

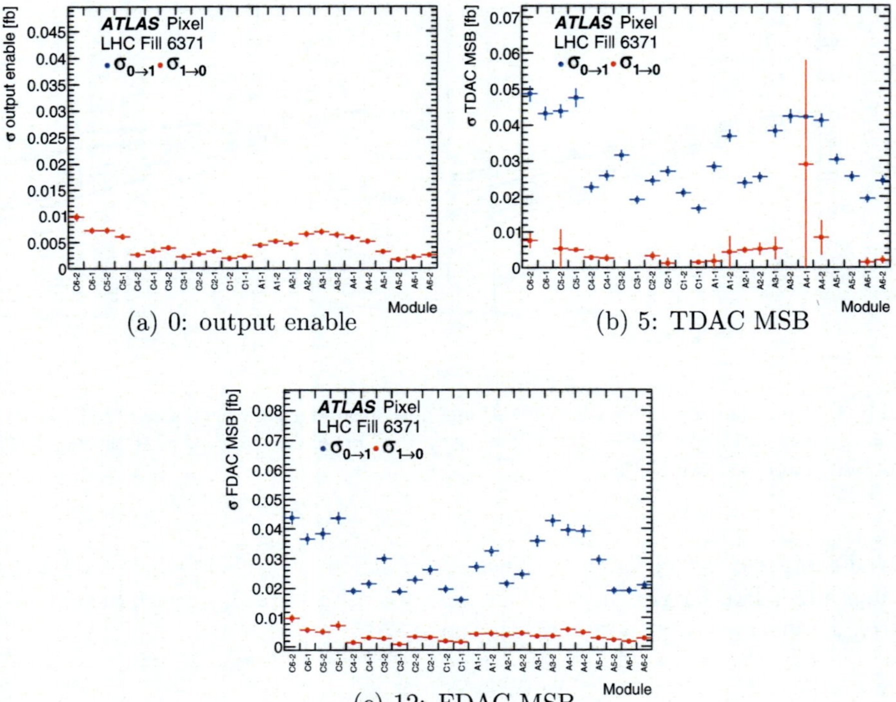

(a) 0: output enable

(b) 5: TDAC MSB

(c) 12: FDAC MSB

Fig. C.3 Probability of bit flip due to SEU in the three DICE latches most important for IBL operations: (**a**) the output enable bit, (**b**) the most significant TDAC bit, and (**c**) the most significant FDAC bit. The probability of the $0 \rightarrow 1$ ($1 \rightarrow 0$) transition is shown in blue (red)

C.2 Measurement of Leakage Current

The **leakage current** is a small current that is present in a reverse-biased semiconductor sensor, even when no particles are traversing the sensor (see Fig. C.4). The leakage current increases linearly with non-ionizing energy loss (NIEL), or **fluence**, which measures the radiation that damages the silicon pixel sensors. Fluence is measured in units of MeV neutron-equivalent radiation per cm^2 (abbreviated MeV n_{eq}/cm^2) and is proportional to integrated luminosity, which is well measured in ATLAS.

The Hamburg model [9] provides a phenomenological model for predicting the leakage current after n time steps:

$$I_{\text{leak}} = \frac{\Phi V}{L_{\text{tot}}} \times \sum_{i=1}^{n} L_i \left[\alpha_I \exp\left(-\sum_{j=1}^{n} \frac{t_j}{\tau(T_j)} \right) + \alpha_0^* - \beta \log\left(\sum_{j=1}^{n} \frac{\Theta(T_j) \times t_j}{t_0} \right) \right] \tag{C.3}$$

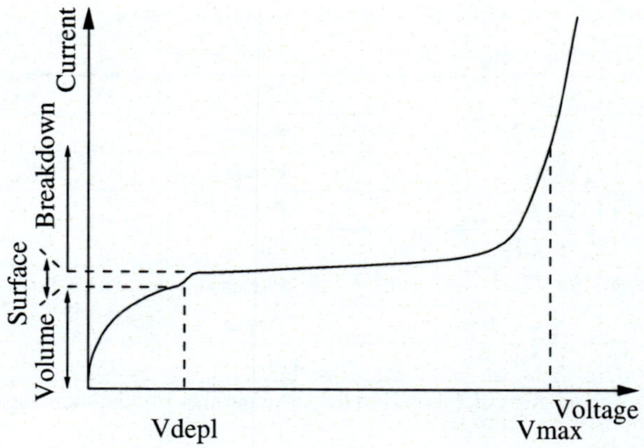

Fig. C.4 Sample IV curve of a silicon pixel sensor [8]. The leakage current is the small current measured at voltages between Vdepl and Vmax

Here, L_i is the integrated luminosity, t_i is the duration, and T_i is the temperature during the ith time step. L_{tot} is the total integrated luminosity, Φ is the fluence, and V is the sensor volume. The measured values of α_I, α_0^*, and β are

$$\alpha_I = (1.23 \pm 0.06) \times 10^{-17}\ \text{A/cm} \tag{C.4}$$

$$\alpha_0^* = 7.07 \times 10^{-17}\ \text{A/cm} \tag{C.5}$$

$$\beta = (3.29 \pm 0.18) \times 10^{-18}\ \text{A/cm} \tag{C.6}$$

The factors τ and Θ are

$$\tau^{-1}(T_j) = 1.2^{+5.3}_{-1.0} \times 10^{13}\ \text{s}^{-1} \times \exp\left[(-1.1 \pm 0.05)\frac{\text{eV}}{k_B T_j}\right] \tag{C.7}$$

$$\Theta(T_j) = \exp\left[-\frac{(1.30 \pm 0.14)\ \text{eV}}{k_B}\left(\frac{1}{T} - \frac{1}{T_R}\right)\right] \tag{C.8}$$

The reference temperature T_R is taken to be 0°C for this study.

Given a prediction for the fluence Φ and measurements of temperature, duration, and integrated luminosity at each time step, the Hamburg model can be used to calculate the leakage current as a function of integrated luminosity. The Hamburg model predictions are overlaid on all measurements below.

To measure the leakage current, readings of high voltage, temperature, and uncorrected leakage current are compiled for each IBL module group listed in Table C.1. Each module group in this table corresponds to four FE-I4B chips in the IBL. The high voltage, temperature, and leakage current readings are averaged over time steps of one **luminosity block (LB)**. The duration of a luminosity block is usually around one minute, and the experimental conditions are assumed to be constant during this period.

The high voltage of the planar sensors (M1-M3) began at 80 V in 2015, increased to 150 V during 2016, to 300 V at the start of 2017, and finally to 400 V at the start of

Table C.1 Summary of the IBL module groups and their positions in ATLAS

Module group	Module type	z (cm)
LI_Sx_A4	3D	> 24
LI_Sx_A3	Planar	[16, 24]
LI_Sx_A2	Planar	[8, 16]
LI_Sx_A1	Planar	[0, 8]
LI_Sx_C1	Planar	[−8, 0]
LI_Sx_C2	Planar	[−16, −8]
LI_Sx_C3	Planar	[−24, −16]
LI_Sx_C4	3D	< -24

The label x corresponds to the stave number of the module group

2018. The high voltage of the 3D sensors began at 20 V in 2015, and increased to 40 V for the remainder of the run. The increase in the high voltage settings is necessary to ensure that the sensors are fully depleted after experiencing more and more radiation damage. Measurements of the leakage current recorded during an LB when the high voltage differs from the setting by $\geq$ 1V are vetoed: most vetoed points correspond to times when the high voltage is off because ATLAS is not taking data.

The temperature of the IBL sensors varies with time and data taking conditions. However, the leakage current can be corrected to a constant reference temperature of $T_R = 0°$C by applying a correction factor:

$$I_{\text{leak}}(T_R) = I_{\text{leak}}(T)\exp\left(-\frac{E_g}{2k_B}\left[\frac{1}{T_R} - \frac{1}{T}\right]\right) \tag{C.9}$$

The silicon band gap energy E_g is taken to be 1.21 eV.

The measured leakage current as a function of Run 2 luminosity is shown in Fig. C.5. Measurements are averaged over all staves at the same z position and over the A and C sides (which exhibit very similar behavior). The Hamburg model prediction is overlaid for comparison, with a z-dependent scale factor applied to normalize the predicted fluence/luminosity (Φ/L) factor to data. The predicted Φ/L from FLUKA and the fitted scale factors are reported in Table C.2. The simulation is found to over-predict Φ/L at high $|z|$.

The sudden drops in the measured leakage current correspond to technical stops and year-end shutdowns. The IBL is warm (room temperature) during this time, and substantial annealing takes place. The level of annealing predicted by the Hamburg model is slightly less than what is observed in data.

The ratio of two leakage current measurements is predicted to be the ratio of the fluences multiplied by the depleted volume. The measured ratio to the 3D modules and corresponding Hamburg model predictions are shown in Fig. C.6. The discontinuity near 35 fb^{-1} corresponds to the increase in high voltage in 2016: before the voltage change, the IBL sensors were running under-depleted. Following this change, the ratio is approximately flat for the remainder of Run 2.

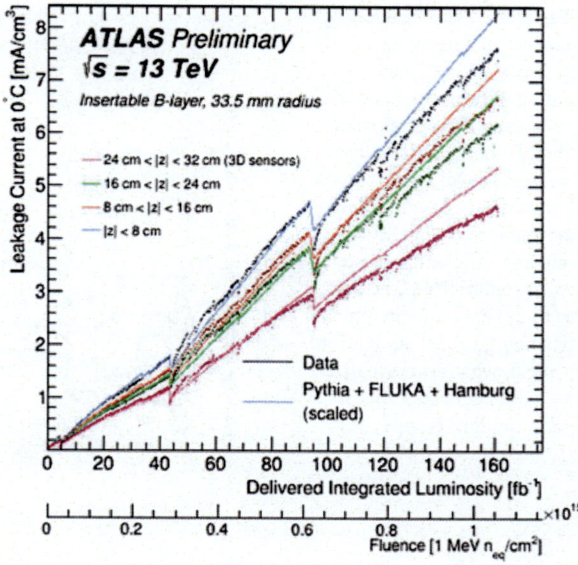

Fig. C.5 Measured leakage current in IBL modules as a function of integrated luminosity. The leakage current is averaged over modules at similar $|z|$ and normalized by the sensor volume. The temperature is corrected to 0°C. The data (points) are compared to a Hamburg model simulation (lines), which is normalized to data independently for each module group

Table C.2 Predicted Φ/L factors from FLUKA [(1 MeV n_{eq}/cm^2)$\times$ fb] and fitted scale factors for each group of IBL modules

Modules	Predicted Φ/L	Scale factor
A4	58.45×10^{11}	0.75
A3	59.90×10^{11}	0.93
A2	62.40×10^{11}	0.96
A1	65.45×10^{11}	1.04
C1	65.45×10^{11}	1.04
C2	62.40×10^{11}	0.94
C3	59.90×10^{11}	0.91
C4	58.45×10^{11}	0.76

Using the scale factors in Table C.2, the Hamburg model predictions for the IBL leakage current can be extended to the end of LHC Run 3. Multiple temperature and luminosity scenarios are under consideration for Run 3, but the maximum leakage current reached is similar in all scenarios.

Figure C.7 shows the Run 2 measured and Run 2–3 predicted leakage current in each module group. This projection corresponds to a total Run 3 luminosity of 320 fb^{-1}, with shutdowns (and therefore warm time for the IBL) after each 80 fb^{-1} period. Since Fig. C.5 has demonstrated that the Hamburg model tends to under-predict the amount of annealing, this projection is considered conservative.

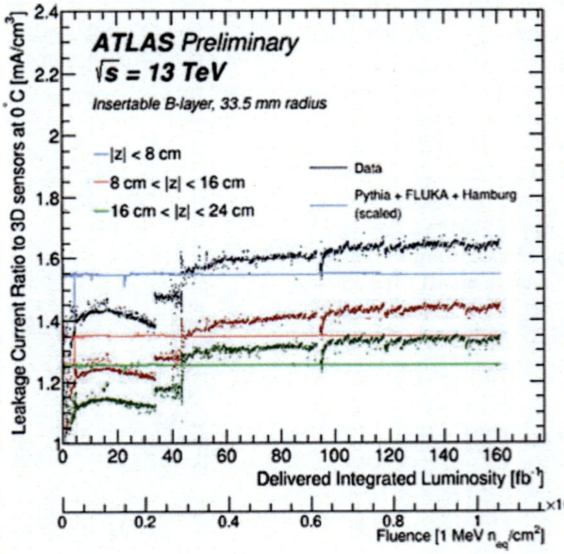

Fig. C.6 Measured leakage current in IBL modules as a function of integrated luminosity, normalized to the measured leakage current in the 3D modules. The leakage current is averaged over modules at similar $|z|$ and normalized by the sensor volume. The temperature is corrected to 0°C. The data (points) are compared to a Hamburg model simulation (lines), which is normalized to data independently for each module group

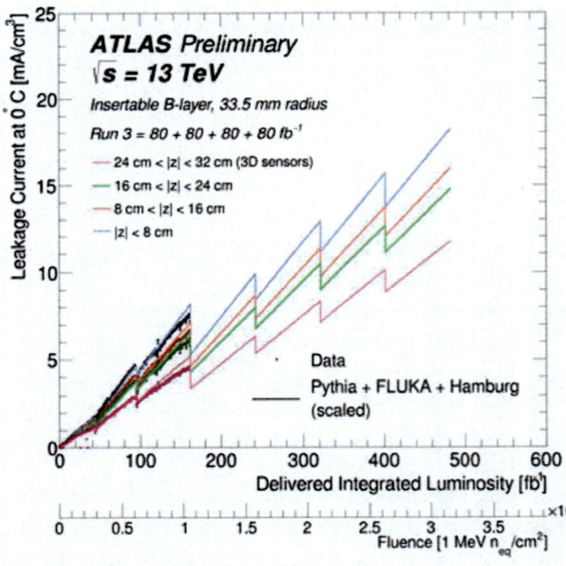

Fig. C.7 Measured leakage current in IBL modules as a function of integrated luminosity. The leakage current is averaged over modules at similar $|z|$ and normalized by the sensor volume. The temperature is corrected to 0°C. The data (points) are shown through the end of Run 2, and the Hamburg model simulation (lines) is projected through the end of Run 3

References

1. Abbott B et al (2018) Production and integration of the ATLAS insertable B-layer. JINST13(05):T05008. https://doi.org/10.1088/1748-0221/13/05/T05008. arXiv:1803.00844 [physics.ins-det].
2. Garcia-Sciveres M et al (2011) The FE-I4 pixel readout integrated circuit. Nucl Instrum Methods Phys Res A 636:S155–S159.
3. FE-I4 Collaboration (2012) The FE-I4B integrated circuit guide.
4. Balbi et al G (2020) Measurements of single event upset in ATLAS IBL.. arXiv:2004.14116 [physics.ins-det].
5. Garcia-Sciveres M, Wermes N (2018) A review of advances in pixel detectors for experiments with high rate and radiation. Rep Progress Phys 81, nbr. 6.
6. Calin T, Nicolaidis M, Velazco R (1996) Upset hardened memory by design for sub- micron CMOS technology SEU. IEEE Trans Nucl Sci 43:2874.
7. Menouni M et al (2006) SEU tolerant memory design for the ATLAS pixel readout chip. JINST 8:C02026.
8. Wermes N et al (2006) Pixel detectors, from fundamentals to applications. Springer-Verlag, 2006. https://doi.org/10.1007/3-540-28333-1. ISBN: 3-540-28332-3
9. Moll M (1999) Radiation damage in silicon particle detectors: microscopic defects and macroscopic properties'. PhD thesis. Hamburg U.